Lecture Notes in Mathematics 1882

T0202524

S. Attal · A. Joye · C.-A. Pillet (Eds.)

Open Quantum Systems III

Recent Developments

Springer

Editors

Stéphane Attal
Institut Camille Jordan
Université Claude Bernard Lyon 1
21 av. Claude Bernard
69622 Villeurbanne Cedex
France
e-mail: attal@math.univ-lyon1.fr

Alain Joye
Institut Fourier
Université de Grenoble 1
BP 74
38402 Saint-Martin d'Hères Cedex
France
e-mail: alain.joye@ujf-grenoble.fr

Claude-Alain Pillet
CPT-CNRS, UMR 6207
Université du Sud Toulon-Var
BP 20132
83957 La Garde Cedex
France
e-mail: pillet@univ-tln.fr

Library of Congress Control Number: 2006923432

Mathematics Subject Classification (2000): 37A60, 37A30, 47A05, 47D06, 47L30, 47L90, 60H10, 60J25, 81Q10, 81S25, 82C10, 82C70

ISSN print edition: 0075-8434
ISSN 1617-9692 (eBook)
ISBN 3-540-30993-4 Springer Berlin Heidelberg New York
ISBN 978-3-540-30993-2 Springer Berlin Heidelberg New York

DOI 10.1007/b128453

Springer is a part of Springer Science+Business Media
springer.com
© Springer-Verlag Berlin Heidelberg 2006

Typesetting: by the authors and SPI Publisher Services using a Springer LaTeX package
Cover design: *design & production* GmbH, Heidelberg

Printed on acid-free paper   SPIN: 11602668   VA41/3100/SPI   5 4 3 2 1 0

Preface

This volume is the third and last of a series devoted to the lecture notes of the Grenoble Summer School on "Open Quantum Systems" which took place at the Institut Fourier from June 16th to July 4th 2003. The contributions presented in this volume correspond to expanded versions of the lecture notes provided by the authors to the students of the Summer School. The corresponding lectures were scheduled in the last part of the School devoted to recent developments in the study of Open Quantum Systems.

Whereas the first two volumes were dedicated to a detailed exposition of the mathematical techniques and physical concepts relevant in the study of Open Systems with no *a priori* pre requisites, the contributions presented in this volume request from the reader some familiarity with these aspects. Indeed, the material presented here aims at leading the reader already acquainted with the basics in quantum statistical mechanics, spectral theory of linear operators, C^*-dynamical systems, and quantum stochastic differential equations to the front of the current research done on various aspects of Open Quantum Systems. Nevertheless, pedagogical efforts have been made by the various authors of these notes so that this volume should be essentially self-contained for a reader with minimal previous exposure to the themes listed above. In any case, the reader in need of complements can always turn to these first two volumes.

The topics covered in these lectures notes start with an introduction to non-equilibrium quantum statistical mechanics. The definitions of the physical concepts as well as the necessary mathematical framework suitable for their description are developed in a general setup. A simple non-trivial physically relevant example of independent electrons in a device connected to several reservoirs is treated in details in the second part of these notes in order to illustrate the notions of non-equilibrium steady states, entropy production and other thermodynamical notions introduced earlier.

The next contribution is devoted to the many aspects of the Fermi Golden Rule used within the Hamiltonian approach of Open Quantum Systems in order to derive a Markovian approximation of the dynamics. In particular, the weak coupling or van Hove limit in both a time-dependent and stationary setting are discussed in an abstract framework. These results are then applied to the case of small systems interacting with reservoirs, within different algebraic representations of the relevant models. The links between the Fermi Golden Rule and the Detailed Balance Condition as well as explicit formulas are also discussed in different physical situations.

The third text of this volume is concerned with the notion of decoherence, relevant, in particular, for a discussion of the measurement theory in Quantum Mechanics. The properties of the large time behavior of the dynamics reduced to a subsystem, which is not Markovian in general, are first reviewed. Then, the so-called isometric-sweeping decomposition of a dynamical semigroup is presented in an general setup and its links with decoherence phenomena are exposed. Applications to physical models such as spin systems or to the unravelling of the classical dynamics in certain regimes are then provided. The properties of dynamical semigroups on CCR algebras are discussed in details in the final section.

The following contribution is devoted to a systematic study of the long time behavior of quantum dynamical semigroups, as they arise in Markovian approximations. More precisely, the key notions for applications of stationary states, convergence towards equilibrium as well as transience and recurrence of such quantum Markov semigroups are developed in an abstract framework. In particular, conditions on unbounded operators defined in the sense of forms to generate a *bona fide* quantum dynamical semigroup are formulated, as well as general criteria insuring the existence of stationary states for a given quantum dynamical semigroup. The relations between return to equilibrium for a quantum dynamical semigroup and the properties of its generator are also discussed. All these concepts are then illustrated by applications to concrete physical models used in quantum optics.

The last notes of this volume provide a detailed account of the process of continual measurements in quantum optics, considered as an application of quantum stochastic calculus. The basics of this quantum stochastic calculus and the modelization of system-field interactions constructed on it are first explained. Then, indirect and continual measurement processes and the corresponding master equations are introduced and discussed. Physical interpretations of computations performed within this quantum stochastic modelization framework are spelled out for various specific processes in quantum optics.

As revealed by this outline, the treatment of the different physical models proposed in this volume makes use of several tools and approximations discussed from a mathematical point of view, both in the Hamiltonian and Markovian approach. At the same time, the different mathematical topics addressed here are illustrated by physically relevant applications in the theory of Open Quantum Systems. We believe the contact made between the practicians of the Markovian and Hamiltonian during the School itself and within the contributions of these volumes is useful and will prove to be even more fruitful for the future developments of the field.

Let us close this introduction by pointing out that some recent results in the theory of Open Quantum Systems are not discussed in these notes. These include notably the descriptions of return to equilibrium by means of renormalization analysis and scattering techniques. These demanding approaches were not addressed in the Grenoble Summer School, because a reasonably complete treatment would simply have required too much time.

We hope the reader will benefit from the pedagogical efforts provided by all authors of these notes in order to introduce the concepts and problems, as well as recent developments in the theory of Open Quantum Systems.

Lyon, Grenoble, Toulon,   *Stéphane Attal*
September 2005   *Alain Joye*
                                                        Claude-Alain Pillet

Contents

List of Contributors

Walter Aschbacher
Zentrum Mathematik M5
Technische Universität München
D-85747 Garching, Germany
e-mail: hbar@ma.tum.de

Alberto Barchielli
Politecnico di Milano
Dipartimento di Matematica
Piazza Leonardo da Vinci 32
20133 Milano, Italy
e-mail: Alberto.Barchielli@polimi.it

Philippe Blanchard
Physics Faculty and BiBoS
University of Bielefeld
Universitätsstrasse 25
33615 Bielefeld, Germany
e-mail: blanchard@
physik.uni-bielefeld.de

Jan Dereziński
Department of Mathematical
Methods in Physics
Warsaw University, Hoża 74
00-682, Warsaw, Poland
e-mail: Jan.Derezinski@fuw.edu.pl

Franco Fagnola
Politecnico di Milano
Dipartmento di Matematica
"F. Brioschi"
Piazza Leonardo da Vinci 32
20133 Milano, Italy
e-mail: franco.fagnola@polimi.it

Rafał Früboes
Department of Mathematical
Methods in Physics
Warsaw University, Hoża 74
00-682, Warsaw, Poland
e-mail: fruboes@fuw.edu.pl

Vojkan Jakšić
Department of Mathematics and
Statistics
McGill University
805 Sherbrooke Street West
Montreal, QC, H3A 2K6, Canada
e-mail: jaksic@math.mcgill.ca

Robert Olkiewicz
Institute of Theoretical Physics
University of Wrocław
pl. M. Borna 9
50-204 Wrocław, Poland
e-mail: rolek@ift.uni.wroc.pl

Yan Pautrat
Laboratoire de Mathématiques
Université Paris-Sud
91405 Orsay cedex, France
e-mail: Yan.Pautrat@math.u-psud.fr

Claude-Alain Pillet
CPT-CNRS, UMR 6207
Université du Sud Toulon-Var
B.P. 20132
83957 La Garde Cedex, France
e-mail: pillet@univ-tln.fr

Rolando Rebolledo
Facultad de Matemáticas
Universidad Católica de Chile
Casilla 306 Santiago 22, Chile
e-mail: rrebolle@uc.cl

Topics in Non-Equilibrium Quantum Statistical Mechanics

Walter Aschbacher[1], Vojkan Jakšić[2], Yan Pautrat[3], and Claude-Alain Pillet[4]

[1] Zentrum Mathematik M5, Technische Universität München,
D-85747 Garching, Germany
e-mail: hbar@ma.tum.de

[2] Department of Mathematics and Statistics, McGill University,
805 Sherbrooke Street West, Montreal, QC, H3A 2K6, Canada
e-mail: jaksic@math.mcgill.co

[3] Laboratoire de Mathématiques, Université Paris-Sud,
91405 Orsay cedex, France
e-mail: yan.pautrat@math.u-psud.fr

[4] CPT-CNRS, UMR 6207, Université du Sud,
Toulon-Var, B.P. 20132, 83957 La Garde Cedex, France
e-mail: pillet@univ-tln.fr

1 Introduction

These lecture notes are an expanded version of the lectures given by the second and the fourth author in the summer school "Open Quantum Systems" held in Grenoble, June 16–July 4, 2003. We are grateful to Stéphane Attal and Alain Joye for their hospitality and invitation to speak.

The lecture notes have their root in the recent review article [JP4] and our goal has been to extend and complement certain topics covered in [JP4]. In particular, we will discuss the scattering theory of non-equilibrium steady states (NESS) (this topic has been only quickly reviewed in [JP4]). On the other hand, we will not discuss the spectral theory of NESS which has been covered in detail in [JP4]. Although the lecture notes are self-contained, the reader would benefit from reading them in parallel with [JP4].

Concerning preliminaries, we will assume that the reader is familiar with the material covered in the lecture notes [At, Jo, Pi]. On occasion, we will mention or use some material covered in the lectures [D1, Ja].

As in [JP4], we will work in the mathematical framework of algebraic quantum statistical mechanics. The basic notions of this formalism are reviewed in Section 3. In Section 4 we introduce open quantum systems and describe their basic properties. The linear response theory (this topic has not been discussed in [JP4]) is described in Subsection 4.4. The linear response theory of open quantum systems (Kubo formulas, Onsager relations, Central Limit Theorem) has been studied in the recent papers [FMU, FMSU, AJPP, JPR2].

The second part of the lecture notes (Sections 6–8) is devoted to an example. The model we will discuss is the simplest non-trivial example of the Electronic Black Box Model studied in [AJPP] and we will refer to it as the *Simple Electronic Black Box Model* (SEBB). The SEBB model is to a large extent exactly solvable— its NESS and entropy production can be exactly computed and Kubo formulas can be verified by an explicit computation. For reasons of space, however, we will not discuss two important topics covered in [AJPP]—the stability theory (which is essentially based on [AM, BM]) and the proof of the Central Limit Theorem. The interested reader may complement Sections 6–8 with the original paper [AJPP] and the recent lecture notes [JKP].

Section 5, in which we discuss statistical mechanics of a free Fermi gas, is the bridge between the two parts of the lecture notes.

Acknowledgment. The research of V.J. was partly supported by NSERC. Part of this work was done while Y.P. was a CRM-ISM postdoc at McGill University and Centre de Recherches Mathématiques in Montreal.

2 Conceptual Framework

The concept of reference state will play an important role in our discussion of non-equilibrium statistical mechanics. To clarify this notion, let us consider first a classical dynamical system with finitely many degrees of freedom and compact phase space $X \subset \mathbb{R}^n$. The normalized Lebesgue measure dx on X provides a physically natural statistics on the phase space in the sense that initial configurations sampled according to it can be considered typical (see [Ru4]). Note that this has nothing to do with the fact that dx is invariant under the flow of the system—any measure of the form $\rho(x)dx$ with a strictly positive density ρ would serve the same purpose. The situation is completely different if the system has infinitely many degrees of freedom. In this case, there is no natural replacement for the Lebesgue dx. In fact, a measure on an infinite-dimensional phase space physically describes a thermodynamic state of the system. Suppose for example that the system is Hamiltonian and is in thermal equilibrium at inverse temperature β and chemical potential μ. The statistics of such a system is described by the Gibbs measure (grand canonical ensemble). Since two Gibbs measures with different values of the intensive thermodynamic parameters β, μ are mutually singular, initial points sampled according to one of them will be atypical relative to the other. In conclusion, if a system has infinitely many degrees of freedom, we need to specify its initial thermodynamic state by choosing an appropriate reference measure. As in the finite-dimensional case, this measure may not be invariant under the flow. It also may not be uniquely determined by the physical situation we wish to describe.

The situation in quantum mechanics is very similar. The Schrödinger representation of a system with finitely many degrees of freedom is (essentially) uniquely determined and the natural statistics is provided by any strictly positive density matrix on the Hilbert space of the system. For systems with infinitely many degrees of freedom there is no such natural choice. The consequences of this fact are however more drastic than in the classical case. There is no natural choice of a Hilbert space in which the system can be represented. To induce a representation, we must specify the thermodynamic state of the system by choosing an appropriate reference state. The algebraic formulation of quantum statistical mechanics provides a mathematical framework to study such infinite system in a representation independent way.

One may object that no real physical system has an infinite number of degrees of freedom and that, therefore, a unique natural reference state always exists. There are however serious methodological reasons to consider this mathematical idealization. Already in equilibrium statistical mechanics the fundamental phenomena of phase transition can only be characterized in a mathematically precise way within such an idealization: A quantum system with finitely many degrees of freedom has a unique thermal equilibrium state. Out of equilibrium, relaxation towards a stationary state and emergence of steady currents can not be expected from the quasi-periodic time evolution of a finite system.

In classical non-equilibrium statistical mechanics there exists an alternative approach to this idealization. A system forced by a non-Hamiltonian or time-dependent force can be driven towards a non-equilibrium steady state, provided the energy supplied by the external source is removed by some thermostat. This *micro-canonical* point of view has a number of advantages over the *canonical*, infinite system idealization. A dynamical system with a relatively small number of degrees of freedom can easily be explored on a computer (numerical integration, iteration of Poincaré sections, ...). A large body of "experimental facts" is currently available from the results of such investigations (see [EM, Do] for an introduction to the techniques and a lucid exposition of the results). From a more theoretical perspective, the full machinery of finite-dimensional dynamical system theory becomes available in the micro-canonical approach. The *Chaotic Hypothesis* introduced in [CG1, CG2] is an attempt to exploit this fact. It justifies phenomenological thermodynamics (Onsager relations, linear response theory, fluctuation-dissipation formulas,...) and has lead to more unexpected results like the Gallavotti-Cohen Fluctuation Theorem. The major drawback of the micro-canonical point of view is the non-Hamiltonian nature of the dynamics, which makes it inappropriate to quantum-mechanical treatment.

The two approaches described above are not completely unrelated. For example, we shall see that the signature of a non-equilibrium steady state in quantum mechanics is its singularity with respect to the reference state, a fact which is well understood in the classical, micro-canonical approach (see Chapter 10 of [EM]). More speculatively, one can expect a general *equivalence principle* for dynamical (micro-canonical and canonical) ensembles (see [Ru5]). The results in this direction are quite scarce and much work remains to be done.

3 Mathematical Framework

In this section we describe the mathematical formalism of algebraic quantum statistical mechanics. Our presentation follows [JP4] and is suited for applications to non-equilibrium statistical mechanics. Most of the material in this section is well known and the proofs can be found, for example, in [BR1, BR2, DJP, Ha, OP, Ta]. The proofs of the results described in Subsection 3.3 are given in Appendix 9.1.

3.1 Basic Concepts

The starting point of our discussion is a pair $(\mathcal{O}, \tau)$, where $\mathcal{O}$ is a C^*-algebra with a unit I and τ is a C^*-dynamics (a strongly continuous group $\mathbb{R} \ni t \mapsto \tau^t$ of $*$-automorphisms of $\mathcal{O}$). The elements of $\mathcal{O}$ describe physical observables of the quantum system under consideration and the group τ specifies their time evolution. The pair $(\mathcal{O}, \tau)$ is sometimes called a C^*-dynamical system.

In the sequel, by the strong topology on $\mathcal{O}$ we will always mean the usual norm topology of $\mathcal{O}$ as Banach space. The C^*-algebra of all bounded operators on a Hilbert space $\mathcal{H}$ is denoted by $\mathcal{B}(\mathcal{H})$.

A state ω on the C^*-algebra $\mathcal{O}$ is a normalized ($\omega(I) = 1$), positive ($\omega(A^*A) \geq 0$), linear functional on $\mathcal{O}$. It specifies a possible *physical state* of the quantum mechanical system. If the system is in the state ω at time zero, the quantum mechanical expectation value of the observable A at time t is given by $\omega(\tau^t(A))$. Thus, states evolve in the Schrödinger picture according to $\omega_t = \omega \circ \tau^t$. The set $E(\mathcal{O})$ of all states on $\mathcal{O}$ is a convex, weak-$*$ compact subset of the Banach space dual $\mathcal{O}^*$ of $\mathcal{O}$.

A linear functional $\eta \in \mathcal{O}^*$ is called τ-invariant if $\eta \circ \tau^t = \eta$ for all t. The set of all τ-invariant states is denoted by $E(\mathcal{O}, \tau)$. This set is always non-empty. A state $\omega \in E(\mathcal{O}, \tau)$ is called ergodic if

$$\lim_{T \to \infty} \frac{1}{2T} \int_{-T}^{T} \omega(B^* \tau^t(A) B) \, \mathrm{d}t = \omega(A)\omega(B^*B),$$

and mixing if

$$\lim_{|t| \to \infty} \omega(B^* \tau^t(A) B) = \omega(A)\omega(B^*B),$$

for all $A, B \in \mathcal{O}$.

Let $(\mathcal{H}_\eta, \pi_\eta, \Omega_\eta)$ be the GNS representation associated to a positive linear functional $\eta \in \mathcal{O}^*$. The enveloping von Neumann algebra of $\mathcal{O}$ associated to η is $\mathfrak{M}_\eta \equiv \pi_\eta(\mathcal{O})'' \subset \mathcal{B}(\mathcal{H}_\eta)$. A linear functional $\mu \in \mathcal{O}^*$ is normal relative to η or η-normal, denoted $\mu \ll \eta$, if there exists a trace class operator ρ_μ on $\mathcal{H}_\eta$ such that $\mu(\cdot) = \mathrm{Tr}(\rho_\mu \pi_\eta(\cdot))$. Any η-normal linear functional μ has a unique normal extension to $\mathfrak{M}_\eta$. We denote by $\mathcal{N}_\eta$ the set of all η-normal states. $\mu \ll \eta$ iff $\mathcal{N}_\mu \subset \mathcal{N}_\eta$.

A state ω is ergodic iff, for all $\mu \in \mathcal{N}_\omega$ and $A \in \mathcal{O}$,

$$\lim_{T \to \infty} \frac{1}{2T} \int_{-T}^{T} \mu(\tau^t(A)) \, \mathrm{d}t = \omega(A).$$

For this reason ergodicity is sometimes called return to equilibrium in mean; see [Ro1, Ro2]. Similarly, ω is mixing (or returns to equilibrium) iff

$$\lim_{|t|\to\infty} \mu(\tau^t(A)) = \omega(A),$$

for all $\mu \in \mathcal{N}_\omega$ and $A \in \mathcal{O}$.

Let η and μ be two positive linear functionals in $\mathcal{O}^*$, and suppose that $\eta \geq \phi \geq 0$ for some μ-normal ϕ implies $\phi = 0$. We then say that η and μ are mutually singular (or orthogonal), and write $\eta \perp \mu$. An equivalent (more symmetric) definition is: $\eta \perp \mu$ iff $\eta \geq \phi \geq 0$ and $\mu \geq \phi \geq 0$ imply $\phi = 0$.

Two positive linear functionals η and μ in $\mathcal{O}^*$ are called disjoint if $\mathcal{N}_\eta \cap \mathcal{N}_\mu = \emptyset$. If η and μ are disjoint, then $\eta \perp \mu$. The converse does not hold— it is possible that η and μ are mutually singular but not disjoint.

To elucidate further these important notions, we recall the following well-known results; see Lemmas 4.1.19 and 4.2.8 in [BR1].

Proposition 3.1. *Let $\mu_1, \mu_2 \in \mathcal{O}^*$ be two positive linear functionals and $\mu = \mu_1 + \mu_2$. Then the following statements are equivalent:*

(i) $\mu_1 \perp \mu_2$.
(ii) There exists a projection P in $\pi_\mu(\mathcal{O})'$ such that

$$\mu_1(A) = (P\Omega_\mu, \pi_\mu(A)\Omega_\mu), \qquad \mu_2(A) = ((I-P)\Omega_\mu, \pi_\mu(A)\Omega_\mu).$$

(iii) The GNS representation $(\mathcal{H}_\mu, \pi_\mu, \Omega_\mu)$ is a direct sum of the two GNS representations $(\mathcal{H}_{\mu_1}, \pi_{\mu_1}, \Omega_{\mu_1})$ and $(\mathcal{H}_{\mu_2}, \pi_{\mu_2}, \Omega_{\mu_2})$, i.e.,

$$\mathcal{H}_\mu = \mathcal{H}_{\mu_1} \oplus \mathcal{H}_{\mu_2}, \qquad \pi_\mu = \pi_{\mu_1} \oplus \pi_{\mu_2}, \qquad \Omega_\mu = \Omega_{\mu_1} \oplus \Omega_{\mu_2}.$$

Proposition 3.2. *Let $\mu_1, \mu_2 \in \mathcal{O}^*$ be two positive linear functionals and $\mu = \mu_1 + \mu_2$. Then the following statements are equivalent:*

(i) μ_1 and μ_2 are disjoint.
(ii) There exists a projection P in $\pi_\mu(\mathcal{O})' \cap \pi_\mu(\mathcal{O})''$ such that

$$\mu_1(A) = (P\Omega_\mu, \pi_\mu(A)\Omega_\mu), \qquad \mu_2(A) = ((I-P)\Omega_\mu, \pi_\mu(A)\Omega_\mu).$$

Let $\eta, \mu \in \mathcal{O}^*$ be two positive linear functionals. The functional η has a unique decomposition $\eta = \eta_n + \eta_s$, where η_n, η_s are positive, $\eta_n \ll \mu$, and $\eta_s \perp \mu$. The uniqueness of the decomposition implies that if η is τ-invariant, then so are η_n and η_s.

To elucidate the nature of this decomposition we need to recall the notions of the universal representation and the universal enveloping von Neumann algebra of $\mathcal{O}$; see Section III.2 in [Ta] and Section 10.1 in [KR].

Set

$$\mathcal{H}_{\mathrm{un}} \equiv \bigoplus_{\omega \in E(\mathcal{O})} \mathcal{H}_\omega, \qquad \pi_{\mathrm{un}} \equiv \bigoplus_{\omega \in E(\mathcal{O})} \pi_\omega, \qquad \mathfrak{M}_{\mathrm{un}} \equiv \pi_{\mathrm{un}}(\mathcal{O})''.$$

$(\mathcal{H}_{\mathrm{un}}, \pi_{\mathrm{un}})$ is a faithful representation. It is called *the universal representation* of $\mathcal{O}$. $\mathfrak{M}_{\mathrm{un}} \subset \mathcal{B}(\mathcal{H}_{\mathrm{un}})$ is its universal enveloping von Neumann algebra. For any $\omega \in E(\mathcal{O})$ the map

$$\pi_{\mathrm{un}}(\mathcal{O}) \to \pi_\omega(\mathcal{O})$$
$$\pi_{\mathrm{un}}(A) \mapsto \pi_\omega(A),$$

extends to a surjective $*$-morphism $\tilde{\pi}_\omega : \mathfrak{M}_{\mathrm{un}} \to \mathfrak{M}_\omega$. It follows that ω uniquely extends to a normal state $\tilde{\omega}(\cdot) \equiv (\Omega_\omega, \tilde{\pi}_\omega(\cdot)\Omega_\omega)$ on $\mathfrak{M}_{\mathrm{un}}$. Moreover, one easily shows that

$$\mathrm{Ker}\, \tilde{\pi}_\omega = \{A \in \mathfrak{M}_{\mathrm{un}} \mid \tilde{\nu}(A) = 0 \text{ for any } \nu \in \mathcal{N}_\omega\}. \tag{1}$$

Since $\mathrm{Ker}\, \tilde{\pi}_\omega$ is a σ-weakly closed two sided ideal in $\mathfrak{M}_{\mathrm{un}}$, there exists an orthogonal projection $p_\omega \in \mathfrak{M}_{\mathrm{un}} \cap \mathfrak{M}'_{\mathrm{un}}$ such that $\mathrm{Ker}\, \tilde{\pi}_\omega = p_\omega \mathfrak{M}_{\mathrm{un}}$. The orthogonal projection $z_\omega \equiv I - p_\omega \in \mathfrak{M}_{\mathrm{un}} \cap \mathfrak{M}'_{\mathrm{un}}$ is called the *support projection* of the state ω. The restriction of $\tilde{\pi}_\omega$ to $z_\omega \mathfrak{M}_{\mathrm{un}}$ is an isomorphism between the von Neumann algebras $z_\omega \mathfrak{M}_{\mathrm{un}}$ and $\mathfrak{M}_\omega$. We shall denote by ϕ_ω the inverse isomorphism.

Let now $\eta, \mu \in \mathcal{O}^*$ be two positive linear functionals. By scaling, without loss of generality we may assume that they are states. Since $\tilde{\eta}$ is a normal state on $\mathfrak{M}_{\mathrm{un}}$ it follows that $\tilde{\eta} \circ \phi_\mu$ is a normal state on $\mathfrak{M}_\mu$ and hence that $\eta_n \equiv \tilde{\eta} \circ \phi_\mu \circ \pi_\mu$ defines a μ-normal positive linear functional on $\mathcal{O}$. Moreover, from the relation $\phi_\mu \circ \pi_\mu(A) = z_\mu \pi_{\mathrm{un}}(A)$ it follows that

$$\eta_n(A) = (\Omega_\eta, \tilde{\pi}_\eta(z_\mu)\pi_\eta(A)\Omega_\eta).$$

Setting

$$\eta_s(A) \equiv (\Omega_\eta, \tilde{\pi}_\eta(p_\mu)\pi_\eta(A)\Omega_\eta),$$

we obtain a decomposition $\eta = \eta_n + \eta_s$. To show that $\eta_s \perp \mu$ let ω be a μ-normal positive linear functional on $\mathcal{O}$ such that $\eta_s \geq \omega$. By the unicity of the normal extension $\tilde{\eta}_s$ one has $\tilde{\eta}_s(A) = \tilde{\eta}(p_\mu A)$ for $A \in \mathfrak{M}_{\mathrm{un}}$. Since $\pi_{\mathrm{un}}(\mathcal{O})$ is σ-strongly dense in $\mathfrak{M}_{\mathrm{un}}$ it follows from the inequality $\tilde{\eta}_s \circ \pi_{\mathrm{un}} \geq \tilde{\omega} \circ \pi_{\mathrm{un}}$ that $\tilde{\eta}(p_\mu A) \geq \tilde{\omega}(A)$ for any positive $A \in \mathfrak{M}_{\mathrm{un}}$. Since ω is μ-normal, it further follows from Equ. (1) that $\omega(A) = \tilde{\omega}(\pi_{\mathrm{un}}(A)) = \tilde{\omega}(z_\mu \pi_{\mathrm{un}}(A)) \leq \tilde{\eta}(p_\mu z_\mu \pi_{\mathrm{un}}(A)) = 0$ for any positive $A \in \mathcal{O}$, i.e., $\omega = 0$. Since $\tilde{\pi}_\eta$ is surjective, one has $\tilde{\pi}_\eta(z_\mu) \in \mathfrak{M}_\eta \cap \mathfrak{M}'_\eta$ and, by Proposition 3.2, the functionals η_n and η_s are disjoint.

Two states ω_1 and ω_2 are called *quasi-equivalent* if $\mathcal{N}_{\omega_1} = \mathcal{N}_{\omega_2}$. They are called unitarily equivalent if their GNS representations $(\mathcal{H}_{\omega_j}, \pi_{\omega_j}, \Omega_{\omega_j})$ are unitarily equivalent, namely if there is a unitary $U : \mathcal{H}_{\omega_1} \to \mathcal{H}_{\omega_2}$ such that $U\Omega_{\omega_1} = \Omega_{\omega_2}$ and $U\pi_{\omega_1}(\cdot) = \pi_{\omega_2}(\cdot)U$. Clearly, unitarily equivalent states are quasi-equivalent.

If ω is τ-invariant, then there exists a unique self-adjoint operator L on $\mathcal{H}_\omega$ such that

$$L\Omega_\omega = 0, \qquad \pi_\omega(\tau^t(A)) = e^{itL}\pi_\omega(A)e^{-itL}.$$

We will call L the ω-Liouvillean of τ.

The state ω is called factor state (or primary state) if its enveloping von Neumann algebra $\mathfrak{M}_\omega$ is a factor, namely if $\mathfrak{M}_\omega \cap \mathfrak{M}'_\omega = \mathbb{C}I$. By Proposition 3.2 ω is a factor state iff it cannot be written as a nontrivial convex combination of disjoint states. This implies that if ω is a factor state and μ is a positive linear functional in $\mathcal{O}^*$, then either $\omega \ll \mu$ or $\omega \perp \mu$.

Two factor states ω_1 and ω_2 are either quasi-equivalent or disjoint. They are quasi-equivalent iff $(\omega_1 + \omega_2)/2$ is also a factor state (this follows from Theorem 4.3.19 in [BR1]).

The state ω is called modular if there exists a C^*-dynamics σ_ω on $\mathcal{O}$ such that ω is a $(\sigma_\omega, -1)$-KMS state. If ω is modular, then Ω_ω is a separating vector for $\mathfrak{M}_\omega$, and we denote by Δ_ω, J and $\mathcal{P}$ the modular operator, the modular conjugation and the natural cone associated to Ω_ω. To any C^*-dynamics τ on $\mathcal{O}$ one can associate a unique self-adjoint operator L on $\mathcal{H}_\omega$ such that for all t

$$\pi_\omega(\tau^t(A)) = e^{itL}\pi_\omega(A)e^{-itL}, \qquad e^{-itL}\mathcal{P} = \mathcal{P}.$$

The operator L is called standard Liouvillean of τ associated to ω. If ω is τ-invariant, then $L\Omega_\omega = 0$, and the standard Liouvillean is equal to the ω-Liouvillean of τ.

The importance of the standard Liouvillean L stems from the fact that if a state η is ω-normal and τ-invariant, then there exists a unique vector $\Omega_\eta \in \operatorname{Ker} L \cap \mathcal{P}$ such that $\eta(\cdot) = (\Omega_\eta, \pi_\omega(\cdot)\Omega_\eta)$. This fact has two important consequences. On one hand, if η is ω-normal and τ-invariant, then some ergodic properties of the quantum dynamical system $(\mathcal{O}, \tau, \eta)$ can be described in terms of the spectral properties of L; see [JP2, Pi]. On the other hand, if $\operatorname{Ker} L = \{0\}$, then the C^*-dynamics τ has no ω-normal invariant states. The papers [BFS, DJ2, FM1, FM2, FMS, JP1, JP2, JP3, Me1, Me2, Og] are centered around this set of ideas.

In quantum statistical mechanics one also encounters L^p-Liouvilleans, for $p \in [1, \infty]$ (the standard Liouvillean is equal to the L^2-Liouvillean). The L^p-Liouvilleans are closely related to the Araki-Masuda L^p-spaces [ArM]. L^1 and L^∞-Liouvilleans have played a central role in the spectral theory of NESS developed in [JP5]. The use of other L^p-Liouvilleans is more recent (see [JPR2]) and they will not be discussed in this lecture.

3.2 Non-Equilibrium Steady States (NESS) and Entropy Production

The central notions of non-equilibrium statistical mechanics are non-equilibrium steady states (NESS) and entropy production. Our definition of NESS follows closely the idea of Ruelle that a "natural" steady state should provide the statistics, over large time intervals $[0, t]$, of initial configurations of the system which are typical with respect to the reference state [Ru3]. The definition of entropy production is more problematic since there is no physically satisfactory definition of the entropy itself out of equilibrium; see [Ga1, Ru2, Ru5, Ru7] for a discussion. Our definition of entropy production is motivated by classical dynamics where the rate of change of thermodynamic (Clausius) entropy can sometimes be related to the phase space contraction rate [Ga2, RC]. The latter is related to the Gibbs entropy (as shown for example in [Ru3]) which is nothing else but the relative entropy with respect to the natural reference state; see [JPR1] for a detailed discussion in a more general context. Thus, it seems reasonable to define the entropy production as the rate of change of the relative entropy with respect to the reference state ω.

Let $(\mathcal{O}, \tau)$ be a C^*-dynamical system and ω a given reference state. The NESS associated to ω and τ are the weak-$*$ limit points of the time averages along the trajectory $\omega \circ \tau^t$. In other words, if

$$\langle \omega \rangle_t \equiv \frac{1}{t} \int_0^t \omega \circ \tau^s \, ds,$$

then ω_+ is a NESS associated to ω and τ if there exists a net $t_\alpha \to \infty$ such that $\langle \omega \rangle_{t_\alpha}(A) \to \omega_+(A)$ for all $A \in \mathcal{O}$. We denote by $\Sigma_+(\omega, \tau)$ the set of such NESS. One easily sees that $\Sigma_+(\omega, \tau) \subset E(\mathcal{O}, \tau)$. Moreover, since $E(\mathcal{O})$ is weak-$*$ compact, $\Sigma_+(\omega, \tau)$ is non-empty.

As already mentioned, our definition of entropy production is based on the concept of relative entropy. The relative entropy of two density matrices ρ and ω is defined, by analogy with the relative entropy of two measures, by the formula

$$\mathrm{Ent}(\rho|\omega) \equiv \mathrm{Tr}(\rho(\log \omega - \log \rho)). \tag{2}$$

It is easy to show that $\mathrm{Ent}(\rho|\omega) \leq 0$. Let φ_i an orthonormal eigenbasis of ρ and by p_i the corresponding eigenvalues. Then $p_i \in [0, 1]$ and $\sum_i p_i = 1$. Let $q_i \equiv (\varphi_i, \omega \varphi_i)$. Clearly, $q_i \in [0, 1]$ and $\sum_i q_i = \mathrm{Tr}\,\omega = 1$. Applying Jensen's inequality twice we derive

$$\mathrm{Ent}(\rho|\omega) = \sum_i p_i((\varphi_i, \log \omega \, \varphi_i) - \log p_i)$$

$$\leq \sum_i p_i(\log q_i - \log p_i) \leq \log \sum_i q_i = 0.$$

Hence $\mathrm{Ent}(\rho|\omega) \leq 0$. It is also not difficult to show that $\mathrm{Ent}(\rho|\omega) = 0$ iff $\rho = \omega$; see [OP]. Using the concept of relative modular operators, Araki has extended the notion of relative entropy to two arbitrary states on a C^*-algebra [Ar1, Ar2]. We refer the reader to [Ar1, Ar2, DJP, OP] for the definition of the Araki relative entropy and its basic properties. Of particular interest to us is that $\mathrm{Ent}(\rho|\omega) \leq 0$ still holds, with equality if and only if $\rho = \omega$.

In these lecture notes we will define entropy production only in a perturbative context (for a more general approach see [JPR2]). Denote by δ the generator of the group τ i.e., $\tau^t = e^{t\delta}$, and assume that the reference state ω is invariant under τ. For $V = V^* \in \mathcal{O}$ we set $\delta_V \equiv \delta + i[V, \cdot]$ and denote by $\tau_V^t \equiv e^{t\delta_V}$ the corresponding perturbed C^*-dynamics (such perturbations are often called *local*, see [Pi]). Starting with a state $\rho \in \mathcal{N}_\omega$, the entropy is pumped out of the system by the perturbation V at a mean rate

$$-\frac{1}{t}\left(\mathrm{Ent}(\rho\circ\tau_V^t|\omega)-\mathrm{Ent}(\rho|\omega)\right).$$

Suppose that ω is a modular state for a C^*-dynamics σ_ω^t and denote by δ_ω the generator of σ_ω. If $V\in\mathrm{Dom}\,(\delta_\omega)$, then one can prove the following entropy balance equation

$$\mathrm{Ent}(\rho\circ\tau_V^t|\omega)=\mathrm{Ent}(\rho|\omega)-\int_0^t\rho(\tau_V^s(\sigma_V))\,\mathrm{d}s, \tag{3}$$

where

$$\sigma_V\equiv\delta_\omega(V),$$

is the entropy production observable (see [JP6, JP7]). In quantum mechanics σ_V plays the role of the phase space contraction rate of classical dynamical systems (see [JPR1]). We define the entropy production rate of a NESS

$$\rho_+=\mathrm{w}^*-\lim_\alpha\frac{1}{t_\alpha}\int_0^{t_\alpha}\rho\circ\tau_V^s\,\mathrm{d}s\in\Sigma_+(\rho,\tau_V),$$

by

$$\mathrm{Ep}(\rho_+)\equiv-\lim_\alpha\frac{1}{t_\alpha}\left(\mathrm{Ent}(\rho\circ\tau_V^{t_\alpha}|\omega)-\mathrm{Ent}(\rho|\omega)\right)=\rho_+(\sigma_V).$$

Since $\mathrm{Ent}(\rho\circ\tau_V^t|\omega)\le 0$, an immediate consequence of this equation is that, for $\rho_+\in\Sigma_+(\rho,\tau_V)$,

$$\mathrm{Ep}(\rho_+)\ge 0. \tag{4}$$

We emphasize that the observable σ_V depends both on the reference state ω and on the perturbation V. As we shall see in the next section, σ_V is related to the thermodynamic fluxes across the system produced by the perturbation V and the positivity of entropy production is the statement of the second law of thermodynamics.

3.3 Structural Properties

In this subsection we shall discuss structural properties of NESS and entropy production following [JP4]. The proofs are given in Appendix 9.1.

First, we will discuss the dependence of $\Sigma_+(\omega,\tau_V)$ on the reference state ω. On physical grounds, one may expect that if ω is sufficiently regular and η is ω-normal, then $\Sigma_+(\eta,\tau_V)=\Sigma_+(\omega,\tau_V)$.

Theorem 3.1. *Assume that ω is a factor state on the C^*-algebra $\mathcal{O}$ and that, for all $\eta\in\mathcal{N}_\omega$ and $A,B\in\mathcal{O}$,*

$$\lim_{T\to\infty}\frac{1}{T}\int_0^T\eta([\tau_V^t(A),B])\,\mathrm{d}t=0,$$

holds (weak asymptotic abelianness in mean). Then $\Sigma_+(\eta,\tau_V)=\Sigma_+(\omega,\tau_V)$ for all $\eta\in\mathcal{N}_\omega$.

The second structural property we would like to mention is:

Theorem 3.2. *Let $\eta \in \mathcal{O}^*$ be ω-normal and τ_V-invariant. Then $\eta(\sigma_V) = 0$. In particular, the entropy production of the normal part of any NESS is equal to zero.*

If $\mathrm{Ent}(\eta|\omega) > -\infty$, then Theorem 3.2 is an immediate consequence of the entropy balance equation (3). The case $\mathrm{Ent}(\eta|\omega) = -\infty$ has been treated in [JP7] and the proof requires the full machinery of Araki's perturbation theory. We will not reproduce it here.

If ω_+ is a factor state, then either $\omega_+ \ll \omega$ or $\omega_+ \perp \omega$. Hence, Theorem 3.2 yields:

Corollary 3.1. *If ω_+ is a factor state and $\mathrm{Ep}(\omega_+) > 0$, then $\omega_+ \perp \omega$. If ω is also a factor state, then ω_+ and ω are disjoint.*

Certain structural properties can be characterized in terms of the standard Liouvillean. Let L be the standard Liouvillean associated to τ and L_V the standard Liouvillean associated to τ_V. By the well-known Araki's perturbation formula, one has $L_V = L + V - JVJ$ (see [DJP, Pi]).

Theorem 3.3. *Assume that ω is modular.*

(i) *Under the assumptions of Theorem 3.1, if $\mathrm{Ker}\, L_V \neq \{0\}$, then it is one-dimensional and there exists a unique normal, τ_V-invariant state ω_V such that*

$$\Sigma_+(\omega, \tau_V) = \{\omega_V\}.$$

(ii) *If $\mathrm{Ker}\, L_V = \{0\}$, then any NESS in $\Sigma_+(\omega, \tau_V)$ is purely singular.*

(iii) *If $\mathrm{Ker}\, L_V$ contains a separating vector for $\mathfrak{M}_\omega$, then $\Sigma_+(\omega, \tau_V)$ contains a unique state ω_+ and this state is ω-normal.*

3.4 C^*-Scattering and NESS

Let $(\mathcal{O}, \tau)$ be a C^*-dynamical system and V a local perturbation. The abstract C^*-scattering approach to the study of NESS is based on the following assumption:

Assumption (S) The strong limit

$$\alpha_V^+ \equiv \mathrm{s} - \lim_{t \to \infty} \tau^{-t} \circ \tau_V^t,$$

exists.

The map α_V^+ is an isometric $*$-endomorphism of $\mathcal{O}$, and is often called Møller morphism. α_V^+ is one-to-one but it is generally not onto, namely

$$\mathcal{O}_+ \equiv \mathrm{Ran}\, \alpha_V^+ \neq \mathcal{O}.$$

Since $\alpha_V^+ \circ \tau_V^t = \tau^t \circ \alpha_V^+$, the pair $(\mathcal{O}_+, \tau)$ is a C^*-dynamical system and α_V^+ is an isomorphism between the dynamical systems $(\mathcal{O}, \tau_V)$ and $(\mathcal{O}_+, \tau)$.

If the reference state ω is τ-invariant, then $\omega_+ = \omega \circ \alpha_V^+$ is the unique NESS associated to ω and τ_V and

$$\text{w}^* - \lim_{t \to \infty} \omega \circ \tau_V^t = \omega_+.$$

Note in particular that if ω is a (τ, β)-KMS state, then ω_+ is a (τ_V, β)-KMS state.

The map α_V^+ is the algebraic analog of the wave operator in Hilbert space scattering theory. A simple and useful result in Hilbert space scattering theory is the Cook criterion for the existence of the wave operator. Its algebraic analog is:

Proposition 3.3.   *(i) Assume that there exists a dense subset $\mathcal{O}_0 \subset \mathcal{O}$ such that for all $A \in \mathcal{O}_0$,*

$$\int_0^\infty \|[V, \tau_V^t(A)]\| \, dt < \infty. \tag{5}$$

Then Assumption (S) holds.
(ii) Assume that there exists a dense subset $\mathcal{O}_1 \subset \mathcal{O}$ such that for all $A \in \mathcal{O}_1$,

$$\int_0^\infty \|[V, \tau^t(A)]\| \, dt < \infty. \tag{6}$$

Then $\mathcal{O}_+ = \mathcal{O}$ and α_V^+ is a $*$-automorphism of $\mathcal{O}$.

Proof. For all $A \in \mathcal{O}$ we have

$$\tau^{-t_2} \circ \tau_V^{t_2}(A) - \tau^{-t_1} \circ \tau_V^{t_1}(A) = i \int_{t_1}^{t_2} \tau^{-t}([V, \tau_V^t(A)]) \, dt,$$

$$\tau_V^{-t_2} \circ \tau^{t_2}(A) - \tau_V^{-t_1} \circ \tau^{t_1}(A) = -i \int_{t_1}^{t_2} \tau_V^{-t}([V, \tau^t(A)]) \, dt, \tag{7}$$

and so

$$\|\tau^{-t_2} \circ \tau_V^{t_2}(A) - \tau^{-t_1} \circ \tau_V^{t_1}(A)\| \leq \int_{t_1}^{t_2} \|[V, \tau_V^t(A)]\| \, dt,$$

$$\|\tau_V^{-t_2} \circ \tau^{t_2}(A) - \tau_V^{-t_1} \circ \tau^{t_1}(A)\| \leq \int_{t_1}^{t_2} \|[V, \tau^t(A)]\| \, dt. \tag{8}$$

To prove Part (i), note that (5) and the first estimate in (8) imply that for $A \in \mathcal{O}_0$ the norm limit

$$\alpha_V^+(A) \equiv \lim_{t \to \infty} \tau^{-t} \circ \tau_V^t(A),$$

exists. Since $\mathcal{O}_0$ is dense and $\tau^{-t} \circ \tau_V^t$ is isometric, the limit exists for all $A \in \mathcal{O}$, and α_V^+ is a $*$-morphism of $\mathcal{O}$. To prove Part (ii) note that the second estimate in (8) and (6) imply that the norm limit

$$\beta_V^+(A) \equiv \lim_{t\to\infty} \tau_V^{-t} \circ \tau^t(A),$$

also exists for all $A \in \mathcal{O}$. Since $\alpha_V^+ \circ \beta_V^+(A) = A$, α_V^+ is a $*$-automorphism of $\mathcal{O}$.
$\square$

Until the end of this subsection we will assume that the Assumption (S) holds and that ω is τ-invariant.

Let $\tilde{\omega} \equiv \omega \upharpoonright \mathcal{O}_+$ and let $(\mathcal{H}_{\tilde{\omega}}, \pi_{\tilde{\omega}}, \Omega_{\tilde{\omega}})$ be the GNS-representation of $\mathcal{O}_+$ associated to $\tilde{\omega}$. Obviously, if α_V^+ is an automorphism, then $\tilde{\omega} = \omega$. We denote by $(\mathcal{H}_{\omega_+}, \pi_{\omega_+}, \Omega_{\omega_+})$ the GNS representation of $\mathcal{O}$ associated to ω_+. Let $L_{\tilde{\omega}}$ and L_{ω_+} be the standard Liouvilleans associated, respectively, to $(\mathcal{O}_+, \tau, \tilde{\omega})$ and $(\mathcal{O}, \tau_V, \omega_+)$. Recall that $L_{\tilde{\omega}}$ is the unique self-adjoint operator on $\mathcal{H}_{\tilde{\omega}}$ such that for $A \in \mathcal{O}_+$,

$$L_{\tilde{\omega}}\Omega_{\tilde{\omega}} = 0, \qquad \pi_{\tilde{\omega}}(\tau^t(A)) = e^{itL_{\tilde{\omega}}}\pi_{\tilde{\omega}}(A)e^{-itL_{\tilde{\omega}}},$$

and similarly for L_{ω_+}.

Proposition 3.4. *The map*

$$U\pi_{\tilde{\omega}}(\alpha_V^+(A))\Omega_{\tilde{\omega}} = \pi_{\omega_+}(A)\Omega_{\omega_+},$$

extends to a unitary $U : \mathcal{H}_{\tilde{\omega}} \to \mathcal{H}_{\omega_+}$ *which intertwines* $L_{\tilde{\omega}}$ *and* L_{ω_+}, *i.e.*,

$$UL_{\tilde{\omega}} = L_{\omega_+}U.$$

Proof. Set $\pi'_{\tilde{\omega}}(A) \equiv \pi_{\tilde{\omega}}(\alpha_V^+(A))$ and note that $\pi'_{\tilde{\omega}}(\mathcal{O})\Omega_{\tilde{\omega}} = \pi_{\tilde{\omega}}(\mathcal{O}_+)\Omega_{\tilde{\omega}}$, so that $\Omega_{\tilde{\omega}}$ is cyclic for $\pi'_{\tilde{\omega}}(\mathcal{O})$. Since

$$\omega_+(A) = \omega(\alpha_V^+(A)) = \tilde{\omega}(\alpha_V^+(A)) = (\Omega_{\tilde{\omega}}, \pi_{\tilde{\omega}}(\alpha_V^+(A))\Omega_{\tilde{\omega}}) = (\Omega_{\tilde{\omega}}, \pi'_{\tilde{\omega}}(A)\Omega_{\tilde{\omega}}),$$

$(\mathcal{H}_{\tilde{\omega}}, \pi'_{\tilde{\omega}}, \Omega_{\tilde{\omega}})$ is also a GNS representation of $\mathcal{O}$ associated to ω_+. Since GNS representations associated to the same state are unitarily equivalent, there is a unitary $U : \mathcal{H}_{\tilde{\omega}} \to \mathcal{H}_{\omega_+}$ such that $U\Omega_{\tilde{\omega}} = \Omega_{\omega_+}$ and

$$U\pi'_{\tilde{\omega}}(A) = \pi_{\omega_+}(A)U.$$

Finally, the identities

$$Ue^{itL_{\tilde{\omega}}}\pi'_{\tilde{\omega}}(A)\Omega_{\tilde{\omega}} = U\pi_{\tilde{\omega}}(\tau^t(\alpha_V^+(A)))\Omega_{\tilde{\omega}} = U\pi_{\tilde{\omega}}(\alpha_V^+(\tau_V^t(A)))\Omega_{\tilde{\omega}}$$

$$= \pi_{\omega_+}(\tau_V^t(A))\Omega_{\omega_+} = e^{itL_{\omega_+}}\pi_{\omega_+}(A)\Omega_{\omega_+}$$

$$= e^{itL_{\omega_+}}U\pi'_{\tilde{\omega}}(A)\Omega_{\tilde{\omega}},$$

yield that U intertwines $L_{\tilde{\omega}}$ and $L_{\omega+}$. $\square$

We finish this subsection with:

Proposition 3.5. *(i) Assume that $\tilde{\omega} \in E(\mathcal{O}_+, \tau)$ is τ-ergodic. Then*

$$\Sigma_+(\eta, \tau_V) = \{\omega_+\},$$

for all $\eta \in \mathcal{N}_\omega$.
(ii) If $\tilde{\omega}$ is τ-mixing, then

$$\lim_{t\to\infty} \eta \circ \tau_V^t = \omega_+,$$

for all $\eta \in \mathcal{N}_\omega$.

Proof. We will prove the Part (i); the proof of the Part (ii) is similar. If $\eta \in \mathcal{N}_\omega$, then $\eta \upharpoonright \mathcal{O}_+ \in \mathcal{N}_{\tilde{\omega}}$, and the ergodicity of $\tilde{\omega}$ yields

$$\lim_{T\to\infty} \frac{1}{T} \int_0^T \eta(\tau^t(\alpha_V^+(A))) \, dt = \tilde{\omega}(\alpha_V^+(A)) = \omega_+(A).$$

This fact, the estimate

$$\|\eta(\tau_V^t(A)) - \eta(\tau^t(\alpha_V^+(A)))\| \le \|\tau^{-t} \circ \tau_V^t(A) - \alpha_V^+(A)\|,$$

and Assumption (S) yield the statement. □

4 Open Quantum Systems

4.1 Definition

Open quantum systems are the basic paradigms of non-equilibrium quantum statistical mechanics. An open system consists of a "small" system $\mathcal{S}$ interacting with a large "environment" or "reservoir" $\mathcal{R}$.

In these lecture notes the small system will be a "quantum dot"—a quantum mechanical system with finitely many energy levels and no internal structure. The system $\mathcal{S}$ is described by a finite-dimensional Hilbert space $\mathcal{H}_\mathcal{S} = \mathbb{C}^N$ and a Hamiltonian $H_\mathcal{S}$. Its algebra of observables $\mathcal{O}_\mathcal{S}$ is the full matrix algebra $M_N(\mathbb{C})$ and its dynamics is given by

$$\tau_\mathcal{S}^t(A) = e^{itH_\mathcal{S}} A e^{-itH_\mathcal{S}} = e^{t\delta_\mathcal{S}}(A),$$

where $\delta_\mathcal{S}(\cdot) = i[H_\mathcal{S}, \cdot]$. The states of $\mathcal{S}$ are density matrices on $\mathcal{H}_\mathcal{S}$. A convenient reference state is the tracial state, $\omega_\mathcal{S}(\cdot) = \mathrm{Tr}(\cdot)/\dim \mathcal{H}_\mathcal{S}$. In the physics literature $\omega_\mathcal{S}$ is sometimes called the chaotic state since it is of maximal entropy, giving the same probability $1/\dim \mathcal{H}_\mathcal{S}$ to any one-dimensional projection in $\mathcal{H}_\mathcal{S}$.

The reservoir is described by a C^*-dynamical system $(\mathcal{O}_\mathcal{R}, \tau_\mathcal{R})$ and a reference state $\omega_\mathcal{R}$. We denote by $\delta_\mathcal{R}$ the generator of $\tau_\mathcal{R}$.

The algebra of observables of the joint system $\mathcal{S} + \mathcal{R}$ is $\mathcal{O} = \mathcal{O}_\mathcal{S} \otimes \mathcal{O}_\mathcal{R}$ and its reference state is $\omega \equiv \omega_\mathcal{S} \otimes \omega_\mathcal{R}$. Its dynamics, still decoupled, is given by $\tau^t = \tau_\mathcal{S}^t \otimes \tau_\mathcal{R}^t$. Let $V = V^* \in \mathcal{O}$ be a local perturbation which couples $\mathcal{S}$ to the reservoir

$\mathcal{R}$. The $*$-derivation $\delta_V \equiv \delta_\mathcal{R} + \delta_\mathcal{S} + \mathrm{i}[V, \cdot]$ generates the coupled dynamics τ_V^t on $\mathcal{O}$. The coupled joint system $\mathcal{S} + \mathcal{R}$ is described by the C^*-dynamical system $(\mathcal{O}, \tau_V)$ and the reference state ω. Whenever the meaning is clear within the context, we will identify $\mathcal{O}_\mathcal{S}$ and $\mathcal{O}_\mathcal{R}$ with subalgebras of $\mathcal{O}$ via $A \otimes I_{\mathcal{O}_\mathcal{R}}, I_{\mathcal{O}_\mathcal{S}} \otimes A$. With a slight abuse of notation, in the sequel we denote $I_{\mathcal{O}_\mathcal{R}}$ and $I_{\mathcal{O}_\mathcal{S}}$ by I.

We will suppose that the reservoir $\mathcal{R}$ has additional structure, namely that it consists of M parts $\mathcal{R}_1, \cdots, \mathcal{R}_M$, which are interpreted as subreservoirs. The subreservoirs are assumed to be independent—they interact only through the small system which allows for the flow of energy and matter between various subreservoirs.

The subreservoir structure of $\mathcal{R}$ can be chosen in a number of different ways and the choice ultimately depends on the class of examples one wishes to describe. One obvious choice is the following: the j-th reservoir is described by the C^*-dynamical system $(\mathcal{O}_{\mathcal{R}_j}, \tau_{\mathcal{R}_j})$ and the reference state $\omega_{\mathcal{R}_j}$, and $\mathcal{O}_\mathcal{R} = \otimes \mathcal{O}_{\mathcal{R}_j}, \tau_\mathcal{R} = \otimes \tau_{\mathcal{R}_j}$, $\omega = \otimes \omega_{\mathcal{R}_j}$ [JP4, Ru1]. In view of the examples we plan to cover, we will choose a more general subreservoir structure.

We will assume that the j-th reservoir is described by a C^*-subalgebra $\mathcal{O}_{\mathcal{R}_j} \subset \mathcal{O}_\mathcal{R}$ which is preserved by $\tau_\mathcal{R}$. We denote the restrictions of $\tau_\mathcal{R}$ and $\omega_\mathcal{R}$ to $\mathcal{O}_{\mathcal{R}_j}$ by $\tau_{\mathcal{R}_j}$ and $\omega_{\mathcal{R}_j}$. Different algebras $\mathcal{O}_{\mathcal{R}_j}$ may not commute. However, we will assume that $\mathcal{O}_{\mathcal{R}_i} \cap \mathcal{O}_{\mathcal{R}_j} = \mathbb{C}I$ for $i \neq j$. If $\mathcal{A}_k, 1 \leq k \leq N$, are subsets of $\mathcal{O}_\mathcal{R}$, we denote by $\langle \mathcal{A}_1, \cdots, \mathcal{A}_N \rangle$ the minimal C^*-subalgebra of $\mathcal{O}_\mathcal{R}$ that contains all $\mathcal{A}_k$. Without loss of generality, we may assume that $\mathcal{O}_\mathcal{R} = \langle \mathcal{O}_{\mathcal{R}_1}, \cdots, \mathcal{O}_{\mathcal{R}_M} \rangle$.

The system $\mathcal{S}$ is coupled to the reservoir $\mathcal{R}_j$ through a *junction* described by a self-adjoint perturbation $V_j \in \mathcal{O}_\mathcal{S} \otimes \mathcal{O}_{\mathcal{R}_j}$ (see Fig. 1). The complete interaction is given by

$$V \equiv \sum_{j=1}^M V_j. \tag{9}$$

An anti-linear, involutive, $*$-automorphism $\mathfrak{r}: \mathcal{O} \to \mathcal{O}$ is called a *time reversal* if it satisfies $\mathfrak{r}(H_\mathcal{S}) = H_\mathcal{S}, \mathfrak{r}(V_j) = V_j$ and $\mathfrak{r} \circ \tau_{\mathcal{R}_j}^t = \tau_{\mathcal{R}_j}^{-t} \circ \mathfrak{r}$. If $\mathfrak{r}$ is a time reversal, then

$$\mathfrak{r} \circ \tau^t = \tau^{-t} \circ \mathfrak{r}, \qquad \mathfrak{r} \circ \tau_V^t = \tau_V^{-t} \circ \mathfrak{r},$$

and a state ω on $\mathcal{O}$ is time reversal invariant if $\omega \circ \mathfrak{r}(A) = \omega(A^*)$ for all $A \in \mathcal{O}$. An open quantum system described by $(\mathcal{O}, \tau_V)$ and the reference state ω is called time reversal invariant (TRI) if there exists a time reversal $\mathfrak{r}$ such that ω is time reversal invariant.

4.2 C^*-Scattering for Open Quantum Systems

Except for Part (ii) of Proposition 3.3, the scattering approach to the study of NESS, described in Subsection 3.4, is directly applicable to open quantum systems. Concerning Part (ii) of Proposition 3.3, note that in the case of open quantum systems the Møller morphism α_V^+ cannot be onto (except in trivial cases). The best one may hope for is that $\mathcal{O}_+ = \mathcal{O}_\mathcal{R}$, namely that α_V^+ is an isomorphism between the C^*-dynamical systems $(\mathcal{O}, \tau_V)$ and $(\mathcal{O}_\mathcal{R}, \tau_\mathcal{R})$. The next theorem was proved in [Ru1].

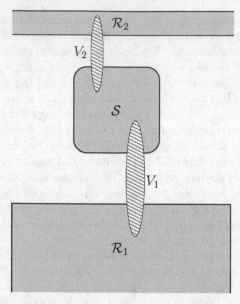

Fig. 1. Junctions V_1, V_2 between the system S and subreservoirs.

Theorem 4.1. *Suppose that Assumption (S) holds.*

(i) *If there exists a dense set $\mathcal{O}_{\mathcal{R}0} \subset \mathcal{O}_{\mathcal{R}}$ such that for all $A \in \mathcal{O}_{\mathcal{R}0}$,*

$$\int_0^\infty \|[V, \tau^t(A)]\| \, dt < \infty, \tag{10}$$

then $\mathcal{O}_{\mathcal{R}} \subset \mathcal{O}_+$.

(ii) *If there exists a dense set $\mathcal{O}_0 \subset \mathcal{O}$ such that for all $X \in \mathcal{O}_S$ and $A \in \mathcal{O}_0$,*

$$\lim_{t \to +\infty} \|[X, \tau_V^t(A)]\| = 0, \tag{11}$$

then $\mathcal{O}_+ \subset \mathcal{O}_{\mathcal{R}}$.

(iii) *If both (10) and (11) hold then α_V^+ is an isomorphism between the C^*-dynamical systems $(\mathcal{O}, \tau_V)$ and $(\mathcal{O}_{\mathcal{R}}, \tau_{\mathcal{R}})$. In particular, if $\omega_{\mathcal{R}}$ is a $(\tau_{\mathcal{R}}, \beta)$-KMS for some inverse temperature β, then ω_+ is a (τ_V, β)-KMS state.*

Proof. The proof of Part (i) is similar to the proof of the Part (i) of Proposition 3.3. The assumption (10) ensures that the limits

$$\beta_V^+(A) = \lim_{t \to \infty} \tau_V^t \circ \tau^{-t}(A),$$

exist for all $A \in \mathcal{O}_{\mathcal{R}}$. Clearly, $\alpha_V^+ \circ \beta_V^+(A) = A$ for all $A \in \mathcal{O}_{\mathcal{R}}$ and so $\mathcal{O}_{\mathcal{R}} \subset$ Ran α_V^+.

To prove Part (ii) recall that $\mathcal{O}_S$ is a N^2-dimensional matrix algebra. It has a basis $\{E_k \,|\, k = 1, \cdots, N^2\}$ such that $\tau^t(E_k) = e^{it\theta_k} E_k$ for some $\theta_k \in \mathbb{R}$. From Assumption (S) and (11) we can conclude that

$$0 = \lim_{t \to +\infty} e^{it\theta_k} \tau^{-t}([E_k, \tau_V^t(A)]) = \lim_{t \to +\infty} [E_k, \tau^{-t} \circ \tau_V^t(A)] = [E_k, \alpha_V^+(A)],$$

for all $A \in \mathcal{O}_0$ and hence, by continuity, for all $A \in \mathcal{O}$. It follows that Ran α_V^+ belongs to the commutant of $\mathcal{O}_S$ in $\mathcal{O}$. Since $\mathcal{O}$ can be seen as the algebra $M_N(\mathcal{O}_{\mathcal{R}})$ of $N \times N$-matrices with entries in $\mathcal{O}_{\mathcal{R}}$, one easily checks that this commutant is precisely $\mathcal{O}_{\mathcal{R}}$.

Part (iii) is a direct consequence of the first two parts. $\square$

4.3 The First and Second Law of Thermodynamics

Let us denote by δ_j the generator of the dynamical group $\tau_{\mathcal{R}_j}$. (Recall that this dynamical group is the restriction of the decoupled dynamics to the subreservoir $\mathcal{R}_j$). Assume that $V_j \in \text{Dom}\,(\delta_j)$. The generator of τ_V is $\delta_V = \delta_{\mathcal{R}} + i[H_S + V, \cdot]$ and it follows from (9) that the total energy flux out of the reservoir is given by

$$\frac{d}{dt} \tau_V^t(H_S + V) = \tau_V^t(\delta_V(H_S + V)) = \tau_V^t(\delta_{\mathcal{R}}(V)) = \sum_{j=1}^{M} \tau_V^t(\delta_j(V_j)).$$

Thus, we can identify the observable describing the heat flux out of the j-th reservoir as

$$\Phi_j = \delta_j(V) = \delta_j(V_j) = \delta_{\mathcal{R}}(V_j).$$

We note that if $\mathfrak{r}$ is a time-reversal, then $\mathfrak{r}(\Phi_j) = -\Phi_j$. The energy balance equation

$$\sum_{j=1}^{M} \Phi_j = \delta_V(H_S + V),$$

yields the conservation of energy (the first law of thermodynamics): for any τ_V-invariant state η,

$$\sum_{j=1}^{M} \eta(\Phi_j) = 0. \tag{12}$$

Besides heat fluxes, there might be other fluxes across the system $\mathcal{S} + \mathcal{R}$ (for example, matter and charge currents). We will not discuss here the general theory of such fluxes (the related information can be found in [FMU, FMSU, TM]). In the rest of this section we will focus on the thermodynamics of heat fluxes. Charge currents will be discussed in the context of a concrete model in the second part of this lecture.

We now turn to the entropy production. Assume that there exists a C^*-dynamics $\sigma_{\mathcal{R}}^t$ on $\mathcal{O}_{\mathcal{R}}$ such that $\omega_{\mathcal{R}}$ is $(\sigma_{\mathcal{R}}, -1)$-KMS state and such that $\sigma_{\mathcal{R}}$ preserves each subalgebra $\mathcal{O}_{\mathcal{R}_j}$. Let $\tilde{\delta}_j$ be the generator of the restriction of $\sigma_{\mathcal{R}}$ to $\mathcal{O}_{\mathcal{R}_j}$ and assume that $V_j \in \mathrm{Dom}\,(\tilde{\delta}_j)$. The entropy production observable associated to the perturbation V and the reference state $\omega = \omega_\mathcal{S} \otimes \omega_\mathcal{R}$, where $\omega_\mathcal{S}(\cdot) = \mathrm{Tr}(\cdot)/\dim \mathcal{H}_\mathcal{S}$, is

$$\sigma_V = \sum_{j=1}^M \tilde{\delta}_j(V_j).$$

Until the end of this section we shall assume that the reservoirs $\mathcal{O}_{\mathcal{R}_j}$ are in thermal equilibrium at inverse temperatures β_j. More precisely, we will assume that $\omega_{\mathcal{R}_j}$ is the *unique* $(\tau_{\mathcal{R}_j}, \beta_j)$-KMS state on $\mathcal{O}_{\mathcal{R}_j}$. Then $\tilde{\delta}_j = -\beta_j \delta_j$, and

$$\sigma_V = -\sum_{j=1}^M \beta_j \Phi_j.$$

In particular, for any NESS $\omega_+ \in \Sigma_+(\omega, \tau_V)$, the second law of thermodynamics holds:

$$\sum_{j=1}^M \beta_j \omega_+(\Phi_j) = -\mathrm{Ep}(\omega_+) \leq 0. \tag{13}$$

In fact, it is not difficult to show that $\mathrm{Ep}(\omega_+)$ is independent of the choice of the reference state of the small system as long as $\omega_\mathcal{S} > 0$; see Proposition 5.3 in [JP4]. In the case of two reservoirs, the relation

$$(\beta_1 - \beta_2)\omega_+(\Phi_1) = \beta_1\,\omega_+(\Phi_1) + \beta_2\,\omega_+(\Phi_2) \leq 0,$$

yields that the heat flows from the hot to the cold reservoir.

4.4 Linear Response Theory

Linear response theory describes thermodynamics in the regime where the "forces" driving the system out of equilibrium are weak. In such a regime, to a very good approximation, the non-equilibrium currents depend linearly on the forces. The ultimate purpose of linear response theory is to justify well known phenomenological laws like Ohm's law for charge currents or Fick's law for heat currents. We are still far from a satisfactory derivation of these laws, even in the framework of classical mechanics; see [BLR] for a recent review on this matter. We also refer to [GVV6] for a rigorous discussion of linear response theory at the macroscopic level.

A less ambitious application of linear response theory concerns transport properties of microscopic and mesoscopic quantum devices (the advances in nanotechnologies during the last decade have triggered a strong interest in the transport properties of such devices). Linear response theory of such systems is much better understood, as we shall try to illustrate.

In our current setting, the forces that drive the system $\mathcal{S} + \mathcal{R}$ out of equilibrium are the different inverse temperatures $\beta_1, \cdots, \beta_M$ of the reservoirs attached to $\mathcal{S}$. If all inverse temperatures β_j are sufficiently close to some value β_{eq}, we expect linear response theory to give a good account of the thermodynamics of the system near thermal equilibrium at inverse temperature β_{eq}.

To emphasize the fact that the reference state $\omega = \omega_{\mathcal{S}} \otimes \omega_{\mathcal{R}}$ depends on the β_j we set $X = (X_1, \cdots, X_M)$ with $X_j \equiv \beta_{\text{eq}} - \beta_j$ and denote by ω_X this reference state. We assume that for some $\epsilon > 0$ and all $|X| < \epsilon$ there exists a unique NESS $\omega_{X+} \in \Sigma_+(\omega_X, \tau_V)$ and that the functions $X \mapsto \omega_{X+}(\Phi_j)$ are C^2. Note that ω_{0+} is the (unique) $(\tau_V, \beta_{\text{eq}})$-KMS state on $\mathcal{O}$. We will denote it simply by $\omega_{\beta_{\text{eq}}}$.

In phenomenological non-equilibrium thermodynamics, the duality between the driving forces F_α, also called *affinities*, and the steady currents ϕ_α they induce is expressed by the entropy production formula

$$\text{Ep} = \sum_\alpha F_\alpha \phi_\alpha,$$

(see [DGM]). The steady currents are themselves functions of the affinities $\phi_\alpha = \phi_\alpha(F_1, \cdots)$. In the linear response regime, these functions are given by the relations

$$\phi_\alpha = \sum_\gamma L_{\alpha\gamma} F_\gamma,$$

which define the *kinetic coefficients* $L_{\alpha\gamma}$.

Comparing with Equ. (13) and using energy conservation (12) we obtain in our case

$$\text{Ep}(\omega_{X+}) = \sum_{j=1}^{M} X_j \, \omega_{X+}(\Phi_j).$$

Thus X_j is the affinity conjugated to the steady heat flux $\phi_j(X) = \omega_{X+}(\Phi_j)$ out of $\mathcal{R}_j$. We note in particular that the equilibrium entropy production vanishes. The kinetic coefficients L_{ji} are given by

$$L_{ji} \equiv \left(\frac{\partial \phi_j}{\partial X_i} \right)_{X=0} = \partial_{X_i} \omega_{X+}(\Phi_j)|_{X=0}.$$

Taylor's formula yields

$$\phi_j(X) = \omega_{X+}(\Phi_j) = \sum_{i=1}^{M} L_{ji} X_i + O(\epsilon^2), \tag{14}$$

$$\text{Ep}(\omega_{X+}) = \sum_{i,j=1}^{M} L_{ji} X_i X_j + o(\epsilon^2). \tag{15}$$

Combining (14) with the first law of thermodynamics (recall (12)) we obtain that for all i,

$$\sum_{j=1}^{M} L_{ji} = 0. \tag{16}$$

Similarly, (15) and the second law (13) imply that the quadratic form

$$\sum_{i,j=1}^{M} L_{ji} X_i X_j,$$

on $\mathbb{R}^M$ is non-negative. Note that this does not imply that the $M \times M$-matrix L is symmetric !

Linear response theory goes far beyond the above elementary relations. Its true cornerstones are the *Onsager reciprocity relations* (ORR), the Kubo *fluctuation-dissipation* formula (KF) and the *Central Limit Theorem* (CLT). All three of them deal with the kinetic coefficients. The Onsager reciprocity relations assert that the matrix L_{ji} of a time reversal invariant (TRI) system is symmetric,

$$L_{ji} = L_{ij}. \tag{17}$$

For non-TRI systems, similar relations hold between the transport coefficients of the system and those of the time reversed one. For example, if time reversal invariance is broken by the action of an external magnetic field B, then the Onsager-Casimir relations

$$L_{ji}(B) = L_{ij}(-B),$$

hold.

The Kubo fluctuation-dissipation formula expresses the transport coefficients of a TRI system in terms of the *equilibrium* current-current correlation function

$$C_{ji}(t) \equiv \frac{1}{2} \omega_{\beta_{eq}} (\tau_V^t(\Phi_j)\Phi_i + \Phi_i \tau_V^t(\Phi_j)), \tag{18}$$

namely

$$L_{ji} = \frac{1}{2} \int_{-\infty}^{\infty} C_{ji}(t)\, dt. \tag{19}$$

The Central Limit Theorem further relates L_{ji} to the statistics of the current fluctuations in equilibrium. In term of characteristic function, the CLT for open quantum systems in thermal equilibrium asserts that

$$\lim_{t \to \infty} \omega_{\beta_{eq}} \left(e^{i\left(\sum_{j=1}^{M} \xi_j \int_0^t \tau_V^s(\Phi_j)\, ds\right)/\sqrt{t}} \right) = e^{-\frac{1}{2} \sum_{i,j=1}^{M} D_{ji} \xi_j \xi_i}, \tag{20}$$

where the covariance matrix D_{ji} is given by

$$D_{ji} = 2 L_{ji}.$$

If, for a self-adjoint $A \in \mathcal{O}$, we denote by $1_{[a,b]}(A)$ the spectral projection on the interval $[a, b]$ of $\pi_{\omega_{\beta_{eq}}}(A)$, the probability of measuring a value of A in $[a, b]$ when the system is in the state $\omega_{\beta_{eq}}$ is given by

$$\text{Prob}_{\omega_{\beta_{eq}}}\{A \in [a, b]\} = (\Omega_{\omega_{\beta_{eq}}}, 1_{[a,b]}(A)\, \Omega_{\omega_{\beta_{eq}}}).$$

It then follows from (20) that

$$\lim_{t\to\infty} \mathrm{Prob}_{\omega_{\beta_{\mathrm{eq}}}} \left\{ \frac{1}{t} \int_0^t \tau_V^s(\Phi_j)\,\mathrm{d}s \in \left[\frac{a}{\sqrt{t}}, \frac{b}{\sqrt{t}} \right] \right\} = \frac{1}{\sqrt{2\pi}L_{jj}} \int_a^b e^{-x^2/2L_{jj}^2}\,\mathrm{d}x.$$

(21)

This is a direct translation to quantum mechanics of the classical central limit theorem. Because fluxes do not commute, $[\Phi_j, \Phi_i] \neq 0$ for $j \neq i$, they can not be measured simultaneously and a simple classical probabilistic interpretation of (20) for the vector variable $\Phi = (\Phi_1, \cdots, \Phi_M)$ is not possible. Instead, the quantum fluctuations of the vector variable Φ are described by the so-called *fluctuation algebra* [GVV1, GVV2, GVV3, GVV4, GVV5, Ma]. The description and study of the fluctuation algebra involve somewhat advanced technical tools and for this reason we will not discuss the quantum CLT theorem in this lecture.

The mathematical theory of ORR, KF, and CLT is reasonably well understood in classical statistical mechanics (see the lecture [Re]). In the context of open quantum systems these important notions are still not completely understood (see however [AJPP, JPR2] for some recent results).

We close this subsection with some general comments about ORR and KF.

The definition (18) of the current-current correlation function involves a symmetrized product in order to ensure that the function $C_{ji}(t)$ is real-valued. The corresponding imaginary part, given by

$$\frac{1}{2}\,\mathrm{i}[\Phi_i, \tau_V^t(\Phi_j)],$$

is usually non-zero. However, since $\omega_{\beta_{\mathrm{eq}}}$ is a KMS state, the stability condition (see [BR2]) yields

$$\int_{-\infty}^{\infty} \omega_{\beta_{\mathrm{eq}}}(\mathrm{i}[\Phi_i, \tau_V^t(\Phi_j)])\,\mathrm{d}t = 0,$$

(22)

so that, in this case, the symmetrization is not necessary and one can rewrite KF as

$$L_{ji} = \frac{1}{2}\int_{-\infty}^{\infty} \omega_{\beta_{\mathrm{eq}}}(\Phi_i \tau_V^t(\Phi_j))\,\mathrm{d}t.$$

Finally, we note that ORR follow directly from KF under the TRI assumption. Indeed, if our system is TRI with time reversal $\mathfrak{r}$ we have

$$\mathfrak{r}(\Phi_i) = -\Phi_i, \quad \mathfrak{r}(\tau_V^t(\Phi_j)) = -\tau_V^{-t}(\Phi_j), \quad \omega_{\beta_{\mathrm{eq}}} \circ \mathfrak{r} = \omega_{\beta_{\mathrm{eq}}},$$

and therefore

$$C_{ji}(t) = \frac{1}{2}\omega_{\beta_{\mathrm{eq}}}(\tau_V^{-t}(\Phi_j)\Phi_i + \Phi_i\tau_V^{-t}(\Phi_j)) = C_{ji}(-t).$$

Since $\omega_{\beta_{\mathrm{eq}}}$ is τ_V-invariant, this implies

$$C_{ji}(t) = \frac{1}{2}\omega_{\beta_{\mathrm{eq}}}(\Phi_j\tau_V^t(\Phi_i) + \tau_V^t(\Phi_i)\Phi_j) = C_{ij}(t),$$

and ORR (17) follows from KF (19).

In the second part of the lecture we will show that the Onsager relations and the Kubo formula hold for the SEBB model. The proof of the Central Limit Theorem for this model is somewhat technically involved and can be found in [AJPP].

4.5 Fermi Golden Rule (FGR) Thermodynamics

Let $\lambda \in \mathbb{R}$ be a control parameter. We consider an open quantum system with coupling λV and write τ_λ for $\tau_{\lambda V}$, $\omega_{\lambda+}$ for ω_+, etc.

The NESS and thermodynamics of the system can be described, to second order of perturbation theory in λ, using the weak coupling (or van Hove) limit. This approach is much older than the "microscopic" Hamiltonian approach discussed so far, and has played an important role in the development of the subject. The classical references are [Da1, Da2, Haa, VH1, VH2, VH3]. The weak coupling limit is also discussed in the lecture notes [D1].

In the weak coupling limit one "integrates" the degrees of freedom of the reservoirs and follows the reduced dynamics of $\mathcal{S}$ on a large time scale t/λ^2. In the limit $\lambda \to 0$ the dynamics of $\mathcal{S}$ becomes irreversible and is described by a semigroup, often called the *quantum Markovian semigroup* (QMS). The generator of this QMS describes the thermodynamics of the open quantum system to second order of perturbation theory.

The "integration" of the reservoir variables is performed as follows. As usual, we use the injection $A \mapsto A \otimes I$ to identify $\mathcal{O}_\mathcal{S}$ with a subalgebra of $\mathcal{O}$. For $A \in \mathcal{O}_\mathcal{S}$ and $B \in \mathcal{O}_\mathcal{R}$ we set

$$P_\mathcal{S}(A \otimes B) = A\omega_\mathcal{R}(B). \tag{23}$$

The map $P_\mathcal{S}$ extends to a projection $P_\mathcal{S} : \mathcal{O} \to \mathcal{O}_\mathcal{S}$. The reduced dynamics of the system $\mathcal{S}$ is described by the family of maps $T_\lambda^t : \mathcal{O}_\mathcal{S} \to \mathcal{O}_\mathcal{S}$ defined by

$$T_\lambda^t(A) \equiv P_\mathcal{S}\left(\tau_0^{-t} \circ \tau_\lambda^t(A \otimes I)\right).$$

Obviously, T_λ^t is neither a group nor a semigroup. Let $\omega_\mathcal{S}$ be an arbitrary reference state (density matrix) of the small system and $\omega = \omega_\mathcal{S} \otimes \omega_\mathcal{R}$. Then for any $A \in \mathcal{O}_\mathcal{S}$,

$$\omega(\tau_0^{-t} \circ \tau_\lambda^t(A \otimes I)) = \mathrm{Tr}_{\mathcal{H}_\mathcal{S}}(\omega_\mathcal{S}\, T_\lambda^t(A)).$$

In [Da1, Da2] Davies proved that under very general conditions there exists a linear map $K_\mathrm{H} : \mathcal{O}_\mathcal{S} \to \mathcal{O}_\mathcal{S}$ such that

$$\lim_{\lambda \to 0} T_\lambda^{t/\lambda^2}(A) = \mathrm{e}^{tK_\mathrm{H}}(A).$$

The operator K_H is the QMS generator (sometimes called the Davies generator) in the *Heisenberg* picture. A substantial body of literature has been devoted to the study of the operator K_H (see the lecture notes [D1]). Here we recall only a few basic results concerning thermodynamics in the weak coupling limit (for additional information see [LeSp]). We will assume that the general conditions described in the lecture notes [D1] are satisfied.

The operator K_H generates a positivity preserving contraction semigroup on $\mathcal{O}_S$. Obviously, $K_H(I) = 0$. We will assume that zero is the only purely imaginary eigenvalue of K_H and that $\mathrm{Ker}\, K_H = \mathbb{C}I$. This non-degeneracy condition can be naturally characterized in algebraic terms, see [D1,Sp]. It implies that the eigenvalue 0 of K_H is semi-simple, that the corresponding eigenprojection has the form $A \mapsto \mathrm{Tr}(\omega_{S+}A)I$, where ω_{S+} is a density matrix, and that for any initial density matrix ω_S,

$$\lim_{t \to \infty} \mathrm{Tr}(\omega_S e^{tK_H}(A)) = \mathrm{Tr}(\omega_{S+}A) \equiv \omega_{S+}(A).$$

The density matrix ω_{S+} describes the NESS of the open quantum system in the weak coupling limit. One further shows that the operator K_H has the form

$$K_H = \sum_{j=1}^{M} K_{H,j},$$

where $K_{H,j}$ is the QMS generator obtained by considering the weak coupling limit of the coupled system $S + \mathcal{R}_j$, i.e.,

$$e^{tK_{H,j}}(A) = \lim_{\lambda \to 0} P_S\left(\tau_0^{-t/\lambda^2} \circ \tau_{\lambda,j}^{t/\lambda^2}(A \otimes I)\right), \tag{24}$$

where $\tau_{\lambda,j}$ is generated by $\delta_j + \mathrm{i}[H_S + \lambda V_j, \cdot]$.

One often considers the QMS generator in the Schrödinger picture, denoted K_S. The operator K_S is the adjoint of K_H with respect to the inner product $(X, Y) = \mathrm{Tr}(X^*Y)$. The semigroup e^{tK_S} is positivity and trace preserving. One similarly defines $K_{S,j}$. Obviously,

$$K_S(\omega_{S+}) = 0, \qquad K_S = \sum_{j=1}^{M} K_{S,j}.$$

Recall our standing assumption that the reservoirs $\mathcal{O}_{\mathcal{R}_j}$ are in thermal equilibrium at inverse temperature β_j. We denote by

$$\omega_\beta = e^{-\beta H_S}/\mathrm{Tr}(e^{-\beta H_S}),$$

the canonical density matrix of S at inverse temperature β (the unique (τ_S, β)-KMS state on $\mathcal{O}_S$). Araki's perturbation theory of KMS-states (see [DJP,BR2]) yields that for $A \in \mathcal{O}_S$,

$$\omega_{\beta_j} \otimes \omega_{\mathcal{R}_j}(\tau_0^{-t} \circ \tau_{\lambda,j}^t(A \otimes I)) = \omega_{\beta_j}(A) + O(\lambda),$$

uniformly in t. Hence, for all $t \geq 0$,

$$\omega_{\beta_j}(e^{tK_{H,j}}(A)) = \omega_{\beta_j}(A),$$

and so $K_{S,j}(\omega_{\beta_j}) = 0$. In particular, if all β_j's are the same and equal to β, then $\omega_{S+} = \omega_\beta$.

Let $\mathcal{O}_d \subset \mathcal{O}_S$ be the $*$-algebra spanned by the eigenprojections of H_S. $\mathcal{O}_d$ is commutative and preserved by K_H, $K_{H,j}$, K_S and $K_{S,j}$ [D1]. The NESS ω_{S+} commutes with H_S. If the eigenvalues of H_S are simple, then the restriction $K_H \upharpoonright \mathcal{O}_d$ is a generator of a Markov process whose state space is the spectrum of H_S. This process has played an important role in the early development of quantum field theory (more on this in Subsection 8.2).

We now turn to the thermodynamics in the weak coupling limit, which we will call *Fermi Golden Rule (FGR) thermodynamics*. The observable describing the heat flux out of the j-th reservoir is

$$\Phi_{\text{fgr},j} = K_{H,j}(H_S).$$

Note that $\Phi_{\text{fgr},j} \in \mathcal{O}_d$. Since $K_S(\omega_{S+}) = 0$ we have

$$\sum_{j=1}^{M} \omega_{S+}(\Phi_{\text{fgr},j}) = \omega_{S+}(K_H(H_S)) = 0,$$

which is the first law of FGR thermodynamics.

The entropy production observable is

$$\sigma_{\text{fgr}} = -\sum_{j=1}^{M} \beta_j \Phi_{\text{fgr},j}, \tag{25}$$

and the entropy production of the NESS ω_{S+} is

$$\text{Ep}_{\text{fgr}}(\omega_{S+}) = \omega_{S+}(\sigma_{\text{fgr}}).$$

Since the semigroup generated by $K_{S,j}$ is trace-preserving we have

$$\frac{d}{dt}\,\text{Ent}(e^{tK_{S,j}}\omega_{S+}|\omega_{\beta_j})|_{t=0} = -\beta_j\,\omega_{S+}(\Phi_{\text{fgr},j}) - \text{Tr}(K_{S,j}(\omega_{S+})\log\omega_{S+}),$$

where the relative entropy is defined by (2). The function

$$t \mapsto \text{Ent}(e^{tK_{S,j}}\omega_{S+}|\omega_{\beta_j}),$$

is non-decreasing (see [Li]), and so

$$\text{Ep}_{\text{fgr}}(\omega_{S+}) = \sum_{j=1}^{M} \frac{d}{dt}\,\text{Ent}(e^{tK_{S,j}}\omega_{S+}|\omega_{\beta_j})|_{t=0} \geq 0,$$

which is the second law of FGR thermodynamics. Moreover, under the usual non-degeneracy assumptions, $\text{Ep}_{\text{fgr}}(\omega_{S+}) = 0$ if and only if $\beta_1 = \cdots = \beta_M$ (see [LeSp] for details).

Let us briefly discuss linear response theory in FGR thermodynamics using the same notational conventions as in Subsection 4.4. The kinetic coefficients are given by

$$L_{\text{fgr},ji} = \partial_{X_i} \omega_{\mathcal{S}+}(\Phi_{\text{fgr},j})|_{X=0}.$$

For $|X| < \epsilon$ one has

$$\omega_{\mathcal{S}+}(\Phi_{\text{fgr},j}) = \sum_{i=1}^{M} L_{\text{fgr},ji} X_i + O(\epsilon^2),$$

$$\text{Ep}_{\text{fgr}}(\omega_{\mathcal{S}+}) = \sum_{i,j=1}^{M} L_{\text{fgr},ji} X_i X_j + o(\epsilon^2).$$

The first and the second law yield that for all i,

$$\sum_{j=1}^{M} L_{\text{fgr},ji} = 0,$$

and that the quadratic form

$$\sum_{i,j=1}^{M} L_{\text{fgr},ji} X_i X_j,$$

is non-negative. The Kubo formula

$$L_{\text{fgr},ji} = \int_0^{\infty} \omega_{\beta_{\text{oq}}}(e^{tK_{\text{H}}}(\Phi_j)\,\Phi_i)\,dt, \tag{26}$$

and the Onsager reciprocity relations

$$L_{\text{fgr},ji} = L_{\text{fgr},ij}, \tag{27}$$

are proven in [LeSp].

Finally, we wish to comment on the relation between microscopic and FGR thermodynamics. One naturally expects FGR thermodynamics to produce the first non-trivial contribution (in λ) to the microscopic thermodynamics. For example, the following relations are expected to hold for small λ:

$$\omega_{\lambda+} = \omega_{\mathcal{S}+} + O(\lambda),$$
$$\omega_{\lambda+}(\Phi_j) = \lambda^2 \omega_{\mathcal{S}+}(\Phi_{\text{fgr},j}) + O(\lambda^3). \tag{28}$$

Indeed, it is possible to prove that if the microscopic thermodynamics exists and is sufficiently regular, then (28) hold. On the other hand, establishing existence and regularity of the microscopic thermodynamics is a formidable task which has been so far carried out only for a few models. FGR thermodynamics is very robust and the weak coupling limit is an effective tool in the study of the models whose microscopic thermodynamics appears beyond reach of the existing techniques.

We will return to this topic in Section 8 where we will discuss the FGR thermodynamics of the SEBB model.

5 Free Fermi Gas Reservoir

In the SEBB model, which we shall study in the second part of this lecture, the reservoir will be described by an infinitely extended free Fermi gas. Our description of the free Fermi gas in this section is suited to this application.

The basic properties of the free Fermi gas are discussed in the lecture [Me3] and in Examples 4.6 and 5.6 of the lecture [Pi] and we will assume that the reader is familiar with the terminology and results described there. A more detailed exposition can be found in [BR2] and in the recent lecture notes [D2].

The free Fermi gas is described by the so called CAR (canonical anticommutation relations) algebra. The mathematical structure of this algebra is well understood (see [D2] for example). In Subsection 5.1 we will review the results we need. Subsection 5.2 contains a few useful examples.

5.1 General Description

Let $\mathfrak{h}$ and h be the Hilbert space and the Hamiltonian of a single Fermion. We will always assume that h is bounded below. Let $\Gamma_-(\mathfrak{h})$ be the anti-symmetric Fock space over $\mathfrak{h}$ and denote by $a^*(f)$, $a(f)$ the creation and annihilation operators for a single Fermion in the state $f \in \mathfrak{h}$. The corresponding self-adjoint field operator

$$\varphi(f) \equiv \frac{1}{\sqrt{2}} \left(a(f) + a^*(f) \right),$$

satisfies the anticommutation relation

$$\varphi(f)\varphi(g) + \varphi(g)\varphi(f) = \mathrm{Re}(f,g)I.$$

In the sequel $a^\#$ stands for either a or a^*. Let $\mathrm{CAR}(\mathfrak{h})$ be the C^*-algebra generated by $\{a^\#(f) \mid f \in \mathfrak{h}\}$. We will refer to $\mathrm{CAR}(\mathfrak{h})$ as the Fermi algebra. The C^*-dynamics induced by h is

$$\tau^t(A) \equiv e^{\mathrm{i}td\Gamma(h)} A e^{-\mathrm{i}td\Gamma(h)}.$$

The pair $(\mathrm{CAR}(\mathfrak{h}), \tau)$ is a C^*-dynamical system. It preserves the Fermion number in the sense that τ^t commutes with the gauge group

$$\vartheta^t(A) \equiv e^{\mathrm{i}td\Gamma(I)} A e^{-\mathrm{i}td\Gamma(I)}.$$

Recall that $N \equiv d\Gamma(I)$ is the Fermion number operator on $\Gamma_-(\mathfrak{h})$ and that τ and ϑ are the groups of Bogoliubov automorphisms

$$\tau^t(a^\#(f)) = a^\#(e^{\mathrm{i}th}f), \qquad \vartheta^t(a^\#(f)) = a^\#(e^{\mathrm{i}t}f).$$

To every self-adjoint operator T on $\mathfrak{h}$ such that $0 \le T \le I$ one can associate a state ω_T on $\mathrm{CAR}(\mathfrak{h})$ satisfying

$$\omega_T(a^*(f_n) \cdots a^*(f_1)a(g_1) \cdots a(g_m)) = \delta_{n,m}\det\{(g_i, Tf_j)\}. \tag{29}$$

This ϑ-invariant state is usually called the quasi-free gauge-invariant state generated by T. It is completely determined by its two point function

$$\omega_T(a^*(f)a(g)) = (g, Tf).$$

We will often call T the *density operator* or simply the *generator* of the state ω_T. Alternatively, quasi-free gauge-invariant states can be described by their action on the field operators. For any integer n we define $\mathcal{P}_n$ as the set of all permutations π of $\{1, \ldots, 2n\}$ such that

$$\pi(2j-1) < \pi(2j), \quad \text{and} \quad \pi(2j-1) < \pi(2j+1),$$

for every $j \in \{1, \ldots, n\}$. Denote by $\epsilon(\pi)$ the signature of $\pi \in \mathcal{P}_n$. ω_T is the unique state on $\mathrm{CAR}(\mathfrak{h})$ with the following properties:

$$\omega_T(\varphi(f_1)\varphi(f_2)) = \frac{1}{2}(f_1, f_2) - i\,\mathrm{Im}(f_1, Tf_2),$$

$$\omega_T(\varphi(f_1)\cdots\varphi(f_{2n})) = \sum_{\pi \in \mathcal{P}_n} \epsilon(\pi) \prod_{j=1}^{n} \omega_T(\varphi(f_{\pi(2j-1)})\varphi(f_{\pi(2j)})),$$

$$\omega_T(\varphi(f_1)\cdots\varphi(f_{2n+1})) = 0.$$

If $\mathfrak{h} = \mathfrak{h}_1 \oplus \mathfrak{h}_2$ and $T = T_1 \oplus T_2$, then for $A \in \mathrm{CAR}(\mathfrak{h}_1)$ and $B \in \mathrm{CAR}(\mathfrak{h}_2)$ one has

$$\omega_T(AB) = \omega_{T_1}(A)\,\omega_{T_2}(B). \tag{30}$$

ω_T is a factor state. It is modular iff $\mathrm{Ker}\,T = \mathrm{Ker}\,(I - T) = \{0\}$. Two states ω_{T_1} and ω_{T_2} are quasi-equivalent iff the operators

$$T_1^{1/2} - T_2^{1/2} \quad \text{and} \quad (I - T_1)^{1/2} - (I - T_2)^{1/2}, \tag{31}$$

are Hilbert-Schmidt; see [De, PoSt, Ri]. Assume that $\mathrm{Ker}\,T_i = \mathrm{Ker}\,(I - T_i) = \{0\}$. Then the states ω_{T_1} and ω_{T_2} are unitarily equivalent iff (31) holds.

If $T = F(h)$ for some function $F\colon \sigma(h) \to [0, 1]$, then ω_T describes a free Fermi gas with energy density per unit volume $F(\varepsilon)$.

The state ω_T is τ-invariant iff T commutes with e^{ith} for all t. If the spectrum of h is simple this means that $T = F(h)$ for some function $F\colon \sigma(h) \to [0, 1]$.

For any $\beta, \mu \in \mathbb{R}$, the Fermi-Dirac distribution $\rho_{\beta\mu}(\varepsilon) \equiv (1 + \mathrm{e}^{\beta(\varepsilon - \mu)})^{-1}$ induces the unique β-KMS state on $\mathrm{CAR}(\mathfrak{h})$ for the dynamics $\tau^t \circ \vartheta^{-\mu t}$. This state, which we denote by $\omega_{\beta\mu}$, describes the free Fermi gas at thermal equilibrium in the grand canonical ensemble with inverse temperature β and chemical potential μ.

The GNS representation of $\mathrm{CAR}(\mathfrak{h})$ associated to ω_T can be explicitly computed as follows. Fix a complex conjugation $f \mapsto \bar{f}$ on $\mathfrak{h}$ and extend it to $\Gamma_-(\mathfrak{h})$. Denote by Ω the vacuum vector and N the number operator in $\Gamma_-(\mathfrak{h})$. Set

$$\mathcal{H}_{\omega_T} = \Gamma_-(\mathfrak{h}) \otimes \Gamma_-(\mathfrak{h}),$$

$$\Omega_{\omega_T} = \Omega \otimes \Omega,$$

$$\pi_{\omega_T}(a(f)) = a((I - T)^{1/2}f) \otimes I + (-I)^N \otimes a^*(\bar{T}^{1/2}\bar{f}).$$

The triple $(\mathcal{H}_{\omega_T}, \pi_{\omega_T}, \Omega_{\omega_T})$ is the GNS representation of the algebra $\mathrm{CAR}(\mathfrak{h})$ associated to ω_T. (This representation was constructed in [AW] and if often called Araki-Wyss representation.) If ω_T is τ-invariant, the corresponding ω_T-Liouvillean is

$$L = \mathrm{d}\Gamma(h) \otimes I - I \otimes \mathrm{d}\Gamma(\bar{h}).$$

If h has purely (absolutely) continuous spectrum so does L, except for the simple eigenvalue 0 corresponding to the vector Ω_{ω_T}. On the other hand, 0 becomes a degenerate eigenvalue as soon as h has some point spectrum. Thus (see the lecture notes [Pi]) the ergodic properties of τ-invariant, gauge-invariant quasi-free states can be described in terms of the spectrum of h. The state ω_T is ergodic iff h has no eigenvalues. If h has purely absolutely continuous spectrum, then ω_T is mixing.

If ω_T is modular, then its modular operator is

$$\log \Delta_{\omega_T} = \mathrm{d}\Gamma(s) \otimes I - I \otimes \mathrm{d}\Gamma(\bar{s}),$$

where $s = \log T(I-T)^{-1}$. The corresponding modular conjugation is $J(\Phi \otimes \Psi) = u\bar{\Psi} \otimes u\bar{\Phi}$, where $u = (-I)^{N(N+I)/2}$.

Let θ be the $*$-automorphism of $\mathrm{CAR}(\mathfrak{h})$ defined by

$$\theta(a(f)) = -a(f). \tag{32}$$

$A \in \mathrm{CAR}(\mathfrak{h})$ is called even if $\theta(A) = A$ and odd if $\theta(A) = -A$. Every element $A \in \mathrm{CAR}(\mathfrak{h})$ can be written in a unique way as a sum $A = A^+ + A^-$ where $A^\pm = (A \pm \theta(A))/2$ is even/odd. The set of all even/odd elements is a vector subspace of $\mathrm{CAR}(\mathfrak{h})$ and $\mathrm{CAR}(\mathfrak{h})$ is a direct sum of these two subspaces. It follows from (29) that $\omega_T(A) = 0$ if A is odd. Therefore one has $\omega_T(A) = \omega_T(A^+)$ and

$$\omega_T \circ \theta = \omega_T. \tag{33}$$

The subspace of even elements is a C^*-subalgebra of $\mathrm{CAR}(\mathfrak{h})$. This subalgebra is called even CAR algebra and is denoted by $\mathrm{CAR}^+(\mathfrak{h})$. It is generated by

$$\{a^\#(f_1) \cdots a^\#(f_{2n}) \mid n \in \mathbb{N}, f_j \in \mathfrak{h}\}.$$

The even CAR algebra plays an important role in physics. It is preserved by τ and ϑ and the pair $(\mathrm{CAR}^+(\mathfrak{h}), \tau)$ is a C^*-dynamical system.

We denote the restriction of ω_T to $\mathrm{CAR}^+(\mathfrak{h})$ by the same letter. In particular, $\omega_{\beta\mu}$ is the unique β-KMS state on $\mathrm{CAR}^+(\mathfrak{h})$ for the dynamics $\tau^t \circ \vartheta^{-\mu t}$.

Let

$$A = a^\#(f_1) \cdots a^\#(f_n), \qquad B = a^\#(g_1) \cdots a^\#(g_m),$$

be two elements of $\mathrm{CAR}(\mathfrak{h})$, where m is *even*. It follows from CAR that

$$\|[A, \tau^t(B)]\| \le C \sum_{i,j} |(f_i, \mathrm{e}^{ith} g_j)|,$$

where one can take $C = (\max(\|f_i\|, \|g_j\|))^{n+m-2}$. If the functions $|(f_i, \mathrm{e}^{ith} g_j)|$ belong to $L^1(\mathbb{R}, \mathrm{d}t)$, then

$$\int_{-\infty}^{\infty} \|[A, \tau^t(B)]\| \, dt < \infty. \tag{34}$$

Let $\mathfrak{h}_0 \subset \mathfrak{h}$ be a subspace such that for any $f, g \in \mathfrak{h}_0$ the function $t \mapsto (f, e^{ith}g)$ is integrable. Let $\mathcal{O}_0 = \{a^\#(f_1) \cdots a^\#(f_n) \,|\, n \in \mathbb{N}, f_j \in \mathfrak{h}_0\}$ and let $\mathcal{O}_0^+$ be the even subalgebra of $\mathcal{O}_0$. Then for $A \in \mathcal{O}_0$ and $B \in \mathcal{O}_0^+$ (34) holds. If $\mathfrak{h}_0$ is dense in $\mathfrak{h}$, then $\mathcal{O}_0$ is dense in $\mathrm{CAR}(\mathfrak{h})$ and $\mathcal{O}_0^+$ is dense in $\mathrm{CAR}^+(\mathfrak{h})$.

Let $\mathfrak{h}_1$ and $\mathfrak{h}_2$ be two Hilbert spaces, and let $\Omega_{\mathfrak{h}_1}, \Omega_{\mathfrak{h}_2}$ be the vaccua in $\Gamma_-(\mathfrak{h}_1)$ and $\Gamma_-(\mathfrak{h}_2)$. The exponential law for Fermions (see [BSZ] and [BR2], Example 5.2.20) states that there exists a unique unitary map $U : \Gamma_-(\mathfrak{h}_1 \oplus \mathfrak{h}_2) \to \Gamma_-(\mathfrak{h}_1) \otimes \Gamma_-(\mathfrak{h}_2)$ such that

$$U\Omega_{\mathfrak{h}_1 \oplus \mathfrak{h}_2} = \Omega_{\mathfrak{h}_1} \otimes \Omega_{\mathfrak{h}_2},$$

$$Ua(f \oplus g)U^{-1} = a(f) \otimes I + (-I)^N \otimes a(g), \tag{35}$$

$$Ua^*(f \oplus g)U^{-1} = a^*(f) \otimes I + (-I)^N \otimes a^*(g),$$

$$U d\Gamma(h_1 \oplus h_2)U^{-1} = d\Gamma(h_1) \otimes I + I \otimes d\Gamma(h_2).$$

The presence of the factors $(-I)^N$ in the above formulas complicates the description of a system containing several reservoirs. The following discussion should help the reader to understand its physical origin.

Consider two boxes $\mathcal{R}_1$, $\mathcal{R}_2$ with one particle Hilbert spaces $\mathfrak{h}_i = L^2(\mathcal{R}_i)$. Denote by $\mathcal{R}$ the combined box i.e., the disjoint union of $\mathcal{R}_1$ and $\mathcal{R}_2$. The corresponding one particle Hilbert space is $\mathfrak{h} \equiv L^2(\mathcal{R})$. Identifying the wave function Ψ_1 of an electron in $\mathcal{R}_1$ with $\Psi_1 \oplus 0$ and similarly for an electron in $\mathcal{R}_2$ we can replace $\mathfrak{h}$ with the direct sum $\mathfrak{h}_1 \oplus \mathfrak{h}_2$.

Assume that each box $\mathcal{R}_i$ contains a single electron with wave functions Ψ_i (see Fig. 2). If the boxes are in thermal contact, the two electrons can exchange energy, but the first one will always stay in $\mathcal{R}_1$ and the second one in $\mathcal{R}_2$. Thus they are distinguishable and the total wave function is just $\Psi_1 \otimes \Psi_2$. The situation is completely different if the electrons are free to move from one box into the other. In this case, the electrons are indistinguishable and Pauli's principle requires the total wave function to be antisymmetric—the total wave function is $\Psi_1 \wedge \Psi_2$. Generalizing this argument to many electrons states we conclude that the second quantized Hilbert

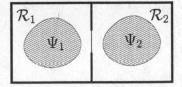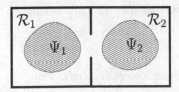

Fig. 2. Thermal contact and open gate between $\mathcal{R}_1$ and $\mathcal{R}_2$.

space is $\Gamma_-(\mathfrak{h}_1) \otimes \Gamma_-(\mathfrak{h}_2)$ in the case of thermal contact and $\Gamma_-(\mathfrak{h}_1 \oplus \mathfrak{h}_2)$ in the other case. The exponential law provides a unitary map U between these two Hilbert and one easily checks that

$$U\Psi_1 \wedge \Psi_2 = Ua^*(\Psi_1 \oplus 0)a^*(0 \oplus \Psi_2)\Omega_{\mathfrak{h}_1 \oplus \mathfrak{h}_2}$$
$$= (a^*(\Psi_1)(-I)^N \otimes a^*(\Psi_2))\Omega_{\mathfrak{h}_1} \otimes \Omega_{\mathfrak{h}_2}$$
$$= \Psi_1 \otimes \Psi_2.$$

Denoting by $\mathcal{O}_{\mathcal{R}_1}, \mathcal{O}_{\mathcal{R}_2}$ and $\mathcal{O}_{\mathcal{R}}$ the CAR (or more appropriately the CAR$^+$) algebras of the boxes $\mathcal{R}_1, \mathcal{R}_2$ and $\mathcal{R}$, the algebra of the combined system in the case of thermal contact is $\mathcal{O}_{\mathcal{R}_1} \otimes \mathcal{O}_{\mathcal{R}_2}$, while it is $\mathcal{O}_{\mathcal{R}}$ in the other case. We emphasize that the unitary map U does not yield an isomorphism between these algebras *i.e.*,

$$U\mathcal{O}_{\mathcal{R}}U^* \neq \mathcal{O}_{\mathcal{R}_1} \otimes \mathcal{O}_{\mathcal{R}_2}.$$

This immediately follows from the observation that $(-I)^N \notin \mathcal{O}_{\mathcal{R}_1}$ (unless, of course, $\mathcal{O}_{\mathcal{R}_1}$ is finite dimensional, see Subsection 6.3), which implies

$$Ua^*(0 \oplus \Psi_2)U^* = (-I)^N \otimes a^*(\Psi_2) \notin \mathcal{O}_{\mathcal{R}_1} \otimes \mathcal{O}_{\mathcal{R}_2}.$$

Note in particular that $a^*(\Psi_1) \otimes I$ and $I \otimes a^*(\Psi_2)$ commute while $a^*(\Psi_1 \oplus 0)$ and $a^*(0 \oplus \Psi_2)$ anticommute. The factor $(-I)^N$ is required in order for $a^*(\Psi_1) \otimes I$ and $(-I)^N \otimes a^*(\Psi_2)$ to anticommute.

5.2 Examples

Recall that the Pauli matrices are defined by

$$\sigma_x \equiv \begin{bmatrix} 0 & 1 \\ 1 & 0 \end{bmatrix}, \qquad \sigma_y \equiv \begin{bmatrix} 0 & -i \\ i & 0 \end{bmatrix}, \qquad \sigma_z \equiv \begin{bmatrix} 1 & 0 \\ 0 & -1 \end{bmatrix}.$$

We set $\sigma_\pm \equiv (\sigma_x \pm i\sigma_y)/2$. Clearly, $\sigma_x^2 = \sigma_y^2 = \sigma_z^2 = I$ and $\sigma_x\sigma_y = -\sigma_y\sigma_x = i\sigma_z$. More generally, with $\vec{\sigma} = (\sigma_x, \sigma_y, \sigma_z)$ and $\vec{u}, \vec{v} \in \mathbb{R}^3$ one has

$$(\vec{u} \cdot \vec{\sigma})(\vec{v} \cdot \vec{\sigma}) = \vec{u} \cdot \vec{v} I + i(\vec{u} \times \vec{v}) \cdot \vec{\sigma}.$$

Example 5.1. Assume that $\dim \mathfrak{h} = 1$, *i.e.*, that $\mathfrak{h} = \mathbb{C}$ and that h is the operator of multiplication by the real constant ω. Then $\Gamma_-(\mathfrak{h}) = \mathbb{C} \oplus \mathbb{C} = \mathbb{C}^2$ and $d\Gamma(h) = \omega N$ with

$$N \equiv d\Gamma(I) = \begin{bmatrix} 0 & 0 \\ 0 & 1 \end{bmatrix} = \frac{1}{2}(I - \sigma_z).$$

Moreover, one easily checks that

$$a(1) = \begin{bmatrix} 0 & 0 \\ 1 & 0 \end{bmatrix}, \qquad a^*(1) = \begin{bmatrix} 0 & 1 \\ 0 & 0 \end{bmatrix},$$

$$a^*(1)a(1) = \begin{bmatrix} 1 & 0 \\ 0 & 0 \end{bmatrix}, \; a(1)a^*(1) = \begin{bmatrix} 0 & 0 \\ 0 & 1 \end{bmatrix}, \tag{36}$$

which shows that $\mathrm{CAR}(\mathfrak{h})$ is the algebra of 2×2 matrices $M_2(\mathbb{C})$ and $\mathrm{CAR}^+(\mathfrak{h})$ its subalgebra of diagonal matrices. A self-adjoint operator $0 \le T \le I$ on $\mathcal{H}$ is multiplication by a constant γ, $0 \le \gamma \le 1$. The associated state ω_T on $\mathrm{CAR}(\mathfrak{h})$ is given by the density matrix

$$\begin{bmatrix} 1 - \gamma & 0 \\ 0 & \gamma \end{bmatrix}.$$

Example 5.2. Assume that $\dim \mathfrak{h} = n$. Without loss of generality we can set $\mathfrak{h} = \mathbb{C}^n$ and assume that $hf_j = \omega_j f_j$ for some $\omega_j \in \mathbb{R}$, where $\{f_j\}$ is the standard basis of $\mathbb{C}^n$. Then,

$$\Gamma_-(\mathfrak{h}) = \mathbb{C} \oplus \mathbb{C}^n \oplus \mathbb{C}^n \wedge \mathbb{C}^n \oplus \cdots \oplus (\mathbb{C}^n)^{\wedge n} \simeq \bigotimes_{i=1}^n \mathbb{C}^2,$$

and $\mathrm{CAR}(\mathfrak{h})$ is isomorphic to the algebra of $2^n \times 2^n$ matrices $M_{2^n}(\mathbb{C})$. This isomorphism is explicitly given by

$$a(f_j) \simeq \left(\otimes_{i=1}^{j-1} \sigma_z \right) \otimes \sigma_+ \otimes \left(\otimes_{i=j+1}^n I \right),$$

for $j = 1, \ldots, n$. It follows that

$$a^*(f_j)a(f_j) \simeq \frac{1}{2} \left(\otimes_{i=1}^{j-1} I \right) \otimes (I - \sigma_z) \otimes \left(\otimes_{i=j+1}^n I \right).$$

The map described by the above formulas is called the Jordan-Wigner transformation. It is a useful tool in the study of quantum spin systems (see [LMS, AB, Ar3]). For $\beta, \mu \in \mathbb{R}$, the quasi-free gauge-invariant state associated to $T = (I + e^{\beta(h-\mu)})^{-1}$ is given by the density matrix

$$\frac{e^{-\beta(H - \mu N)}}{\mathrm{Tr}\, e^{-\beta(H - \mu N)}},$$

with

$$H \equiv \mathrm{d}\Gamma(h) = \sum_{j=1}^n \omega_j \, a^*(f_j)a(f_j), \qquad N \equiv \mathrm{d}\Gamma(I) = \sum_{j=1}^n a^*(f_j)a(f_j).$$

It is an instructive exercise to work out the thermodynamics of the finite dimensional free Fermi gas following Section 3 in [Jo].

Example 5.3. In this example we will briefly discuss the finite dimensional approximation of a free Fermi gas. Assume that $\mathfrak{h}$ is a separable Hilbert space and let $\Lambda_n \subset \operatorname{Dom} h$ be an increasing sequence of finite dimensional subspaces. The algebras $\operatorname{CAR}(\Lambda_n)$ are identified with subalgebras of $\operatorname{CAR}(\mathfrak{h})$. We also assume that $\cup_n \Lambda_n$ is dense in $\mathfrak{h}$. Let p_n be the orthogonal projection on Λ_n. Set $h_n = p_n h p_n$ and let τ_n be the corresponding C^*-dynamics on $\operatorname{CAR}(\Lambda_n)$. Since p_n converges strongly to I one has, for $f \in \mathcal{H}$,

$$\lim_{n \to \infty} \|a^\#(p_n f) - a^\#(f)\| = 0, \qquad \lim_{n \to \infty} \|\tau_n^t(a^\#(p_n f)) - \tau^t(a^\#(f))\| = 0.$$

Let ω_T be the gauge-invariant quasi-free state on $\operatorname{CAR}(\mathfrak{h})$ associated to T. Let $T_n = p_n T p_n$. Then

$$\lim_{n \to \infty} \omega_{T_n}(a^*(p_n f) a(p_n g)) = \omega_T(a^*(f) a(g)).$$

Assume that μ and η are two faithful ω_T-normal states and let $\operatorname{Ent}(\mu|\eta)$ be their Araki relative entropy. Let μ_n and η_n be the restrictions of μ and η to $\operatorname{CAR}^+(\Lambda_n)$. Then the function

$$n \mapsto \operatorname{Ent}(\mu_n|\eta_n) = \operatorname{Tr}_{\Lambda_n}(\mu_n(\log \mu_n - \log \eta_n)),$$

is monotone increasing and

$$\lim_{n \to \infty} \operatorname{Ent}(\mu_n|\eta_n) = \operatorname{Ent}(\mu|\eta).$$

Additional information about the last result can be found in [BR2], Proposition 6.2.33.

Example 5.4. The tight binding approximation for an electron in a single Bloch band of a d-dimensional (cubic) crystal is defined by $\mathfrak{h} \equiv \ell^2(\mathbb{Z}^d)$ with the translation invariant Hamiltonian

$$(h\psi)(x) \equiv \frac{1}{2d} \sum_{|x-y|=1} \psi(y), \tag{37}$$

where $|x| \equiv \sum_i |x_i|$. In the sequel δ_x denotes the Kronecker delta function at $x \in \mathbb{Z}^d$.

Writing $a_x \equiv a(\delta_x)$, the second quantized energy and number operators are given by

$$d\Gamma(h) = \frac{1}{2d} \sum_{|x-y|=1} a_x^* a_y, \qquad d\Gamma(I) = \sum_x a_x^* a_x.$$

The Fourier transform $\hat{\psi}(k) \equiv \sum_x \psi(x) \, e^{-ix \cdot k}$ maps $\mathfrak{h}$ unitarily onto

$$\hat{\mathfrak{h}} \equiv L^2([-\pi, \pi]^d, \frac{dk}{(2\pi)^d}).$$

The set $[-\pi, \pi]^d$ is the Brillouin zone of the crystal and k is the quasi-momentum of the electron. The Fourier transform diagonalizes the Hamiltonian which becomes multiplication by the band function $\varepsilon(k) \equiv \frac{1}{d} \sum_i \cos(k_i)$. Thus h has purely absolutely continuous spectrum $\sigma(h) = [-1, 1]$, and in particular is bounded.

A simple stationary phase argument shows that

$$(f, e^{ith} g) = O(t^{-n}),$$

for arbitrary n provided $\hat{f}$ and $\hat{g}$ are smooth and vanish in a neighborhood of the critical set $\{k \mid |\nabla_k \varepsilon(k)| = 0\}$. Since this set has Lebesgue measure 0, such functions are dense in $\mathfrak{h}$. If f and g have bounded support in $\mathbb{Z}^d$, then

$$(f, e^{ith} g) = O(t^{-d/2}).$$

Example 5.5. The tight binding approximation of a semi-infinite wire is obtained by restricting the Hamiltonian (37), for $d = 1$, to the space of odd functions $\psi \in \ell^2(\mathbb{Z})$ and identifying such ψ with elements of $\ell^2(\mathbb{Z}_+)$, where $\mathbb{Z}_+ \equiv \{1, 2, \cdots\}$. This is clearly equivalent to imposing a Dirichlet boundary condition at $x = 0$ and

$$h = \frac{1}{2} \sum_{x=1}^{\infty} \left((\delta_x, \cdot) \delta_{x+1} + (\delta_{x+1}, \cdot) \delta_x \right).$$

The Fourier-sine transform $\tilde{\psi}(k) \equiv \sum_{x \in \mathbb{Z}_+} \psi(x) \sin(kx)$ maps unitarily $\ell^2(\mathbb{Z}_+)$ onto the space $L^2([0, \pi], \frac{2dk}{\pi})$ and the Hamiltonian becomes multiplication by $\cos k$. By a simple change of variable $r = \cos k$ we obtain the spectral representation of the Hamiltonian h:

$$(h\psi)^\#(r) = r\psi^\#(r),$$

where

$$\psi^\#(r) \equiv \sqrt{\frac{2}{\pi \sqrt{1 - r^2}}} \, \tilde{\psi}(\arccos(r)),$$

maps unitarily the Fourier space $L^2([0, \pi], \frac{2dk}{\pi})$ onto $L^2([-1, 1], dr)$. A straightforward integration by parts shows that

$$(f, e^{ith} g) = O(t^{-n}),$$

if $f^\#, g^\# \in C_0^n((-1, 1))$. A more careful analysis shows that

$$(f, e^{ith} g) = O(t^{-3/2}),$$

if f and g have bounded support in $\mathbb{Z}_+$.

Example 5.6. The non-relativistic spinless Fermion of mass m is described in the position representation by the Hilbert space $L^2(\mathbb{R}^d, dx)$ and the Hamiltonian $h = -\Delta/2m$, where Δ is the usual Laplacian in $\mathbb{R}^d$. The cases of physical interest are $d = 1, 2, 3$. In the momentum representation the Hilbert space of the Fermion is $L^2(\mathbb{R}^d, dk)$ and its Hamiltonian (which we will again denote by h) is the operator of multiplication by $|k|^2/2m$.

The spectrum of h is purely absolutely continuous. Integration by parts yields that

$$(f, e^{ith} g) = O(t^{-n}),$$

for arbitrary n provided $\hat{f}$ and $\hat{g}$ are smooth, compactly supported and vanish in a neighborhood of the origin. Such functions are dense in $\mathfrak{h}$. If $f, g \in \mathfrak{h}$ are compactly supported in the position representation, then

$$(f, e^{ith} g) = O(t^{-d/2}).$$

6 The Simple Electronic Black-Box (SEBB) Model

In the second part of this lecture we shall study in detail the non-equilibrium statistical mechanics of the simplest non-trivial example of the electronic black box model introduced in [AJPP]. The electronic black-box model is a general, independent electron model for a localized quantum device S connected to M electronic reservoirs $\mathcal{R}_1, \cdots, \mathcal{R}_M$. The device is called black-box since, according to the scattering approach introduced in Subsection 4.2, the thermodynamics of the coupled system is largely independent of the internal structure of the device. The NESS and the steady currents are completely determined by the Møller morphism which in our simple model further reduces to the one-particle wave operator.

6.1 The Model

The black-box itself is a two level system. Its Hilbert space is $\mathcal{H}_S \equiv \mathbb{C}^2$, its algebra of observables is $\mathcal{O}_S \equiv M_2(\mathbb{C})$, and its Hamiltonian is

$$H_S \equiv \begin{bmatrix} 0 & 0 \\ 0 & \varepsilon_0 \end{bmatrix}.$$

The associated C^*-dynamics is $\tau_S^t(A) = e^{itH_S} A e^{-itH_S}$. The black-box has a one-parameter family of steady states with density matrices

$$\omega_S \equiv \begin{bmatrix} 1 - \gamma & 0 \\ 0 & \gamma \end{bmatrix}, \qquad \gamma \in [0, 1],$$

which we shall use as the reference states.

According to Example 5.1 of Subsection 5.2, we can also think of S as a free Fermi gas over $\mathbb{C}$, namely $H_S = \Gamma_-(\mathbb{C})$, $H_S = d\Gamma(\varepsilon_0) = \varepsilon_0 a^*(1)a(1)$ and $\mathcal{O}_S =$

CAR($\mathbb{C}$). In this picture, the black-box S can only accommodate a single Fermion of energy ε_0. We denote by $N_S = a^*(1)a(1)$ the corresponding number operator. In physical terms, S is a quantum dot without internal structure. We also note that ω_S is the quasi-free gauge-invariant state generated by $T_S \equiv \gamma$. Therefore, we can interpret γ as the occupation probability of the box.

Let $\mathfrak{h}_\mathcal{R}$ be a Hilbert space and $h_\mathcal{R}$ a self-adjoint operator on $\mathfrak{h}_\mathcal{R}$. We set $\mathcal{O}_\mathcal{R} \equiv$ CAR($\mathfrak{h}_\mathcal{R}$) and

$$\tau_\mathcal{R}^t(A) \equiv \mathrm{e}^{\mathrm{it}\mathrm{d}\Gamma(h_\mathcal{R})} A \, \mathrm{e}^{-\mathrm{it}\mathrm{d}\Gamma(h_\mathcal{R})}.$$

The reference state of the reservoir, $\omega_\mathcal{R}$, is the quasi-free gauge-invariant state associated to the radiation density operator $T_\mathcal{R}$. We assume that $h_\mathcal{R}$ is bounded from below and that $T_\mathcal{R}$ commutes with $h_\mathcal{R}$.

To introduce the subreservoir structure we shall assume that

$$\mathfrak{h}_\mathcal{R} = \oplus_{j=1}^M \mathfrak{h}_{\mathcal{R}_j}, \qquad h_\mathcal{R} = \oplus_{j=1}^M h_{\mathcal{R}_j}, \qquad T_\mathcal{R} = \oplus_{j=1}^M T_{\mathcal{R}_j}.$$

The algebra of observables of the j-th reservoir is $\mathcal{O}_{\mathcal{R}_j} \equiv$ CAR($\mathfrak{h}_{\mathcal{R}_j}$) and its dynamics $\tau_{\mathcal{R}_j} \equiv \tau_\mathcal{R} \restriction \mathcal{O}_{\mathcal{R}_j}$ is generated by the Hamiltonian $\mathrm{d}\Gamma(h_{\mathcal{R}_j})$. The state $\omega_{\mathcal{R}_j} = \omega_\mathcal{R} \restriction \mathcal{O}_{\mathcal{R}_j}$ is the gauge-invariant quasi-free state associated to $T_{\mathcal{R}_j}$. If p_j is the orthogonal projection on $\mathfrak{h}_{\mathcal{R}_j}$, then $N_{\mathcal{R}_j} = \mathrm{d}\Gamma(p_j)$ is the charge (or number) operator associated to the j-th reservoir. The total charge operator of the reservoir is $N_\mathcal{R} = \sum_{j=1}^M N_{\mathcal{R}_j}$.

The algebra of observables of the joint system $S + \mathcal{R}$ is $\mathcal{O} \equiv \mathcal{O}_S \otimes \mathcal{O}_\mathcal{R}$, its reference state is $\omega = \omega_S \otimes \omega_\mathcal{R}$, and its decoupled dynamics is $\tau_0 = \tau_S \otimes \tau_\mathcal{R}$. Note that

$$\tau_0^t(A) = \mathrm{e}^{\mathrm{it}H_0} A \, \mathrm{e}^{-\mathrm{it}H_0},$$

where

$$H_0 \equiv H_S \otimes I + I \otimes \mathrm{d}\Gamma(h_\mathcal{R}).$$

The junction between the box S and the reservoir $\mathcal{R}_j$ works in the following way: The box can make a transition from its ground state to its excited state by absorbing an electron of $\mathcal{R}_j$ in state $f_j/\|f_j\|$. Reciprocally, the excited box can relax to its ground state by emitting an electron in state $f_j/\|f_j\|$ in $\mathcal{R}_j$. These processes have a fixed rate $\lambda^2 \|f_j\|^2$. More precisely, the junction is described by

$$\lambda V_j \equiv \lambda \left(a(1) \otimes a^*(f_j) + a^*(1) \otimes a(f_j) \right),$$

where $\lambda \in \mathbb{R}$ and the $f_j \in \mathfrak{h}_j$. The normalization is fixed by the condition $\sum_j \|f_j\|^2 = 1$. The complete interaction is given by

$$\lambda V \equiv \sum_{j=1}^M \lambda V_j = \lambda(a(1) \otimes a^*(f) + a^*(1) \otimes a(f)),$$

where $f \equiv \oplus_{j=1}^M f_j$. Note that "charge" is conserved at the junction, i.e., V commutes with the total number operator $N \equiv N_S \otimes I + I \otimes N_\mathcal{R}$.

The full Hamiltonian is

$$H_\lambda \equiv H_0 + \lambda V,$$

and the corresponding C^*-dynamics

$$\tau_\lambda^t(A) \equiv e^{itH_\lambda} A e^{-itH_\lambda},$$

is charge-preserving. In other words, τ_λ commutes with the gauge group

$$\vartheta^t(A) \equiv e^{itN} A e^{-itN},$$

and $[H_\lambda, N] = 0$. The C^*-dynamical system $(\mathcal{O}, \tau_\lambda)$ with its decoupled dynamics τ_0^t and the reference state $\omega = \omega_S \otimes \omega_R$ is our *simple electronic black box model (SEBB)*. This model is an example of the class of open quantum systems described in Section 4.

6.2 The Fluxes

The heat flux observables have been defined in Subsection 4.3. The generator of $\tau_{\mathcal{R}_j}$ is given by $\delta_j(\cdot) = i[d\Gamma(h_{\mathcal{R}_j}), \cdot]$. Note that $V_j \in \text{Dom}\,\delta_j$ iff $f_j \in \text{Dom}\,h_{\mathcal{R}_j}$. If $V_j \in \text{Dom}\,\delta_j$, then the observable describing the heat flux out of $\mathcal{R}_j$ is

$$\Phi_j = \lambda\delta_j(V_j) = \lambda(a(1) \otimes a^*(ih_{\mathcal{R}_j}f_j) + a^*(1) \otimes a(ih_{\mathcal{R}_j}f_j)).$$

In a completely similar way we can define the charge current. The rate of change of the charge in the box S is

$$\frac{d}{dt}\tau_\lambda^t(N_S)|_{t=0} = i[d\Gamma(H_\lambda), N_S]$$

$$= -\lambda i[N_S, V] = \lambda i[N_R, V] = \sum_{j=1}^M \lambda i[N_{\mathcal{R}_j}, V], \tag{38}$$

which allows us to identify

$$\mathcal{J}_j \equiv \lambda i[N_{\mathcal{R}_j}, V]$$

$$= \lambda i[N_{\mathcal{R}_j}, V_j] = \lambda i[N_R, V_j] = \lambda(a(1) \otimes a^*(if_j) + a^*(1) \otimes a(if_j)),$$

as the observable describing the charge current out of $\mathcal{R}_j$.

Let us make a brief comment concerning these definitions. If $\mathfrak{h}_{\mathcal{R}_j}$ is finite dimensional, then the energy and the charge of $\mathcal{R}_j$ are observables, given by the Hamiltonian $d\Gamma(h_{\mathcal{R}_j})$ and the number operator $N_{\mathcal{R}_j} = d\Gamma(p_j)$, and

$$-\frac{d}{dt}\tau_\lambda^t(d\Gamma(h_{\mathcal{R}_j}))|_{t=0} = \lambda i[d\Gamma(h_{\mathcal{R}_j}), V_j] = \Phi_j,$$

$$-\frac{d}{dt}\tau_\lambda^t(d\Gamma(p_j))|_{t=0} = \lambda i[d\Gamma(p_j), V_j] = \mathcal{J}_j.$$

When $\mathfrak{h}_{\mathcal{R}_j}$ becomes infinite dimensional (recall Example 5.3 in Subsection 5.2), $d\Gamma(h_{\mathcal{R}_j})$ and $N_{\mathcal{R}_j}$ are no longer observables. However, the flux observables Φ_j and $\mathcal{J}_j$ are still well-defined and they are equal to the limit of the flux observables corresponding to finite-dimensional approximations.

The first law of thermodynamics (energy conservation) has been verified in Subsection 4.3—for any τ_λ-invariant state η one has

$$\sum_{j=1}^{M} \eta(\Phi_j) = 0.$$

The analogous statement for charge currents is proved in a similar way. By (38),

$$\sum_{j=1}^{M} \mathcal{J}_j = \frac{d}{dt} \tau_\lambda^t(N_\mathcal{S})|_{t=0},$$

and so for any τ_λ-invariant state η one has

$$\sum_{j=1}^{M} \eta(\mathcal{J}_j) = 0. \tag{39}$$

6.3 The Equivalent Free Fermi Gas

In this subsection we shall show how to use the exponential law for fermionic systems to map the SEBB model to a free Fermi gas. Let

$$\mathfrak{h} \equiv \mathbb{C} \oplus \mathfrak{h}_\mathcal{R} = \mathbb{C} \oplus \left(\bigoplus_{j=1}^{M} \mathfrak{h}_{\mathcal{R}_j} \right), \qquad \tilde{\mathcal{O}} \equiv \mathrm{CAR}(\mathfrak{h}), \qquad h_0 \equiv \varepsilon_0 \oplus h_\mathcal{R},$$

and, with a slight abuse of notation, denote by $1, f_1, \cdots, f_M$ the elements of $\mathfrak{h}$ canonically associated with $1 \in \mathbb{C}$ and $f_j \in \mathfrak{h}_{\mathcal{R}_j}$. Then

$$v_j \equiv (1, \cdot) f_j + (f_j, \cdot) 1,$$

is a finite rank, self-adjoint operator on $\mathfrak{h}$ and so is the sum $v \equiv \sum_{j=1}^{M} v_j$. We further set

$$h_\lambda \equiv h_0 + \lambda v, \tag{40}$$

and define the dynamical group

$$\tilde{\tau}_\lambda^t(A) \equiv e^{itd\Gamma(h_\lambda)} A\, e^{-itd\Gamma(h_\lambda)},$$

on $\tilde{\mathcal{O}}$. Finally, we set

$$\tilde{T} \equiv T_\mathcal{S} \oplus T_\mathcal{R},$$

and denote by $\tilde{\omega}$ be the quasi-free gauge-invariant state on $\tilde{\mathcal{O}}$ generated by $\tilde{T}$.

Theorem 6.1. *Let $U : \Gamma_-(\mathbb{C} \oplus \mathfrak{h}_\mathcal{R}) \to \Gamma_-(\mathbb{C}) \otimes \Gamma_-(\mathfrak{h}_\mathcal{R})$ be the unitary map defined by the exponential law (35) and set $\phi(A) \equiv U^{-1}AU$.*

(i) $\phi : \mathcal{O} \to \tilde{\mathcal{O}}$ is a $*$-isomorphism.
(ii) For any $\lambda, t \in \mathbb{R}$ one has $\phi \circ \tau_\lambda^t = \tilde{\tau}_{-\lambda}^t \circ \phi$.
(iii) $\omega = \tilde{\omega} \circ \phi$.
(iv) For $j = 1, \cdots, M$, one has

$$\tilde{\Phi}_j \equiv \phi(\Phi_j) = -\lambda \left(a^*(ih_j f_j)a(1) + a^*(1)a(ih_j f_j) \right),$$

and

$$\tilde{\mathcal{J}}_j \equiv \phi(\mathcal{J}_j) = -\lambda(a^*(if_j)a(1) + a^*(1)a(if_j)).$$

Proof. Clearly, ϕ is a $*$-isomorphism from $\mathcal{B}(\Gamma_-(\mathbb{C} \oplus \mathfrak{h}))$ onto $\mathcal{B}(\Gamma_-(\mathbb{C}) \otimes \Gamma_-(\mathfrak{h}))$. Using the canonical injections $\mathbb{C} \to \mathfrak{h}$ and $\mathfrak{h}_\mathcal{R} \to \mathfrak{h}$ we can identify $\mathcal{O}_\mathcal{S}$ and $\mathcal{O}_\mathcal{R}$ with the subalgebras of $\tilde{\mathcal{O}}$ generated by $a(1 \oplus 0)$ and $\{a(0 \oplus f) \mid f \in \mathfrak{h}_\mathcal{R}\}$. With this identification, (35) gives

$$\phi(a(\alpha) \otimes I + (-I)^{N_S} \otimes a(f)) = a(\alpha) + a(f),$$

for $\alpha \in \mathbb{C}$ and $f \in \mathfrak{h}_\mathcal{R}$. We conclude that

$$\phi(A \otimes I) = A, \tag{41}$$

for any $A \in \mathcal{O}_\mathcal{S}$. In particular, since $b \equiv (-I)^{N_S} = [a(1), a^*(1)] \in \mathcal{O}_\mathcal{S}$, we have $\phi(b \otimes I) = b$. Relation $b^2 = I$ yields $\phi(I \otimes a(f)) = b \, a(f)$. Since $[b, a(f)] = 0$, we conclude that for $A \in \mathcal{O}_\mathcal{R}$

$$\phi(I \otimes A) = \begin{cases} A & \text{if } A \in \mathcal{O}_\mathcal{R}^+, \\ b \, A & \text{if } A \in \mathcal{O}_\mathcal{R}^-, \end{cases} \tag{42}$$

where $\mathcal{O}_\mathcal{R}^\pm$ denote the even and odd parts of $\mathcal{O}_\mathcal{R}$. Equ. (41) and (42) show that $\phi(\mathcal{O}) \subset \tilde{\mathcal{O}}$. Since $\tilde{\mathcal{O}} = \langle \mathcal{O}_\mathcal{S}, \mathcal{O}_\mathcal{R}^+, \mathcal{O}_\mathcal{R}^- \rangle$, it follows from $\phi(\mathcal{O}_\mathcal{S} \otimes I) = \mathcal{O}_\mathcal{S}$, $\phi(I \otimes \mathcal{O}_\mathcal{R}^+) = \mathcal{O}_\mathcal{R}^+$ and $\phi(b \otimes \mathcal{O}_\mathcal{R}^-) = \mathcal{O}_\mathcal{R}^-$ that $\phi(\mathcal{O}) \supset \tilde{\mathcal{O}}$. This proves Part (i).

From (35) we can see that $U^{-1}H_0 U = d\Gamma(h_0)$ and from (41) and (42) that

$$U^{-1}V_j U = \phi(V_j) = a(1) \, b \, a^*(f_j) + a^*(1) \, b \, a(f_j).$$

Since it also follows from CAR that

$$a(1) \, b = -a(1), \qquad a^*(1) \, b = a^*(1), \tag{43}$$

we get

$$U^{-1}V_j U = -a(1) \, a^*(f_j) + a^*(1) \, a(f_j) = -a(1) \, a^*(f_j) - a(f_j) \, a^*(1) = -d\Gamma(v_j).$$

Therefore $U^{-1}H_\lambda U = d\Gamma(h_{-\lambda})$ from which Part (ii) follows. A similar computation yields Part (iv).

It remains to prove Part (iii). Using the morphism θ (recall Equ. (32)) to express the even and odd parts of $B \in \mathcal{O}_\mathcal{R}$, we can rewrite (41) and (42) as

$$\phi(A \otimes B) = A(B + \theta(B))/2 + A\,b\,(B - \theta(B))/2,$$

from which we easily get

$$\phi(A \otimes B) = Aa(1)a^*(1)B + Aa^*(1)a(1)\theta(B).$$

It follows from the factorization property (30) and the invariance property (33) of quasi-free states that

$$\begin{aligned}
\tilde{\omega} \circ \phi(A \otimes B) &= \tilde{\omega}(Aa(1)a^*(1))\tilde{\omega}(B) + \tilde{\omega}(Aa^*(1)a(1))\tilde{\omega}(B) \\
&= \tilde{\omega}(Aa(1)a^*(1)B + Aa^*(1)a(1)B) \\
&= \tilde{\omega}(AB) = \tilde{\omega}(A)\tilde{\omega}(B) \\
&= \omega_\mathcal{S}(A)\omega_\mathcal{R}(B) = \omega(A \otimes B).
\end{aligned}$$

$\square$

By Theorem 6.1, the SEBB model can be equivalently described by the C^*-dynamical system $(\tilde{O}, \tilde{\tau}_{-\lambda})$ and the reference state $\omega_{\tilde{T}}$. The heat and charge flux observables are $\tilde{\Phi}_j$ and $\tilde{\mathcal{J}}_j$. Since the change $\lambda \to -\lambda$ affects neither the model nor the results, *in the sequel we will work with the system $(\tilde{O}, \tilde{\tau}_\lambda)$ and we will drop the* $\sim$. Hence, we will use the C^*-algebra $\mathcal{O} = \mathrm{CAR}(\mathbb{C} \oplus \mathfrak{h}_\mathcal{R})$ and C^*-dynamics

$$\tau_\lambda^t(A) = e^{\mathrm{i}td\Gamma(h_\lambda)}A\,e^{-\mathrm{i}td\Gamma(h_\lambda)},$$

with the reference state ω, the quasi-free gauge-invariant state generated by $T = T_\mathcal{S} \oplus T_\mathcal{R}$. The corresponding heat and charge flux observables are

$$\Phi_j \equiv \lambda\,(a^*(\mathrm{i}h_j f_j)a(1) + a^*(1)a(\mathrm{i}h_j f_j)),$$

$$\mathcal{J}_j \equiv \lambda(a^*(\mathrm{i}f_j)a(1) + a^*(1)a(\mathrm{i}f_j)).$$

The entropy production observable associated to ω is computed as follows. Assume that for $j = 1, \cdots, M$ one has $\mathrm{Ker}\,T_{\mathcal{R}_j} = \mathrm{Ker}\,(I - T_{\mathcal{R}_j}) = \{0\}$ and set

$$s_j \equiv -\log T_{\mathcal{R}_j}(I - T_{\mathcal{R}_j})^{-1}, \qquad s_\mathcal{R} = \oplus_{j=1}^M s_j.$$

We also assume that $0 < \gamma < 1$ and set $s_\mathcal{S} = \log \gamma(1 - \gamma)^{-1}$. Let $s \equiv -s_\mathcal{S} \oplus s_\mathcal{R}$. Under the above assumptions, the reference state ω is modular and its modular automorphism group is

$$\sigma_\omega^t(A) = e^{\mathrm{i}td\Gamma(s)}A\,e^{-\mathrm{i}td\Gamma(s)}.$$

If $f_j \in \mathrm{Dom}(s_j)$, then the entropy production observable is

$$\sigma = -\lambda\,(a^*(f)a(\mathrm{i}s_\mathcal{S}) + a^*(\mathrm{i}s_\mathcal{S})a(f)) - \lambda\,(a^*(\mathrm{i}s_\mathcal{R}f)a(1) + a^*(1)a(\mathrm{i}s_\mathcal{R}f)).$$
$$(44)$$

The entropy balance equation

$$\mathrm{Ent}(\omega \circ \tau_\lambda^t | \omega) = -\int_0^t \omega(\tau_\lambda^s(\sigma))\,\mathrm{d}s,$$

holds and so, as in Subsection 3.2, the entropy production of any NESS $\omega_+ \in \Sigma_+(\omega,\tau_\lambda)$ is non-negative. In fact, it is not difficult to show that the entropy production of ω_+ is independent of γ as long as $\gamma \in (0,1)$ (see Proposition 5.3 in [JP4]). *In the sequel, whenever we speak about the entropy production, we will assume that* $\gamma = 1/2$ *and hence that*

$$\sigma = -\lambda\left(a^*(\mathrm{i}s_\mathcal{R}f)a(1) + a^*(1)a(\mathrm{i}s_\mathcal{R}f)\right). \tag{45}$$

In particular, if

$$T_{\mathcal{R}_j} = (I + e^{\beta_j(h_{\mathcal{R}_j}-\mu_j)}),$$

then $s_j = -\beta_j(h_{\mathcal{R}_j} - \mu_j)$, and

$$\sigma = -\sum_{j=1}^M \beta_j(\Phi_j - \mu_j \mathcal{J}_j). \tag{46}$$

We finish with the following remark. In the physics literature, the Hamiltonian (40) is sometimes called the *Wigner-Weisskopf atom* [WW] (see [JKP] for references and additional information). The operators of this type are also often called *Friedrich Hamiltonians* [Fr]. The point we wish to emphasize is that such Hamiltonians are often used as toy models which allow for simple mathematical analysis of physically important phenomena.

6.4 Assumptions

In this subsection we describe a set of assumptions under which we shall study the thermodynamics of the SEBB model.

Assumption (SEBB1) $\mathfrak{h}_{\mathcal{R}_j} = L^2((e_-,e_+),\mathrm{d}r)$ for some $-\infty < e_- < e_+ \leq \infty$ and $h_{\mathcal{R}_j}$ is the operator of multiplication by r.

The assumption (SEBB1) yields that $\mathfrak{h}_\mathcal{R} = L^2((e_-,e_+),\mathrm{d}r;\mathbb{C}^M)$ and that $h_\mathcal{R}$ is the operator of multiplication by r. With a slight abuse of the notation we will sometimes denote $h_{\mathcal{R}_j}$ and $h_\mathcal{R}$ by r. Note that the spectrum of $h_\mathcal{R}$ is purely absolutely continuous and equal to $[e_-,e_+]$ with uniform multiplicity M. With the shorthand $f \equiv (f_1,\cdots,f_M) \in \mathfrak{h}_\mathcal{R}$, the Hamiltonian (40) acts on $\mathbb{C} \oplus \mathfrak{h}_\mathcal{R}$ and has the form

$$h_\lambda = \varepsilon_0 \oplus r + \lambda((1,\cdot)f + (f,\cdot)1). \tag{47}$$

Assumption (SEBB2) The functions

$$g_j(t) \equiv \int_{e_-}^{e_+} e^{itr} |f_j(r)|^2 \, dr,$$

belong to $L^1(\mathbb{R}, dt)$.

Assumption (SEBB2) implies that the function

$$G(z) \equiv \int_{e_-}^{e_+} \frac{|f(r)|^2}{r - z} \, dr = -i \int_0^\infty g(t) \, e^{-itz} \, dt,$$

which is obviously analytic in the lower half-plane $\mathbb{C}_- \equiv \{z \,|\, \mathrm{Im}\, z < 0\}$, is continuous and bounded on its closure $\bar{\mathbb{C}}_-$. We denote by $G(r - io)$ the value of this function at $r \in \mathbb{R}$.

Assumption (SEBB3) For $j = 1, \cdots, M$, the generator $T_{\mathcal{R}_j}$ is the operator of multiplication by a continuous function $\rho_j(r)$ such that $0 < \rho_j(r) < 1$ for $r \in (e_-, e_+)$. Moreover, if

$$s_j(r) \equiv \log \left[\frac{\rho_j(r)}{1 - \rho_j(r)} \right],$$

we assume that $s_j(r) f_j(r) \in L^2((e_-, e_+), dr)$.

Assumption (SEBB3) ensures that the reference state $\omega_\mathcal{R}$ of the reservoir is modular. The function $\rho_j(r)$ is the energy density of the j-th reservoir. The second part of this assumption ensures that the entropy production observable (44) is well defined.

The study of SEBB model depends critically on the spectral and scattering properties of h_λ. Our final assumption will ensure that Assumption (S) of Subsection 3.4 holds and will allow us to use a simple scattering approach to study SEBB.

Assumption (SEBB4) $\varepsilon_0 \in (e_-, e_+)$ and $|f(\varepsilon_0)| \neq 0$.

We set

$$F(r) \equiv \varepsilon_0 - r - \lambda^2 G(r - io) = \varepsilon_0 - r - \lambda^2 \int_{e_-}^{e_+} \frac{|f(r')|^2}{r' - r + io} \, dr'. \qquad (48)$$

By a well-known result in harmonic analysis (see, e.g., [Ja] or any harmonic analysis textbook),

$$\mathrm{Im}\, F(r) = \lambda^2 \pi |f(r)|^2, \qquad (49)$$

for $r \in (e_-, e_+)$. We also mention that for any $g \in \mathfrak{h}_\mathcal{R} = L^2((e_-, e_+), dr; \mathbb{C}^M)$, the function

$$r \mapsto \int_{e_-}^{e_+} \frac{\bar{f}(r') \cdot g(r')}{r' - r + io} \, dr',$$

is also in $\mathfrak{h}_\mathcal{R}$:

The main spectral and scattering theoretic results on h_λ are given in the following Theorem which is an easy consequence of the techniques described in [Ja]. Its proof can be found in [JKP].

Theorem 6.2. *Suppose that Assumptions* (SEBB1), (SEBB2) *and* (SEBB4) *hold. Then there exists a constant* $\Lambda > 0$ *such that, for any* $0 < |\lambda| < \Lambda$:

(i) The spectrum of h_λ *is purely absolutely continuous and equal to* $[e_-, e_+]$.
(ii) The wave operators

$$W_\pm \equiv \mathrm{s} - \lim_{t \to \pm\infty} \mathrm{e}^{\mathrm{i}th_0}\, \mathrm{e}^{-\mathrm{i}th_\lambda},$$

exist and are complete, i.e., $\mathrm{Ran}\, W_\pm = \mathfrak{h}_\mathcal{R}$ *and* $W_\pm : \mathfrak{h} \to \mathfrak{h}_\mathcal{R}$ *are unitary. Moreover, if* $\psi = \alpha \oplus g \in \mathfrak{h}$, *then*

$$(W_-\psi)(r) = g(r) - \lambda F(r)^{-1} \left[\alpha - \lambda \int_{e_-}^{e_+} \frac{\bar{f}(r') \cdot g(r')}{r' - r + \mathrm{io}}\, \mathrm{d}r' \right] f(r). \quad (50)$$

Needless to say, the thermodynamics of the SEBB model can be studied under much more general assumptions than (SEBB1)-(SEBB4). However, these assumptions allow us to describe the results of [AJPP] with the least number of technicalities.

Parenthetically, we note that the SEBB model is obviously time-reversal invariant. Write $f_j(r) = \mathrm{e}^{\mathrm{i}\theta_j(r)}|f_j(r)|$, and let

$$\mathrm{j}(\alpha \oplus (g_1, \cdots, g_M)) = \bar{\alpha} \oplus (\mathrm{e}^{2\mathrm{i}\theta_1}\bar{g}_1, \cdots, \mathrm{e}^{2\mathrm{i}\theta_M}\bar{g}_M),$$

where $\bar{}$ denotes the usual complex conjugation. Then the map

$$\mathfrak{r}(A) = \Gamma(\mathrm{j})A\Gamma(\mathrm{j}^{-1}).$$

is a time reversal and ω is time reversal invariant.

Finally, as an example, consider a concrete SEBB model where each reservoir is a semi-infinite wire in the tight-binding approximation described in Example 5.5 of Subsection 5.2. Thus, for each j, $\mathfrak{h}_{\mathcal{R}_j} = \ell^2(\mathbb{Z}_+)$ and $h_{\mathcal{R}_j}$ is the discrete Laplacian on $\mathbb{Z}_+$ with Dirichlet boundary condition at 0. Choosing $f_j = \delta_1$ we obtain, in the spectral representation of $h_{\mathcal{R}_j}$,

$$\mathfrak{h}_{\mathcal{R}_j} = L^2((-1,1), \mathrm{d}r),$$
$$h_{\mathcal{R}_j} = r,$$
$$f_j^{\#}(r) = \sqrt{\frac{2}{\pi}}(1 - r^2)^{1/4}.$$

Thus, Assumptions (SEBB1) and (SEBB4) hold. Since, as $t \to \infty$, one has

$$\int_{-1}^{1} \mathrm{e}^{\mathrm{i}tr}|f^{\#}(r)|^2\, \mathrm{d}r = \frac{2M}{t} J_1(t) = O(t^{-3/2}),$$

where J_1 denotes a Bessel function of the first kind, Assumption (SEBB2) is also satisfied. Hence, if $\epsilon_0 \in (-1,1)$, then the conclusions of Theorem 6.2 hold. In fact one can show that in this case

$$\Lambda = \sqrt{\frac{1 - |\varepsilon_0|}{2M}}.$$

7 Thermodynamics of the SEBB Model

Throughout this and the next section we will assume that Assumptions (SEBB1)-(SEBB4) hold.

7.1 Non-Equilibrium Steady States

In this subsection we show that the SEBB model has a unique NESS $\omega_{\lambda+}$ which does not depend on the choice of the initial state $\eta \in \mathcal{N}_\omega$. Recall that the reference state ω of the SEBB model is the quasi-free gauge-invariant state generated by $T = T_S \oplus T_\mathcal{R}$, where $T_S = \gamma \in (0,1)$ and $T_\mathcal{R} = \oplus_j \rho_j(r)$.

Theorem 7.1. *Let $\Lambda > 0$ be the constant introduced in Theorem 6.2. Then, for any real λ such that $0 < |\lambda| < \Lambda$ the following hold:*

(i) The limit

$$\alpha_\lambda^+(A) \equiv \lim_{t \to \infty} \tau_0^{-t} \circ \tau_\lambda^t(A), \tag{51}$$

exists for all $A \in \mathcal{O}$. Moreover, $\operatorname{Ran} \alpha_\lambda^+ = \mathcal{O}_\mathcal{R}$ and α_λ^+ is an isomorphism between the $\check{C}^*$-dynamical systems $(\mathcal{O}, \tau_\lambda)$ and $(\mathcal{O}_\mathcal{R}, \tau_\mathcal{R})$.

(ii) Let $\omega_{\lambda+} \equiv \omega_\mathcal{R} \circ \alpha_\lambda^+$. Then

$$\lim_{t \to \infty} \eta \circ \tau_\lambda^t = \omega_{\lambda+},$$

for all $\eta \in \mathcal{N}_\omega$.

(iii) $\omega_{\lambda+}$ is the gauge-invariant quasi-free state on $\mathcal{O}$ generated by

$$T_+ \equiv W_-^* T_\mathcal{R} W_-,$$

where W_- is the wave operator of Theorem 6.2.

Proof. Recall that τ_λ^t is a group of Bogoliubov automorphisms, $\tau_\lambda^t(a^\#(f)) = a^\#(e^{ith_\lambda} f)$. Hence, for any observable of the form

$$A = a^\#(\psi_1) \cdots a^\#(\psi_n), \tag{52}$$

$$\tau_0^{-t} \circ \tau_\lambda^t(A) = a^\#(e^{-ith_0} e^{ith_\lambda} \psi_1) \cdots a^\#(e^{-ith_0} e^{ith_\lambda} \psi_n).$$

It follows from Theorem 6.2 that

$$\lim_{t \to \infty} \tau_0^{-t} \circ \tau_\lambda^t(A) = a^\#(W_- \psi_1) \cdots a^\#(W_- \psi_n).$$

Since the linear span of set of elements of the form (52) is dense in $\mathcal{O}$, the limit (51) exists and is given by the Bogoliubov morphism $\alpha_\lambda^+(a^\#(f)) = a^\#(W_- f)$. Since W_- is a unitary operator between $\mathfrak{h}$ and $\mathfrak{h}_\mathcal{R}$, $\operatorname{Ran} \alpha_\lambda^+ = \operatorname{CAR}(\mathfrak{h}_\mathcal{R}) = \mathcal{O}_\mathcal{R}$, which proves Part (i).

Since $h_\mathcal{R}$ has purely absolutely continuous spectrum, it follows from our discussion of quasi-free states in Subsection 5.1 that $\omega_\mathcal{R}$ is mixing for τ_0^t. Part (ii) is thus a restatement of Proposition 3.5.

If $A = a^*(\psi_n) \cdots a^*(\psi_1) a(\phi_1) \cdots a(\phi_m)$ is an element of $\mathcal{O}$, then

$$\omega^+(A) = \omega_\mathcal{R}(a^*(W_-\psi_n) \cdots a^*(W_-\psi_1) a(W_-\phi_1) \cdots a(W_-\phi_m))$$

$$= \delta_{n,m} \det \{(W_-\phi_i, T_\mathcal{R} W_-\psi_j)\}$$

$$= \delta_{n,m} \det \{(\phi_i, T_+\psi_j)\}.$$

and Part (iii) follows. $\square$

7.2 The Hilbert-Schmidt Condition

Since ω and $\omega_{\lambda+}$ are factor states, they are either quasi-equivalent ($\mathcal{N}_\omega = \mathcal{N}_{\omega_{\lambda+}}$) or disjoint ($\mathcal{N}_\omega \cap \mathcal{N}_{\omega_{\lambda+}} = \emptyset$). Since $\operatorname{Ker} T = \operatorname{Ker}(I - T) = \{0\}$, we also have $\operatorname{Ker} T_+ = \operatorname{Ker}(I - T_+) = \{0\}$, and so ω and $\omega_{\lambda+}$ are quasi-equivalent iff they are unitarily equivalent.

Let $\alpha > 0$. A function $h : (e_-, e_+) \to \mathbb{C}$ is α-Hölder continuous if there exists a constant C such that for all $r, r' \in (e_-, e_+)$, $|h(r) - h(r')| \leq C|r - r'|^\alpha$.

Theorem 7.2. *Assume that all the densities $\rho_j(r)$ are the same and equal to $\rho(r)$. Assume further that the functions $\rho(r)^{1/2}$ and $(1-\rho(r))^{1/2}$ are α-Hölder continuous for some $\alpha > 1/2$. Then the operators*

$$(T_+)^{1/2} - T^{1/2} \qquad and \qquad (I - T_+)^{1/2} - (I - T)^{1/2}$$

are Hilbert-Schmidt. In particular, the reference state ω and the NESS $\omega_{\lambda+}$ are unitarily equivalent and $\operatorname{Ep}(\omega_{\lambda+}) = 0$.

Remark. We will prove this theorem in Appendix 9.2. Although the Hölder continuity assumption is certainly not optimal, it covers most cases of interest and allows for a technically simple proof.

Theorem 7.2 requires a comment. By the general principles of statistical mechanics, one expects that $\operatorname{Ep}(\omega_{\lambda+}) = 0$ if and only if all the reservoirs are in *thermal equilibrium* at the same inverse temperature β and chemical potential μ (see Section 4.3 in [JP4]). This is not the case in the SEBB model because the perturbations V_j are chosen in such a special way that the coupled dynamics is still given by a Bogoliubov automorphism. Following the strategy of [JP4], one can show that the Planck law $\rho(r) = (1 + e^{\beta(r-\mu)})^{-1}$ can be deduced from the stability requirement $\operatorname{Ep}(\omega_{\lambda+}) = 0$ for a more general class of interactions V_j. For reasons of space we will not discuss this subject in detail in these lecture notes (the interested reader may consult [AJPP]).

We will see below that the entropy production of the SEBB model is non-vanishing whenever the density operators of the reservoirs are not identical.

7.3 The Heat and Charge Fluxes

Recall that the observables describing heat and charge currents out of the j-th reservoir are

$$\Phi_j = \lambda(a^*(irf_j)a(1) + a^*(1)a(irf_j)),$$
$$\mathcal{J}_j = \lambda(a^*(if_j)a(1) + a^*(1)a(if_j)).$$

The expectation of the currents in the state $\omega_{\lambda+}$ are thus

$$\begin{aligned}
\omega_{\lambda+}(\Phi_j) &= i\lambda\omega_{\lambda+}\big(a^*(rf_j)a(1) - a^*(1)a(rf_j)\big) \\
&= 2\lambda\mathrm{Im}\,(rf_j, T_+1) \\
&= 2\lambda\mathrm{Im}\,(W_-rf_j, T_{\mathcal{R}}W_-1),
\end{aligned}$$

and

$$\begin{aligned}
\omega_{\lambda+}(\mathcal{J}_j) &= i\lambda\omega_{\lambda+}\big(a^*(f_j)a(1) - a^*(1)a(f_j)\big) \\
&= 2\lambda\mathrm{Im}\,(f_j, T_+1) \\
&= 2\lambda\mathrm{Im}\,(W_-f_j, T_{\mathcal{R}}W_-1).
\end{aligned}$$

Setting

$$G_j(r) \equiv \int_{e_-}^{e_+} \frac{r|f_j(r')|^2}{r' - r + io}\, dr',$$

it easily follows from Formula (50) that for $k = 1, \cdots, M$,

$$(T_{\mathcal{R}}W_-1)_k(r) = -\lambda\frac{\rho_k(r)f_k(r)}{F(r)},$$

$$(W_-rf_j)_k(r) = \delta_{kj}\, rf_j(r) + \lambda^2\frac{G_j(r)f_k(r)}{F(r)},$$

from which we obtain

$$(W_-rf_j, T_{\mathcal{R}}W_-1) = -\lambda\sum_{k=1}^{M}\int_{e_-}^{e_+} \frac{|f_k(r)|^2\rho_k(r)}{|F(r)|^2}\left[r\bar{F}(r)\delta_{kj} + \lambda^2\bar{G}_j(r)\right]\,dr.$$

From Equ. (49) we have $\mathrm{Im}\,\bar{F}(r) = -\lambda^2\pi|f(r)|^2$ and similarly $\mathrm{Im}\,\bar{G}_j(r) = \pi r|f_j(r)|^2$. Hence,

$$\omega_{\lambda+}(\Phi_j) = 2\pi\lambda^4\sum_{k=1}^{M}\int_{e_-}^{e_+} \frac{r|f_k(r)|^2\rho_k(r)}{|F(r)|^2}\left[|f(r)|^2\delta_{kj} - |f_j(r)|^2\right]\,dr.$$

Since $|f|^2 = \sum_k |f_k|^2$, the last formula can be rewritten as

$$\omega_{\lambda+}(\Phi_j) = 2\pi\lambda^4\sum_{k=1}^{M}\int_{e_-}^{e_+} |f_j(r)|^2|f_k(r)|^2(\rho_j(r) - \rho_k(r))\frac{r\,dr}{|F(r)|^2}. \tag{53}$$

In a completely similar way one obtains

$$\omega_{\lambda+}(\mathcal{J}_j) = 2\pi\lambda^4 \sum_{k=1}^{M} \int_{e_-}^{e_+} |f_j(r)|^2 |f_k(r)|^2 (\rho_j(r) - \rho_k(r)) \frac{dr}{|F(r)|^2}. \tag{54}$$

An immediate consequence of Formulas (53) and (54) is that all the fluxes vanish if $\rho_1 = \cdots = \rho_M$. Note also the antisymmetry in k and j of the integrands which ensures that the conservation laws

$$\sum_{j=1}^{M} \omega_{\lambda+}(\Phi_j) = \sum_{j=1}^{M} \omega_{\lambda+}(\mathcal{J}_j) = 0,$$

hold.

7.4 Entropy Production

By the Assumption (SEBB3) the entropy production observable of the SEBB model is well defined and is given by Equ. (45) which we rewrite as

$$\sigma = -\lambda \sum_{j=1}^{M} \left(a^*(\mathrm{i}s_j f_j)a(1) + a^*(1)a(\mathrm{i}s_j f_j) \right). \tag{55}$$

Proceeding as in the previous section we obtain

$$\omega_{\lambda+}(\sigma) = -2\lambda \sum_{j=1}^{M} \mathrm{Im}\, (W_- s_j f_j, T_{\mathcal{R}} W_- 1),$$

which yields

$$\omega_{\lambda+}(\sigma) = 2\pi\lambda^4 \sum_{j,k=1}^{M} \int_{e_-}^{e_+} \frac{|f_j(r)|^2 |f_k(r)|^2}{|F(r)|^2} (s_j(r) - s_k(r)) \rho_k(r)\, dr.$$

Finally, symmetrizing the sum over j and k we get

$$\omega_{\lambda+}(\sigma) = \pi\lambda^4 \sum_{j,k=1}^{M} \int_{e_-}^{e_+} \frac{|f_j(r)|^2 |f_k(r)|^2}{|F(r)|^2} (s_j(r) - s_k(r)) (\rho_k(r) - \rho_j(r))\, dr.$$

Since $\rho_j = (1 + e^{s_j})^{-1}$ is a strictly decreasing function of s_j,

$$(s_j(r) - s_k(r))(\rho_k(r) - \rho_j(r)) \geq 0,$$

with equality if and only if $\rho_k(r) = \rho_j(r)$. We summarize:

Theorem 7.3. *The entropy production of* $\omega_{\lambda+}$ *is*

$$\omega_{\lambda+}(\sigma) = \pi\lambda^4 \sum_{j,k=1}^{M} \int_{e_-}^{e_+} \frac{|f_j(r)|^2 |f_k(r)|^2}{|F(r)|^2} \left(s_j(r) - s_k(r)\right)\left(\rho_k(r) - \rho_j(r)\right) \mathrm{d}r.$$

In particular, $\mathrm{Ep}(\omega_+) \geq 0$ *(something we already know from the general principles)* *and* $\mathrm{Ep}(\omega_+) = 0$ *if and only if* $\rho_1 = \cdots = \rho_M$.

Since ω and $\omega_{\lambda+}$ are factor states, they are either quasi-equivalent or disjoint. By Theorem 3.2, if $\mathrm{Ep}(\omega_{\lambda+}) > 0$, then $\omega_{\lambda+}$ is not ω-normal. Hence, Theorem 7.3 implies that if the densities ρ_j are not all equal, then the reference state ω and the NESS $\omega_{\lambda+}$ are disjoint states.

Until the end of this section we will assume that the energy density of the j-th reservoir is

$$\rho_{\beta_j \mu_j}(r) \equiv \frac{1}{1 + e^{\beta_j(r-\mu_j)}},$$

where β_j is the inverse temperature and $\mu_j \in \mathbb{R}$ is the chemical potential of the j-th reservoir. Then, by (46), $\mathrm{Ep}(\omega_{\lambda+})$ can be written as

$$\mathrm{Ep}(\omega_{\lambda+}) = \mathrm{Ep}_{\mathrm{heat}}(\omega_{\lambda+}) + \mathrm{Ep}_{\mathrm{charge}}(\omega_{\lambda+}),$$

where

$$\mathrm{Ep}_{\mathrm{heat}}(\omega_{\lambda+}) = -\sum_{j=1}^{M} \beta_j \omega_{\lambda+}(\Phi_j),$$

is interpreted as the entropy production due to the heat fluxes and

$$\mathrm{Ep}_{\mathrm{charge}}(\omega_{\lambda+}) = \sum_{j=1}^{M} \beta_j \mu_j \omega_{\lambda+}(\mathcal{J}_j).$$

as the entropy production due to the electric currents.

7.5 Equilibrium Correlation Functions

In this subsection we compute the integrated current-current correlation functions

$$L_\rho(A, B) \equiv \lim_{T \to \infty} \frac{1}{2} \int_{-T}^{T} \omega_{\rho+}(\tau_\lambda^t(A)B)\, \mathrm{d}t,$$

where A and B are heat or charge flux observables and $\omega_{\rho+}$ denotes the NESS $\omega_{\lambda+}$ in the equilibrium case $\rho_1 = \cdots = \rho_M = \rho$. To do this, note that $\Phi_l = \mathrm{d}\Gamma(\varphi_l)$ and $\mathcal{J}_l = \mathrm{d}\Gamma(j_l)$ where

$$\varphi_l = \mathrm{i}[h_{\mathcal{R}_l}, \lambda v] = -\mathrm{i}[h_\lambda, h_{\mathcal{R}_j}],$$
$$j_l = \mathrm{i}[p_j, \lambda v] = -\mathrm{i}[h_\lambda, p_j],$$

are finite rank operators. We will only consider $L_\rho(\Phi_j, \Phi_k)$, the other cases are completely similar.

Using the CAR, Formula (29) and the fact that $\omega_{\rho+}(\Phi_l) = 0$, one easily shows that

$$\omega_{\rho+}(\tau_\lambda^t(\Phi_j)\Phi_k) - \mathrm{Tr}\,(T_+ e^{ith_\lambda}\varphi_j e^{-ith_\lambda}(I - T_+)\varphi_k).$$

Since

$$e^{ith_\lambda}\varphi_j e^{-ith_\lambda} = -\frac{d}{dt}e^{ith_\lambda}h_{\mathcal{R}_j}e^{-ith_\lambda},$$

the integration can be explicitly performed and we have

$$L_\rho(\Phi_j, \Phi_k) = -\lim_{T\to\infty}\frac{1}{2}\mathrm{Tr}\,(T_+ e^{ith_\lambda}h_{\mathcal{R}_j}e^{-ith_\lambda}(I - T_+)\varphi_k)\Big|_{-T}^{T}.$$

Writing $e^{ith_\lambda}h_{\mathcal{R}_j}e^{-ith_\lambda} = e^{ith_\lambda}e^{-ith_0}h_{\mathcal{R}_j}e^{ith_0}e^{-ith_\lambda}$ and using the fact that φ_k is finite rank, we see that the limit exists and can be expressed in terms of the wave operators $W_\pm$ as

$$L_\rho(\Phi_j, \Phi_k) = \frac{1}{2}\Big\{\mathrm{Tr}\,(T_+ W_-^* h_{\mathcal{R}_j}W_-(I - T_+)\varphi_k)$$

$$- \mathrm{Tr}\,(T_+ W_+^* h_{\mathcal{R}_j}W_+(I - T_+)\varphi_k)\Big\}.$$

The intertwining property of the wave operators gives

$$T_+ = W_-^*\rho(h_{\mathcal{R}})W_- = \rho(h_\lambda) = W_+^*\rho(h_{\mathcal{R}})W_+,$$

from which we obtain

$$L_\rho(\Phi_j, \Phi_k) = \frac{1}{2}\mathrm{Tr}\,(T_{\mathcal{R}}(I - T_{\mathcal{R}})h_{\mathcal{R}_j}(W_-\varphi_k W_-^* - W_+\varphi_k W_+^*)),$$

with $T_{\mathcal{R}} = \rho(h_{\mathcal{R}})$. Time reversal invariance further gives

$$W_+ = j W_- j, \qquad j\varphi_k j = -\varphi_k,$$

and so

$$L_\rho(\Phi_j, \Phi_k) = \frac{1}{2}\mathrm{Tr}\,(T_{\mathcal{R}}(I - T_{\mathcal{R}})h_{\mathcal{R}_j}(W_-\varphi_k W_-^* + j W_-\varphi_k W_-^* j))$$

$$= \mathrm{Tr}\,(T_{\mathcal{R}}(I - T_{\mathcal{R}})h_{\mathcal{R}_j}W_-\varphi_k W_-^*).$$

The last trace is easily evaluated (use the formula $\varphi_k = \lambda i[h_{\mathcal{R}_k}, v]$ and follow the steps of the computation in Subsection 7.3). The result is

$$L_\rho(\Phi_j, \Phi_k) = -2\pi\lambda^4 \int_{e_-}^{e_+} |f_j(r)|^2 \left[|f_k(r)|^2 - \delta_{jk}|f(r)|^2\right] \rho(r)(1 - \rho(r)) \frac{r^2 dr}{|F(r)|^2},$$

$$L_\rho(\mathcal{J}_j, \Phi_k) = -2\pi\lambda^4 \int_{e_-}^{e_+} |f_j(r)|^2 \left[|f_k(r)|^2 - \delta_{jk}|f(r)|^2\right] \rho(r)(1 - \rho(r)) \frac{r dr}{|F(r)|^2},$$

$$L_\rho(\Phi_j, \mathcal{J}_k) = -2\pi\lambda^4 \int_{e_-}^{e_+} |f_j(r)|^2 \left[|f_k(r)|^2 - \delta_{jk}|f(r)|^2\right] \rho(r)(1 - \rho(r)) \frac{r dr}{|F(r)|^2},$$

$$L_\rho(\mathcal{J}_j, \mathcal{J}_k) = -2\pi\lambda^4 \int_{e_-}^{e_+} |f_j(r)|^2 \left[|f_k(r)|^2 - \delta_{jk}|f(r)|^2\right] \rho(r)(1 - \rho(r)) \frac{dr}{|F(r)|^2}.$$

$$(56)$$

Note the following symmetries:

$$\begin{aligned} L_\rho(\Phi_j, \Phi_k) &= L_\rho(\Phi_k, \Phi_j), \\ L_\rho(\mathcal{J}_j, \mathcal{J}_k) &= L_\rho(\mathcal{J}_k, \mathcal{J}_j), \\ L_\rho(\Phi_j, \mathcal{J}_k) &= L_\rho(\mathcal{J}_k, \Phi_j). \end{aligned} \qquad (57)$$

Note also that $L_\rho(\Phi_j, \Phi_k) \leq 0$ and $L_\rho(\mathcal{J}_j, \mathcal{J}_k) \leq 0$ for $j \neq k$ while $L_\rho(\Phi_j, \Phi_j) \geq 0$ and $L_\rho(\mathcal{J}_j, \mathcal{J}_j) \geq 0$.

7.6 Onsager Relations. Kubo Formulas.

Let β_{eq} and μ_{eq} be given equilibrium values of the inverse temperature and the chemical potential. The affinities (thermodynamic forces) conjugated to the currents Φ_j and $\mathcal{J}_j$ are

$$X_j = \beta_{eq} - \beta_j, \qquad Y_j = \beta_j\mu_j - \beta_{eq}\mu_{eq}.$$

Indeed, it follows from the conservations laws (12) and (39) that

$$\mathrm{Ep}(\omega_{\lambda+}) = \sum_{j=1}^{M} \left(X_j\, \omega_{\lambda+}(\Phi_j) + Y_j\, \omega_{\lambda+}(\mathcal{J}_j)\right).$$

Since

$$\rho_{\beta_j\mu_j}(r) = \frac{1}{1 + e^{\beta_{eq}(r - \mu_{eq}) - (X_j r + Y_j)}},$$

we have

$$\begin{aligned} \partial_{X_k}\rho_{\beta_j\mu_j}(r)|_{X=Y=0} &= \delta_{kj}\, \rho(r)(1 - \rho(r))\, r, \\ \partial_{Y_k}\rho_{\beta_j\mu_j}(r)|_{X=Y=0} &= \delta_{kj}\, \rho(r)(1 - \rho(r)), \end{aligned}$$

where $\rho \equiv \rho_{\beta_{eq}\mu_{eq}}$. Using these formulas, and explicit differentiation of the steady currents (53) and (54) and comparison with (56) lead to

$$\begin{aligned} \partial_{X_k}\omega_{\lambda+}(\Phi_j)|_{X=Y=0} &= L_\rho(\Phi_j, \Phi_k), \\ \partial_{Y_k}\omega_{\lambda+}(\Phi_j)|_{X=Y=0} &= L_\rho(\Phi_j, \mathcal{J}_k), \\ \partial_{X_k}\omega_{\lambda+}(\mathcal{J}_j)|_{X=Y=0} &= L_\rho(\mathcal{J}_j, \Phi_k), \\ \partial_{Y_k}\omega_{\lambda+}(\mathcal{J}_j)|_{X=Y=0} &= L_\rho(\mathcal{J}_j, \mathcal{J}_k), \end{aligned}$$

which are the *Kubo Fluctuation-Dissipation Formulas*. The symmetry (57) gives the *Onsager reciprocity relations*

$$\partial_{X_j}\omega_{\lambda+}(\Phi_k)|_{X=Y=0} = \partial_{X_k}\omega_{\lambda+}(\Phi_j)|_{X=Y=0},$$
$$\partial_{Y_j}\omega_{\lambda+}(\mathcal{J}_k)|_{X=Y=0} = \partial_{Y_k}\omega_{\lambda+}(\mathcal{J}_j)|_{X=Y=0},$$
$$\partial_{Y_j}\omega_{\lambda+}(\Phi_k)|_{X=Y=0} = \partial_{X_k}\omega_{\lambda+}(\mathcal{J}_j)|_{X=Y=0}.$$

The fact that $L_\rho(\Phi_j,\Phi_j) \geq 0$ and $L_\rho(\mathcal{J}_j,\mathcal{J}_j) \geq 0$ while $L_\rho(\Phi_j,\Phi_k) \leq 0$ and $L_\rho(\mathcal{J}_j,\mathcal{J}_k) \leq 0$ for $j \neq k$ means that increasing a force results in an increase of the conjugated current and a decrease of the other currents. This is not only true in the linear regime. Direct differentiation of (53) and (54) yields

$$\partial_{X_k}\omega_{\lambda+}(\Phi_k) = 2\pi\lambda^4\sum_{j\neq k}\int_{e_-}^{e_+}|f_j(r)|^2|f_k(r)|^2\rho_{\beta_k\mu_k}(r)(1-\rho_{\beta_k\mu_k}(r))\frac{r^2\mathrm{d}r}{|F(r)|^2}\geq 0,$$

$$\partial_{Y_k}\omega_{\lambda+}(\mathcal{J}_k) = 2\pi\lambda^4\sum_{j\neq k}\int_{e_-}^{e_+}|f_j(r)|^2|f_k(r)|^2\rho_{\beta_k\mu_k}(r)(1-\rho_{\beta_k\mu_k}(r))\frac{\mathrm{d}r}{|F(r)|^2}\geq 0,$$

$$\partial_{X_k}\omega_{\lambda+}(\Phi_j) = -2\pi\lambda^4\int_{e_-}^{e_+}|f_j(r)|^2|f_k(r)|^2\rho_{\beta_k\mu_k}(r)(1-\rho_{\beta_k\mu_k}(r))\frac{r^2\mathrm{d}r}{|F(r)|^2}\leq 0,$$

$$\partial_{Y_k}\omega_{\lambda+}(\mathcal{J}_j) = -2\pi\lambda^4\int_{e_-}^{e_+}|f_j(r)|^2|f_k(r)|^2\rho_{\beta_k\mu_k}(r)(1-\rho_{\beta_k\mu_k}(r))\frac{\mathrm{d}r}{|F(r)|^2}\leq 0.$$

Note that these derivatives do not depend on the reference states of the reservoirs $\mathcal{R}_j$ for $j \neq k$.

8 FGR Thermodynamics of the SEBB Model

For $j = 1,\cdots,M$, we set

$$\tilde{g}_j(t) \equiv \int_{e_-}^{e_+} e^{\mathrm{i}tr}\rho_j(r)|f_j(r)|^2\,\mathrm{d}r.$$

In addition to (SEBB1)-(SEBB4) in this section we will assume

Assumption (SEBB5) $\tilde{g}_j(t) \in L^1(\mathbb{R},\mathrm{d}t)$ for $j = 1,\cdots,M$.

8.1 The Weak Coupling Limit

In this subsection we study the dynamics restricted to the small system on the van Hove time scale t/λ^2.

Recall that by Theorem 6.1 the algebra of observables $\mathcal{O}_S$ of the small system is the 4-dimensional subalgebra of $\mathcal{O} = \mathrm{CAR}(\mathbb{C}\oplus\mathfrak{h}_\mathcal{R})$ generated by $a(1)$. It is the full matrix algebra of the subspace $\mathfrak{h}_S \subset \Gamma_-(\mathbb{C} \oplus \mathfrak{h}_\mathcal{R})$ generated by the vectors $\{\Omega, a(1)\Omega\}$. In this basis, the Hamiltonian and the reference state of the small system are

$$H_S = \begin{bmatrix} 0 & 0 \\ 0 & \varepsilon_0 \end{bmatrix}, \qquad \omega_S = \begin{bmatrix} 1 - \gamma & 0 \\ 0 & \gamma \end{bmatrix}.$$

Let $A \in \mathcal{O}_S$ be an observable of the small system. We will study the expectation values

$$\omega(\tau_\lambda^{t/\lambda^2}(A)), \tag{58}$$

as $\lambda \to 0$. If $A = a^\#(1)$, then (58) vanishes, so we need only to consider the Abelian 2-dimensional even subalgebra $\mathcal{O}_S^+ \subset \mathcal{O}_S$. Since $a^*(1)a(1) = N_S$ and $a(1)a^*(1) = I - N_S$, it suffices to consider $A = N_S$. In this case we have

$$\begin{aligned}
\omega \circ \tau_\lambda^{t/\lambda^2}(N_S) &= \omega(a^*(e^{ith_\lambda/\lambda^2}1)a(e^{ith_\lambda/\lambda^2}1)) \\
&= (e^{ith_\lambda/\lambda^2}1, (\gamma \oplus T_\mathcal{R})e^{ith_\lambda/\lambda^2}1).
\end{aligned} \tag{59}$$

Using the projection p_j on the Hilbert space $\mathfrak{h}_{\mathcal{R}_j}$ of the j-th reservoir we can rewrite this expression as

$$\omega \circ \tau_\lambda^{t/\lambda^2}(N_S) = \gamma |(1, e^{ith_\lambda/\lambda^2}1)|^2 + \sum_{j=1}^M (p_j e^{ith_\lambda/\lambda^2}1, T_{\mathcal{R}_j} p_j e^{ith_\lambda/\lambda^2}1).$$

Theorem 8.1. *Assume that Assumptions* (SEBB1)-(SEBB5) *hold.*

(i) For any $t \geq 0$,

$$\lim_{\lambda \to 0} |(1, e^{ith_\lambda/\lambda^2}1)|^2 = e^{-2\pi t |f(\varepsilon_0)|^2}. \tag{60}$$

(ii) For any $t \geq 0$ and $j = 1, \cdots, M$,

$$\lim_{\lambda \to 0} (p_j e^{ith_\lambda/\lambda^2}1, T_{\mathcal{R}_j} p_j e^{ith_\lambda/\lambda^2}1) = \frac{|f_j(\varepsilon_0)|^2}{|f(\varepsilon_0)|^2} \rho_j(\varepsilon_0) \left(1 - e^{-2\pi t |f(\varepsilon_0)|^2}\right). \tag{61}$$

The proof of Theorem 8.1 is not difficult—for Part (i) see [Da1, D1], and for Part (ii) [Da2]. These proofs use the regularity Assumption (SEBB5). An alternative proof of Theorem 8.1, based on the explicit form of the wave operator W_-, can be found in [JKP].

Theorem 8.1 implies that

$$\begin{aligned}
\gamma(t) &\equiv \lim_{\lambda \to 0} \omega \circ \tau_\lambda^{t/\lambda^2}(N_S) \\
&= \gamma e^{-2\pi t |f(\varepsilon_0)|^2} + \left(1 - e^{-2\pi t |f(\varepsilon_0)|^2}\right) \sum_{j=1}^M \frac{|f_j(\varepsilon_0)|^2}{|f(\varepsilon_0)|^2} \rho_j(\varepsilon_0),
\end{aligned}$$

from which we easily conclude that for all $A \in \mathcal{O}_S$ one has

$$\lim_{\lambda \to 0} \omega \circ \tau_\lambda^{t/\lambda^2}(A) = \mathrm{Tr}(\omega_\mathcal{S}(t)A),$$

where

$$\omega_\mathcal{S}(t) = \begin{bmatrix} 1 - \gamma(t) & 0 \\ 0 & \gamma(t) \end{bmatrix}.$$

According to the general theory described in Section 4.5 we also have

$$\omega_\mathcal{S}(t) = \mathrm{e}^{tK_\mathrm{S}}\omega_\mathcal{S},$$

where K_S is the QMS generator in the Schrödinger picture. We shall now discuss its restriction to the algebra of diagonal 2×2-matrices. In the basis

$$\begin{bmatrix} 1 & 0 \\ 0 & 0 \end{bmatrix}, \begin{bmatrix} 0 & 0 \\ 0 & 1 \end{bmatrix}, \tag{62}$$

of this subalgebra we obtain the matrix representation

$$K_\mathrm{S} = 2\pi \sum_{j=1}^{M} |f_j(\varepsilon_0)|^2 \begin{bmatrix} -\rho_j(\varepsilon_0) & 1 - \rho_j(\varepsilon_0) \\ \rho_j(\varepsilon_0) & -(1 - \rho_j(\varepsilon_0)) \end{bmatrix}.$$

In the Heisenberg picture we have

$$\lim_{\lambda \to 0} \omega_\mathcal{S} \circ \tau_\lambda^{t/\lambda^2}(A) = \mathrm{Tr}(\omega_\mathcal{S} \, \mathrm{e}^{tK_\mathrm{H}}A),$$

where K_H is related to K_S by the duality

$$\mathrm{Tr}(K_\mathrm{S}(\omega_\mathcal{S})A) = \mathrm{Tr}(\omega_\mathcal{S} K_\mathrm{H}(A)).$$

The restriction of K_H to the subalgebra of diagonal 2×2-matrices has the following matrix representation relative to the basis (62),

$$K_\mathrm{H} = 2\pi \sum_{j=1}^{M} |f_j(\varepsilon_0)|^2 \begin{bmatrix} -\rho_j(\varepsilon_0) & \rho_j(\varepsilon_0) \\ 1 - \rho_j(\varepsilon_0) & -(1 - \rho_j(\varepsilon_0)) \end{bmatrix}.$$

We stress that K_S and K_H are the diagonal parts of the full Davies generators in the Schrödinger and Heisenberg pictures discussed in the lecture notes [D1].

As we have discussed in Section 4.5, an important property of the generators K_S and K_H is the decomposition

$$K_\mathrm{S} = \sum_{j=1}^{M} K_{\mathrm{S},j}, \qquad K_\mathrm{H} = \sum_{j=1}^{M} K_{\mathrm{H},j},$$

where $K_{\mathrm{S},j}$ and $K_{\mathrm{H},j}$ are the generators describing interaction of $\mathcal{S}$ with the j-th reservoir only. Explicitly,

$$K_{\mathrm{S},j} = 2\pi |f_j(\varepsilon_0)|^2 \begin{bmatrix} -\rho_j(\varepsilon_0) & 1 - \rho_j(\varepsilon_0) \\ \rho_j(\varepsilon_0) & -(1 - \rho_j(\varepsilon_0)) \end{bmatrix},$$

$$K_{\mathrm{H},j} = 2\pi |f_j(\varepsilon_0)|^2 \begin{bmatrix} -\rho_j(\varepsilon_0) & \rho_j(\varepsilon_0) \\ 1 - \rho_j(\varepsilon_0) & -(1 - \rho_j(\varepsilon_0)) \end{bmatrix}.$$

Finally, we note that

$$\omega_{\mathcal{S}+} \equiv \lim_{t \to \infty} \omega_{\mathcal{S}}(t) = \sum_{j=1}^{M} \frac{|f_j(\varepsilon_0)|^2}{|f(\varepsilon_0)|^2} \begin{bmatrix} 1 - \rho_j(\varepsilon_0) & 0 \\ 0 & \rho_j(\varepsilon_0) \end{bmatrix}.$$

$\omega_{\mathcal{S}+}$ is the NESS on the Fermi Golden Rule time scale: for any observable A of the small system,

$$\lim_{t \to \infty} \lim_{\lambda \to 0} \omega \circ \tau_\lambda^{t/\lambda^2}(A) = \mathrm{Tr}(\omega_{\mathcal{S}+} A) = \omega_{\mathcal{S}+}(A).$$

In the sequel we will refer to $\omega_{\mathcal{S}+}$ as the FGR NESS.

8.2 Historical Digression—Einstein's Derivation of the Planck Law

Einstein's paper [Ei], published in 1917, has played an important role in the historical development of quantum mechanics and quantum field theory. In this paper Einstein made some deep insights into the nature of interaction between radiation and matter which have led him to a new derivation of the Planck law. For the history of these early developments the interested reader may consult [Pa].

The original Einstein argument can be paraphrased as follows. Consider a two-level quantum system $\mathcal{S}$ with energy levels 0 and ε_0, which is in equilibrium with a radiation field reservoir with energy density $\rho(r)$. Due to the interaction with the reservoir, the system $\mathcal{S}$ will make constant transitions between the energy levels 0 and ε_0. Einstein *conjectured* that the corresponding transition rates (transition probabilities per unit time) have the form

$$k(\varepsilon_0, 0) = A_{\varepsilon_0}(1 - \rho(\varepsilon_0)), \qquad k(0, \varepsilon_0) = B_{\varepsilon_0} \rho(\varepsilon_0),$$

where A_{ε_0} and B_{ε_0} are the coefficients which depend on the mechanics of the interaction. (Of course, in 1917 Einstein considered the bosonic reservoir (the light)—in this case in the first formula one has $1 + \rho(\varepsilon_0)$ instead of $1 - \rho(\varepsilon_0)$). These formulas are the celebrated Einstein's A and B laws. Let $\bar{p}_0$ and $\bar{p}_{\varepsilon_0}$ be probabilities that in equilibrium the small system has energies 0 and ε_0 respectively. If $\mathcal{S}$ is in thermal equilibrium at inverse temperature β, then by the Gibbs postulate,

$$\bar{p}_0 = (1 + e^{-\beta \varepsilon_0})^{-1}, \qquad \bar{p}_{\varepsilon_0} = e^{-\beta \varepsilon_0}(1 + e^{-\beta \varepsilon_0})^{-1}.$$

The equilibrium condition

$$k(0, \varepsilon_0)\bar{p}_0 = k(\varepsilon_0, 0)\bar{p}_{\varepsilon_0},$$

yields

$$\rho(\varepsilon_0) = \frac{A_{\varepsilon_0}}{B_{\varepsilon_0}}(1 - \rho(\varepsilon_0))e^{-\beta\varepsilon_0}.$$

In 1917 Einstein naturally could not compute the coefficients A_{ε_0} and B_{ε_0}. However, if $A_{\varepsilon_0}/B_{\varepsilon_0} = 1$ for all ε_0, then the above relation yields the Planck law for energy density of the free fermionic reservoir in thermal equilibrium,

$$\rho(\varepsilon_0) = \frac{1}{1 + e^{\beta\varepsilon_0}}.$$

In his paper Einstein points out that to compute the numerical value of A_{ε_0} and B_{ε_0} one would need an exact [quantum] theory of electro-dynamical and mechanical processes.

The quantum theory of mechanical processes was developed in the 1920's by Schrödinger, Heisenberg, Jordan, Dirac and others. In 1928, Dirac extended quantum theory to electrodynamical processes and computed the coefficients A_{ε_0} and B_{ε_0} from the first principles of quantum theory. Dirac's seminal paper [Di] marked the birth of quantum field theory. To compute A_{ε_0} and B_{ε_0} Dirac developed the so-called time-dependent perturbation theory, which has been discussed in lecture notes [D1, JKP] (see also Chapter XXI in [Mes], or any book on quantum mechanics). In his 1949 Chicago lecture notes [Fer] Fermi called the basic formulas of Dirac's theory *the Golden Rule*, and since then they have been called *the Fermi Golden Rule*.

In this section we have described the mathematically rigorous Fermi Golden Rule theory of the SEBB model. In this context Dirac's theory reduces to the computation of K_S and K_H since the matrix elements of these operators give the transition probabilities $k(\varepsilon_0, 0)$ and $k(0, \varepsilon_0)$. In particular, in the case of a single reservoir with energy density $\rho(r)$,

$$A_{\varepsilon_0} = B_{\varepsilon_0} = 2\pi|f(\varepsilon_0)|^2.$$

Einstein's argument can be rephrased as follows: if the energy density ρ is such that

$$\omega_{S+} = e^{-\beta H_S}/\text{Tr}(e^{-\beta H_S}) = (1 + e^{-\beta\varepsilon_0})^{-1}\begin{bmatrix} 1 & 0 \\ 0 & e^{-\beta\varepsilon_0} \end{bmatrix},$$

for all ε_0 (namely H_S), then

$$\rho(\varepsilon_0) = \frac{1}{1 + e^{\beta\varepsilon_0}}.$$

8.3 FGR Fluxes, Entropy Production and Kubo Formulas

Any diagonal observable $A \in \mathcal{O}_S^+$ of the small system is a function of the Hamiltonian H_S. We identify such an observable with a function $g : \{0, \varepsilon_0\} \to \mathbb{R}$. Occasionally, we will write g as a column vector with components $g(0)$ and $g(\varepsilon_0)$. In the sequel we will use such identifications without further comment. A vector ν is called a probability vector if $\nu(0) \geq 0$, $\nu(\varepsilon_0) \geq 0$ and $\nu(0) + \nu(\varepsilon_0) = 1$. The diagonal part of any density matrix defines a probability vector. We denote the probability vector associated to FGR NESS $\omega_{\mathcal{S}+}$ by the same letter. Similarly, to a probability vector one uniquely associates a diagonal density matrix. With these conventions, the Hamiltonian and the number operator of the small system are

$$H_{\mathcal{S}} = \varepsilon_0 a^*(1) a(1) = \begin{bmatrix} 0 \\ \varepsilon_0 \end{bmatrix}, \qquad N_{\mathcal{S}} = a^*(1) a(1) = \begin{bmatrix} 0 \\ 1 \end{bmatrix}.$$

The Fermi Golden Rule (FGR) heat and charge flux observables are

$$\Phi_{\text{fgr},j} = K_{\text{H},j}(H_{\mathcal{S}}) = 2\pi \varepsilon_0 |f_j(\varepsilon_0)|^2 \begin{bmatrix} \rho_j(\varepsilon_0) \\ -(1 - \rho_j(\varepsilon_0)) \end{bmatrix},$$

$$\mathcal{J}_{\text{fgr},j} = K_{\text{H},j}(N_{\mathcal{S}}) = 2\pi |f_j(\varepsilon_0)|^2 \begin{bmatrix} \rho_j(\varepsilon_0) \\ -(1 - \rho_j(\varepsilon_0)) \end{bmatrix}.$$

The steady heat and the charge currents in the FGR NESS are given by

$$\omega_{\mathcal{S}+}(\Phi_{\text{fgr},j}) = 2\pi \sum_{k=1}^{M} \frac{|f_j(\varepsilon_0)|^2 |f_k(\varepsilon_0)|^2}{|f(\varepsilon_0)|^2} \varepsilon_0 (\rho_j(\varepsilon_0) - \rho_k(\varepsilon_0)),$$

$$\omega_{\mathcal{S}+}(\mathcal{J}_{\text{fgr},j}) = 2\pi \sum_{k=1}^{M} \frac{|f_j(\varepsilon_0)|^2 |f_k(\varepsilon_0)|^2}{|f(\varepsilon_0)|^2} (\rho_j(\varepsilon_0) - \rho_k(\varepsilon_0)).$$

(63)

The conservation laws

$$\sum_{j=1}^{M} \omega_{\mathcal{S}+}(\Phi_{\text{fgr},j}) = 0, \qquad \sum_{j=1}^{M} \omega_{\mathcal{S}+}(\mathcal{J}_{\text{fgr},j}) = 0,$$

follow from the definition of the fluxes and the relation $K_{\mathcal{S}}(\omega_{\mathcal{S}+}) = 0$. Of course, they also follow easily from the above explicit formulas.

Until the end of this subsection we will assume that

$$\rho_j(r) = \frac{1}{1 + e^{\beta_j(r - \mu_j)}}.$$

Using Equ. (63), we can also compute the expectation of the entropy production in the FGR NESS $\omega_{\mathcal{S}+}$. The natural extension of the definition (25) is

$$\sigma_{\text{fgr}} \equiv - \sum_{j=1}^{M} \beta_j \left(\Phi_{\text{fgr},j} - \mu_j \mathcal{J}_{\text{fgr},j} \right),$$

from which we get

$$\omega_{\mathcal{S}+}(\sigma_{\text{fgr}}) = 2\pi \sum_{j,k=1}^{M} \frac{|f_j(\varepsilon_0)|^2 |f_k(\varepsilon_0)|^2}{|f(\varepsilon_0)|^2} (\rho_k(\varepsilon_0) - \rho_j(\varepsilon_0)) \beta_j(\varepsilon_0 - \mu_j). \quad (64)$$

Writing

$$s_j \equiv \log \frac{\rho_j(\varepsilon_0)}{1 - \rho_j(\varepsilon_0)} = \beta_j(\varepsilon_0 - \mu_j),$$

and symmetrizing the sum in Equ. (64) we obtain

$$\omega_{\mathcal{S}+}(\sigma_{\text{fgr}}) = \pi \sum_{j,k=1}^{M} \frac{|f_j(\varepsilon_0)|^2 |f_k(\varepsilon_0)|^2}{|f(\varepsilon_0)|^2} (\rho_k(\varepsilon_0) - \rho_j(\varepsilon_0))(s_j - s_k),$$

which is non-negative since $\rho_l(\varepsilon_0)$ is a strictly decreasing function of s_l. The FGR entropy production vanishes iff all s_j's are the same. Note however that this condition does not require that all the β_j's and μ_j's are the same.

Let β_{eq} and μ_{eq} be given equilibrium values of the inverse temperature and chemical potential, and

$$\omega_{\mathcal{S}\text{eq}} = e^{-\beta_{\text{eq}}(H_{\mathcal{S}} - \mu_{\text{eq}})} / \text{Tr}(e^{-\beta_{\text{eq}}(H_{\mathcal{S}} - \mu_{\text{eq}})}) = \begin{bmatrix} (1 + e^{-\beta_{\text{eq}}\varepsilon_0})^{-1} & 0 \\ 0 & (1 + e^{\beta_{\text{eq}}\varepsilon_0})^{-1} \end{bmatrix},$$

the corresponding NESS. As in Subsection 7.6, the affinities (thermodynamic forces) are $X_j = \beta_{\text{eq}} - \beta_j$ and $Y_j = \beta_j \mu_j - \beta_{\text{eq}} \mu_{\text{eq}}$. A simple computation yields the FGR Onsager reciprocity relations

$$\partial_{X_j} \omega_{\mathcal{S}+}(\Phi_{\text{fgr},k})|_{X=Y=0} = \partial_{X_k} \omega_{\mathcal{S}+}(\Phi_{\text{fgr},j})|_{X=Y=0},$$
$$\partial_{Y_j} \omega_{\mathcal{S}+}(\mathcal{J}_{\text{fgr},k})|_{X=Y=0} = \partial_{Y_k} \omega_{\mathcal{S}+}(\mathcal{J}_{\text{fgr},i})|_{X=Y=0}, \quad (65)$$
$$\partial_{Y_j} \omega_{\mathcal{S}+}(\Phi_{\text{fgr},k})|_{X=Y=0} = \partial_{X_k} \omega_{\mathcal{S}+}(\mathcal{J}_{\text{fgr},i})|_{X=Y=0}.$$

We set

$$L_{\text{fgr}}(A, B) = \int_0^\infty \omega_{\mathcal{S}\text{eq}}(e^{tK_{\text{H}}}(A)B) \, dt,$$

where A and B are the FGR heat or charge flux observables. Explicit computations yield the FGR Kubo formulas

$$\partial_{X_k} \omega_{\mathcal{S}+}(\Phi_{\text{fgr},j})|_{X=Y=0} = L_{\text{fgr}}(\Phi_{\text{fgr},j}, \Phi_{\text{fgr},k}),$$
$$\partial_{Y_k} \omega_{\mathcal{S}+}(\Phi_{\text{fgr},j})|_{X=Y=0} = L_{\text{fgr}}(\Phi_{\text{fgr},j}, \mathcal{J}_{\text{fgr},k}),$$
$$\partial_{X_k} \omega_{\mathcal{S}+}(\mathcal{J}_{\text{fgr},j})|_{X=Y=0} = L_{\text{fgr}}(\mathcal{J}_{\text{fgr},j}, \Phi_{\text{fgr},k}), \quad (66)$$
$$\partial_{Y_k} \omega_{\mathcal{S}+}(\mathcal{J}_{\text{fgr},j})|_{X=Y=0} = L_{\text{fgr}}(\mathcal{J}_{\text{fgr},j}, \mathcal{J}_{\text{fgr},k}).$$

8.4 From Microscopic to FGR Thermodynamics

At the end of Subsection 4.5 we have briefly discussed the passage from the microscopic to the FGR thermodynamics. We now return to this subject in the context of the SEBB model. The next theorem is a mathematically rigorous version of the heuristic statement that the FGR thermodynamics is the first non-trivial contribution (in λ) to the microscopic thermodynamics.

Theorem 8.2.   *(i) For any diagonal observable $A \in \mathcal{O}_S$,*

$$\lim_{\lambda \to 0} \omega_{\lambda+}(A) = \omega_{S+}(A).$$

(ii) For $j = 1, \cdots, M$,

$$\lim_{\lambda \to 0} \lambda^{-2} \omega_{\lambda+}(\Phi_j) = \omega_{S+}(\Phi_{\text{fgr},j}), \qquad \lim_{\lambda \to 0} \lambda^{-2} \omega_{\lambda+}(\mathcal{J}_j) = \omega_{S+}(\mathcal{J}_{\text{fgr},j}).$$

(iii) Let $s_j \equiv \log \rho_j(\varepsilon_0)/(1 - \rho_j(\varepsilon_0))$ and define the FGR entropy production by

$$\sigma_{\text{fgr}} \equiv 2\pi \sum_{j=1}^{M} |f_j(\varepsilon_0)|^2 s_j \begin{bmatrix} -\rho_j(\varepsilon_0) \\ 1 - \rho_j(\varepsilon_0) \end{bmatrix}.$$

Then

$$\lim_{\lambda \to 0} \lambda^{-2} \, \text{Ep}(\omega_{\lambda+}) = \omega_{S+}(\sigma_{\text{fgr}}).$$

The proof of this theorem is an integration exercise. We will restrict ourselves to an outline of the proof of Part (i) and several comments. Let $A = N_S = a^*(1)a(1)$. Then

$$\omega_{\lambda+}(A) = (W_{-1}, T_{\mathcal{R}} W_{-1}) = \sum_{j=1}^{M} \lambda^2 \int_{e_-}^{e_+} \frac{|f_j(r)|^2}{|F(r)|^2} \rho_j(r) \, dr,$$

, and

$$\omega_{S+}(A) = \sum_{j=1}^{M} \frac{|f_j(\varepsilon_0)|^2}{|f(\varepsilon_0)|^2} \rho_j(\varepsilon_0).$$

Hence, to prove Part (i) we need to show that

$$\lim_{\lambda \to 0} \lambda^2 \int_{e_-}^{e_+} \frac{|f_j(r)|^2}{|F(r)|^2} \rho_j(r) \, dr = \frac{|f_j(\varepsilon_0)|^2}{|f(\varepsilon_0)|^2} \rho_j(\varepsilon_0).$$

By Assumption (SEBB2), $R(r) \equiv \text{Re}\, G(r - io)$ and $\pi |f(r)|^2 = \text{Im}\, G(r - io)$ are bounded continuous functions. The same is true for $\rho_j(r)$ by Assumption (SEBB3). Since

$$F(r) = \varepsilon_0 - r - \lambda^2 R(r) + i\lambda^2 \pi |f(r)|^2,$$

we have

$$\int_{e_-}^{e_+} \frac{|f_j(r)|^2}{|F(r)|^2} \rho_j(r) \, dr = \int_{e_-}^{e_+} \frac{|f_j(r)|^2 \rho_j(r)}{(r - \varepsilon_0 + \lambda^2 R(r))^2 + \pi^2 \lambda^4 |f(r)|^4} \, dr.$$

Using the above-mentioned continuity and boundedness properties it is not hard to show that

$$\lim_{\lambda \to 0} \lambda^2 \int_{e_-}^{e_+} \frac{|f_j(r)|^2}{|F(r)|^2} \rho_j(r)\, dr$$

$$= \rho_j(\varepsilon_0)|f_j(\varepsilon_0)|^2 \lim_{\lambda \to 0} \lambda^2 \int_{e_-}^{e_+} \frac{dr}{(r - \varepsilon_0 + \lambda^2 R(r))^2 + \pi^2 \lambda^4 |f(r)|^4}$$

$$= \rho_j(\varepsilon_0)|f_j(\varepsilon_0)|^2 \lim_{\lambda \to 0} \lambda^2 \int_{-\infty}^{\infty} \frac{dr}{r^2 + \pi^2 \lambda^4 |f(\varepsilon_0)|^4}$$

$$= \frac{|f_j(\varepsilon_0)|^2}{|f(\varepsilon_0)|^2} \rho_j(\varepsilon_0).$$

The proofs of Parts (ii) and (iii) are similar. Clearly, under additional regularity assumptions one can get information on the rate of convergence in Parts (i)-(iii). Finally, it is not difficult to show, using the Kubo formulas described in Subsection 7.6 and 8.3, that

$$\lim_{\lambda \to 0} \lambda^{-2} L_\rho(A, B) = L_{\mathrm{fgr}}(A_{\mathrm{fgr}}, B_{\mathrm{fgr}}),$$

where A, B are the microscopic heat or charge flux observables and A_{fgr}, B_{fgr} are their FGR counterparts.

9 Appendix

9.1 Structural Theorems

Proof of Theorem 3.1

Recall that $\pi_\omega(\mathcal{O})''$ is the Banach space dual of $\mathcal{N}_\omega$. If $A \in \mathcal{O}$ and $\tilde{A} \in \pi_\omega(\mathcal{O})''$ is a weak-$*$ accumulation point of the net

$$\frac{1}{t} \int_0^t \pi_\omega(\tau_V^s(A))\, ds,$$

$t \geq 0$, it follows from the asymptotic abelianness in mean that $\tilde{A} \in \pi_\omega(\mathcal{O})'$. Since ω is a factor state we have $\pi_\omega(\mathcal{O})' \cap \pi_\omega(\mathcal{O})'' = \mathbb{C}I$ and therefore, for any $\eta \in \mathcal{N}_\omega$, one has

$$\eta(\tilde{A}) = \omega(\tilde{A}). \tag{67}$$

Let $\mu, \nu \in \mathcal{N}_\omega$ and $\mu_+ \in \Sigma_+(\mu, \tau_V)$. Let $t_\alpha \to \infty$ be a net such that

$$\lim_\alpha \frac{1}{t_\alpha} \int_0^{t_\alpha} \mu \circ \tau_V^s(A)\, ds = \mu_+(A),$$

for all $A \in \mathcal{O}$. Passing to a subnet, we may also assume that for all $A \in \mathcal{O}$ and some $\nu_+ \in \Sigma_+(\nu, \tau_V)$,

$$\lim_\alpha \frac{1}{t_\alpha} \int_0^{t_\alpha} \nu \circ \tau_V^s(A)\, ds = \nu_+(A).$$

By the Banach-Alaoglu theorem, for any $A \in \mathcal{O}$ there exists a subnet $t_\gamma(A)$ of the net t_α and $A^\# \in \pi_\omega(\mathcal{O})''$ such that, for all $\eta \in \mathcal{N}_\omega$

$$\lim_\gamma \frac{1}{t_\gamma(A)} \int_0^{t_\gamma(A)} \eta(\pi_\omega(\tau_V^s(A))) \, \mathrm{d}s = \eta(A^\#).$$

Hence, $\mu_+(A) = \mu(A^\#)$ and $\nu_+(A) = \nu(A^\#)$. By (67) we also have $\mu(A^\#) = \omega(A^\#) = \nu(A^\#)$ and so $\mu_+(A) = \nu_+(A)$. We conclude that $\mu_+ = \nu_+$ and that

$$\Sigma_+(\mu, \tau_V) \subset \Sigma_+(\nu, \tau_V).$$

By symmetry, the reverse inclusion also holds and

$$\Sigma_+(\mu, \tau_V) = \Sigma_+(\omega, \tau_V)$$

for all $\mu \in \mathcal{N}_\omega$. $\square$

Proof of Theorem 3.3

To prove this theorem we use the correspondence between ω-normal states and elements of the standard cone $\mathcal{P}$ obtained from ω (see Theorem 4.41 in [Pi]); this is possible since ω is modular by assumption.

Note that if $\mathrm{Ker}\, L_V \neq \{0\}$, then there is an ω-normal, τ_V-invariant state η. By Theorem 3.1, $\Sigma_+(\omega, \tau_V) = \Sigma_+(\eta, \tau_V)$ and obviously $\Sigma_+(\eta, \tau_V) = \{\eta\}$. Two non-zero elements in $\mathrm{Ker}\, L_V$ therefore yield the same vector state and are represented by the same vector in the standard cone, *i.e.*, $\mathrm{Ker}\, L_V \cap \mathcal{P}$ is a one-dimensional half-line. Recall that any $\zeta \in \mathfrak{h}_\omega$ can be uniquely decomposed as

$$\zeta = \zeta_1 - \zeta_2 + \mathrm{i}\zeta_3 - \mathrm{i}\zeta_4,$$

with ζ_i in $\mathcal{P}$. Since $e^{\mathrm{i}tL_V}$ preserves the standard cone, $e^{\mathrm{i}tL_V}\zeta = \zeta$ iff $e^{\mathrm{i}tL_V}\zeta_i = \zeta_i$ for all i (*i.e.*, $\zeta_i \in \mathrm{Ker}\, L_V \cap \mathcal{P}$ for all i). Hence, $\mathrm{Ker}\, L_V$ is one-dimensional and Part (i) follows.

The proof of Part (ii) is simple. Any NESS $\eta \in \Sigma_+(\omega, \tau_V)$ can be uniquely decomposed as $\eta_n + \eta_s$ where $\eta_n \ll \omega$ and $\eta_s \perp \omega$. Since η is τ_V-invariant, η_n and η_s are also τ_V-invariant. Therefore η_n is represented by a vector ζ in $\mathrm{Ker}\, L_V \cap \mathcal{P}$. If $\mathrm{Ker}\, L_V = \{0\}$, then $\eta_n = 0$ and $\eta \perp \omega$.

It remains to prove Part (iii). Let $\varphi \in \mathrm{Ker}\, L_V$ be a separating vector for $\mathfrak{M}_\omega$. Let $B \in \pi_\omega(\mathcal{O})'$ be such that $\|B\varphi\| = 1$ and let ν_B be the vector state associated to $B\varphi$, $\nu_B(\cdot) = (B\varphi, \cdot B\varphi)$. For any $A \in \pi_\omega(\mathcal{O})$,

$$\frac{1}{t}\int_0^t \nu_B(\tau_V^s(A)) \, \mathrm{d}s = \frac{1}{t}\int_0^t \left(B\varphi,\, e^{\mathrm{i}sL_V}\pi_\omega(A)e^{-\mathrm{i}sL_V}B\varphi\right) \mathrm{d}s$$

$$= \left(\frac{1}{t}\int_0^t e^{-\mathrm{i}sL_V}B^*B\,\varphi\,\mathrm{d}s,\, \pi_\omega(A)\varphi\right).$$

Hence, by the von Neumann ergodic theorem,

$$\nu_{B+}(A) \equiv \lim_{t \to \infty} \frac{1}{t} \int_0^t \nu_B\big(\tau_V^s(A)\big) \, \mathrm{d}s = \big(P_{\mathrm{Ker}\, L_V} B^* B \, \varphi, \pi_\omega(A)\varphi\big),$$

where $P_{\mathrm{Ker}\, L_V}$ is the projection on $\mathrm{Ker}\, L_V$. Since φ is cyclic for $\pi_\omega(\mathcal{O})'$, for every $n \in \mathbb{N}$ we can find a B_n such that $\|\omega - \nu_{B_n}\| < 1/n$. The sequence ν_{B_n} is Cauchy in norm and for all $\omega_+ \in \Sigma_+(\omega, \tau_V)$,

$$\|\omega_+ - \nu_{B_n+}\| \le \|\omega - \nu_{B_n}\| < 1/n.$$

This implies that the norm limit of ν_{B_n} is the unique NESS in $\Sigma_+(\omega, \tau_V)$. Since $\nu_{B_n+} \in \mathcal{N}_\omega$ and $\mathcal{N}_\omega$ is a norm closed subset of $\mathcal{O}^*$, this NESS is ω-normal. $\quad\square$

9.2 The Hilbert-Schmidt Condition

Proof of Theorem 7.2

We will prove that $T_+^{1/2} - T^{1/2}$ is Hilbert-Schmidt. The proof that $(I - T_+)^{1/2} - (I - T)^{1/2}$ is also Hilbert-Schmidt is identical. For an elementary introduction to Hilbert-Schmidt operators (which suffices for the proof below) the reader may consult Section VI.6 in [RS].

By our general assumptions, the functions $f(r)$ and $F(r)^{-1}$ are bounded and continuous. By the assumption of Theorem 7.2, all the densities $\rho_j(r)$ are the same and equal to $\rho(r)$. Hence,

$$T_\mathcal{R} = \bigoplus_{j=1}^M \rho_j(r) = \rho(h_\mathcal{R}).$$

Let $p_\mathcal{R}$ be the orthogonal projection on the reservoir Hilbert space $\mathfrak{h}_\mathcal{R}$. Since $T^{1/2} - T_\mathcal{R}^{1/2} = T_\mathcal{S}^{1/2}, T_+^{1/2}(I - p_\mathcal{R}), (I - p_\mathcal{R})T_+^{1/2}$ are obviously Hilbert-Schmidt, it suffices to show that $p_\mathcal{R} T_+^{1/2} p_\mathcal{R} - T_\mathcal{R}^{1/2}$ is a Hilbert-Schmidt operator on the Hilbert space $\mathfrak{h}_\mathcal{R}$. Since

$$p_\mathcal{R} T_+^{1/2} p_\mathcal{R} - T_\mathcal{R}^{1/2} = -p_\mathcal{R} W_-^* [W_- p_\mathcal{R}, T_\mathcal{R}^{1/2}],$$

it suffices to show that $K \equiv [W_- p_\mathcal{R}, T_\mathcal{R}^{1/2}]$ is a Hilbert-Schmidt operator on $\mathfrak{h}_\mathcal{R}$. By Theorem 6.2, for $g \in \mathfrak{h}_\mathcal{R}$,

$$(Kg)(r) = \lambda^2 \frac{f(r)}{F(r)} \int_{e_-}^{e_+} \frac{\rho(r')^{1/2} - \rho(r)^{1/2}}{r' - r + \mathrm{i}o} \bar{f}(r') \cdot g(r') \, \mathrm{d}r'.$$

Let K_{ij} be an operator on $L^2((e_-, e_+), \mathrm{d}r)$ defined by

$$(K_{ij}h)(r) = \lambda^2 \frac{f_i(r)}{F(r)} \int_{e_-}^{e_+} \frac{\rho(r')^{1/2} - \rho(r)^{1/2}}{r' - r + \mathrm{i}o} \bar{f}_j(r') h(r') \, \mathrm{d}r'.$$

To prove that K is Hilbert-Schmidt on $\mathfrak{h}_\mathcal{R}$, it suffices to show that K_{ij} is Hilbert-Schmidt on $L^2((e_-, e_+), \mathrm{d}r)$ for all i, j.

Let $h_1, h_2 \in L^2((e_-, e_+), \mathrm{d}r)$ be bounded continuous functions. Then

$$(h_1, K_{ij}h_2) = \lambda^2 \int_{e_-}^{e_+} \frac{\bar{h}_1(r)f_i(r)g_2(r)}{F(r)}\,\mathrm{d}r, \tag{68}$$

where

$$g_2(r) = \lim_{\epsilon \downarrow 0} \int_{e_-}^{e_+} \frac{\rho(r')^{1/2} - \rho(r)^{1/2}}{r' - r + \mathrm{i}\epsilon} \bar{f}_j(r')h_2(r')\,\mathrm{d}r'.$$

Using the identity

$$\frac{1}{r' - r + \mathrm{i}\epsilon} = \frac{r' - r}{(r' - r)^2 + \epsilon^2} - \frac{\mathrm{i}\epsilon}{(r' - r)^2 + \epsilon^2},$$

and the fact that, for $r \in (e_-, e_+)$, one has

$$\lim_{\epsilon \downarrow 0} \epsilon \int_{e_-}^{e_+} \frac{\rho(r')^{1/2} - \rho(r)^{1/2}}{(r' - r)^2 + \epsilon^2} \bar{f}_j(r')h_2(r')\,\mathrm{d}r' = \pi(\rho(r)^{1/2} - \rho(r)^{1/2})\bar{f}_j(r)h_2(r)$$

$$= 0,$$

(see the Lecture [Ja]), we obtain

$$g_2(r) = \lim_{\epsilon \downarrow 0} \int_{e_-}^{e_+} \frac{(r' - r)(\rho(r')^{1/2} - \rho(r)^{1/2})}{(r' - r)^2 + \epsilon^2} \bar{f}_j(r')h_2(r')\,\mathrm{d}r'.$$

Since f_j and h_2 are bounded and $\rho(r)^{1/2}$ is $\frac{1}{2}$-Hölder continuous, we have

$$\sup_{\epsilon > 0, r \in (e_-, e_+)} \left| \int_{e_-}^{e_+} \frac{(r' - r)(\rho(r')^{1/2} - \rho(r)^{1/2})}{(r' - r)^2 + \epsilon^2} \bar{f}_j(r')h_2(r')\,\mathrm{d}r' \right|$$

$$\leq C \sup_{r \in (e_-, e_+)} \int_{e_-}^{e_+} \frac{\bar{f}_j(r')h_2(r')}{|r' - r|^{1/2}}\,\mathrm{d}r' < \infty.$$

Moreover, since $\bar{h}_1(r)F(r)^{-1}f_i(r) \in L^1((e_-, e_+), \mathrm{d}r)$, we can invoke the dominated convergence theorem to rewrite Equ. (68) as

$$(h_1, K_{ij}h_2) = \lim_{\epsilon \downarrow 0}(h_1, K_{ij,\epsilon}h_2) \tag{69}$$

where $K_{ij,\epsilon}$ is the integral operator on $L^2((e_-, e_+), \mathrm{d}r)$ with kernel

$$k_\epsilon(r, r') = \lambda^2 \frac{f_i(r)\bar{f}_j(r')}{F(r)} \frac{(r' - r)(\rho(r')^{1/2} - \rho(r)^{1/2})}{(r' - r)^2 + \epsilon^2}.$$

We denote by $\| \cdot \|_{\mathrm{HS}}$ the Hilbert-Schmidt norm. Then

$$\|K_{ij,\epsilon}\|_{\mathrm{HS}}^2 = \int |k_\epsilon(r,r')|^2 \, \mathrm{d}r \, \mathrm{d}r'.$$

Since $\rho(r)^{1/2}$ is α-Hölder continuous for $\alpha > 1/2$ and $F(r)^{-1}$ is bounded there exists a constant C such that, for $r, r' \in (e_-, e_+)$ and $\epsilon > 0$, one has the estimate

$$|k_\epsilon(r,r')|^2 \le C \, \frac{|f_i(r)|^2 |f_j(r')|^2}{|r - r'|^{2(1-\alpha)}}.$$

Therefore, since $2(1 - \alpha) < 1$, we conclude that

$$\sup_{\epsilon > 0} \|K_{ij,\epsilon}\|_{\mathrm{HS}}^2 = \sup_{\epsilon > 0} \int |k_\epsilon(r,r')|^2 \, \mathrm{d}r \, \mathrm{d}r' < \infty.$$

The Hilbert-Schmidt class of operators on $L^2((e_-, e_+), \mathrm{d}r)$ is a Hilbert space with the inner product $(X, Y) = \mathrm{Tr}(X^* Y)$. Since $\{K_{ij,\epsilon}\}_{\epsilon > 0}$ is a bounded set in this Hilbert space, there is a sequence $\epsilon_n \to 0$ and a Hilbert-Schmidt operator $\tilde{K}_{ij}$ such that for any Hilbert-Schmidt operator X on $L^2((e_-, e_+), \mathrm{d}r)$,

$$\lim_{n \to \infty} \mathrm{Tr}(X^* K_{ij,\epsilon_n}) = \mathrm{Tr}(X^* \tilde{K}_{ij}).$$

Taking $X = (h_1, \cdot) h_2$, where $h_i \in L^2((e_-, e_+), \mathrm{d}r)$ are bounded and continuous, we derive from (69) that $(h_1, \tilde{K}_{ij} h_2) = (h_1, K_{ij} h_2)$. Since the set of such h's is dense in $L^2((e_-, e_+), \mathrm{d}r)$, $\tilde{K}_{ij} = K_{ij}$ and so K_{ij} is Hilbert-Schmidt. $\quad\square$

References

[AB]   Araki, H., Barouch, E.: On the dynamics and ergodic properties of the XY-model. J. Stat. Phys. **31**, 327 (1983).

[AJPP] Aschbacher, W., Jakšić, V., Pautrat, Y., Pillet, C.-A.: Transport properties of ideal Fermi gases (in preparation).

[AM]   Aizenstadt, V.V., Malyshev, V.A.: Spin interaction with an ideal Fermi gas. J. Stat. Phys. **48**, 51 (1987).

[Ar1]   Araki, H.: Relative entropy of states of von Neumann algebras. Publ. Res. Inst. Math. Sci. Kyoto Univ. **11**, 809 (1975/76).

[Ar2]   Araki, H.: Relative entropy of states of von Neumann algebras II. Publ. Res. Inst. Math. Sci. Kyoto Univ. **13**, 173 (1977/78).

[Ar3]   Araki, H.: On the XY-model on two-sided infinite chain. Publ. Res. Inst. Math. Sci. Kyoto Univ. **20**, 277 (1984).

[ArM]   Araki, H., Masuda, T.: Positive cones and L^p-spaces for von Neumann algebras. Publ. RIMS Kyoto Univ. **18**, 339 (1982).

[At]   Attal, S.: Elements of operator algebras and modular theory. Volume I of this series.

[AW]   Araki, H., Wyss, W.: Representations of canonical anti-commutation relations. Helv. Phys. Acta **37**, 136 (1964).

[BFS]   Bach, V., Fröhlich, J., Sigal, I.: Return to equilibrium. J. Math. Phys. **41**, 3985 (2000).

[BLR]   Bonetto, F., Lebowitz, J.L., Rey-Bellet, L.: Fourier Law: A challenge to theorists. In *Mathematical Physics 2000*. Imp. Coll. Press, London (2000).

[BM]   Botvich, D.D., Malyshev, V.A.: Unitary equivalence of temperature dynamics for ideal and locally perturbed Fermi Gas. Commun. Math. Phys. **61**, 209 (1978).

[BR1]   Bratteli, O, Robinson D. W.. *Operator Algebras and Quantum Statistical Mechanics 1*. Springer, Berlin (1987).

[BR2]   Bratteli, O, Robinson D. W.: *Operator Algebras and Quantum Statistical Mechanics 2*. Springer, Berlin (1996).

[BSZ]   Baez, J.C., Segal, I.E., Zhou, Z.: *Introduction to algebraic and constructive quantum field theory*. Princeton University Press, Princeton NJ, (1991).

[CG1]   Cohen, E.G.D., Gallavotti, G.: Dynamical ensembles in stationary states. J. Stat. Phys. **80**, 931 (1995).

[CG2]   Cohen, E.G.D., Gallavotti, G.: Dynamical ensembles in nonequilibrium statistical mechanics. Phys. Rev. Lett. **74**, 2694 (1995).

[Da1]   Davies, E.B.: Markovian master equations. Commun. Math. Phys. **39**, 91 (1974).

[Da2]   Davies, E.B.: Markovian master equations II. Math. Ann. **219**, 147 (1976).

[De]   Dell'Antonio, G.F.: Structure of the algebra of some free systems. Commun. Math. Phys. **9**, 81 (1968).

[DGM]   De Groot, S.R., Mazur, P.: *Non-Equilibrium Thermodynamics*. North-Holland, Amsterdam (1969).

[D1]   Dereziński, J.: Fermi Golden Rule and open quantum systems. This volume.

[D2]   Dereziński, J.: Inroduction to representations of canonical commutation and anti-commutation relations. Lecture notes of the Nordfjordeid Summer School "Large Coulomb Systems—Quantum Electrodynamics", August 2003.

[Di]   Dirac P.A.M.: The quantum theory of the emission and absorption of radiation. Proc. Roy. Soc. London, Ser. A **114**, 243 (1927).

[DJ]   Dereziński, J., Jakšić, V.: Return to equilibrium for Pauli-Fierz systems. Ann. Henri Poincaré **4**, 739 (2003).

64   Walter Aschbacher et al.

[DJP]   Dereziński, J., Jakšić, V., Pillet, C.-A.: Perturbation theory of W^*-dynamics, KMS-·
        states and Liouvillean. Rev. Math. Phys. **15**, 447 (2003).
[Do]   Dorfman, J.R.: *An Introduction to Chaos in Nonequilibrium Statistical Mechanics.*
        Cambridge University Press, Cambridge (1999)
[Ei]   Einstein, A.: Zur Quantentheorie der Strahlung. Physik. Zeitschr. **18**, 121 (1917).
        This paper is reprinted in: van der Waerden, B.L., *Sources of Quantum Mechanics.*
        Dover, New York (1967).
[EM]   Evans, D.J., Morriss, G.P.: *Statistical Mechanics of Non-Equilibrium Liquids.* Aca-
        demic Press, New York (1990).
[Fer]   Fermi, E.: *Nuclear Physics.* Notes compiled by Orear J., Rosenfeld A.H. and
        Schluter R.A. The University of Chicago Press, Chicago, 1950.
[FM1]   Fröhlich, J., Merkli, M.: Thermal Ionization. Mathematical Physics, Analysis and
        Geometry **7**, 239 (2004).
[FM2]   Fröhlich, J., Merkli, M.: Another return of "return to equilibrium". Commun. Math.
        Phys., **251**, 235 (2004).
[FMS]   Fröhlich, J., Merkli, M., Sigal, I.M.: Ionization of atoms in a thermal field. J. Stat.
        Phys. **116**, 311 (2004).
[FMSU] Fröhlich, J., Merkli, M., Schwarz, S., Ueltschi, D.: Statistical mechanics of thermo-
        dynamic processes. In *A Garden of Quanta*, 345. World Scientific Publishing, River
        Edge NJ (2003).
[FMU]   Fröhlich, J., Merkli, M., Ueltschi, D.: Dissipative transport: thermal contacts and
        tunneling junctions. Ann. Henri Poincaré **4**, 897 (2004).
[Fr]   Friedrichs, K. O.: *Perturbation of Spectra in Hilbert Space*, AMS, Providence
        (1965).
[Ga1]   Gallavotti, G.: Nonequilibrium thermodynamics. Preprint, mp-arc 03-11 (2003).
[Ga2]   Gallavotti, G.: Entropy production in nonequilibrium thermodynamics: a review.
        Chaos **14**, 680 (2004).
[GVV1] Goderis, D., Verbeure, A., Vets, P.: Noncommutative central limits. Probab. Theory
        Related Fields **82** 527 (1989).
[GVV2] Goderis, V., Verbeure, A., Vets, P.: Quantum central limit and coarse graining. In
        Quantum probability and applications, V. Lecture Notes in Math., **1442**, 178 (1988).
[GVV3] Goderis, D., Verbeure, A., Vets, P.: About the mathematical theory of quantum fluc-
        tuations. In *Mathematical Methods in Statistical Mechanics.* Leuven Notes Math.
        Theoret. Phys. Ser. A Math. Phys., **1**, 31. Leuven Univ. Press, Leuven (1989).
[GVV4] Goderis, D., Verbeure, A., Vets, P.: Theory of quantum fluctuations and the Onsager
        relations. J. Stat. Phys. **56**, 721 (1989).
[GVV5] Goderis, D., Verbeure, A., Vets, P.: Dynamics of fluctuations for quantum lattice
        systems. Commun. Math. Phys. **128**, 533 (1990).
[GVV6] Goderis, D., Verbeure, A., Vets, P.: About the exactness of the linear response the-
        ory. Commun. Math. Phys. **136**, 265 (1991).
[Haa]   Haake, F.: *Statistical Treatment of Open Systems by Generalized Master Equation.*
        Springer Tracts in Modern Physics **66**, Springer, Berlin (1973).
[Ha]   Haag, R.: *Local Quantum Physics.* Springer, New York (1993).
[Ja]   Jakšić, V.: Topics in spectral theory. Volume I of this series.
[JKP]   Jakšić, V., Kritchevski, E., Pillet, C.-A.: Mathematical theory of the Wigner-Weiss-
        kopf atom. Lecture notes of the Nordfjordeid Summer School "Large Coulomb
        Systems—Quantum Electrodynamics", August 2003.
[Jo]   Joye, A.: Introduction to quantum statistical mechanics. Volume I of this series.
[JP1]   Jakšić, V., Pillet, C.-A.: On a model for quantum friction II: Fermi's golden rule and
        dynamics at positive temperature. Commun. Math. Phys. **176**, 619 (1996).

[JP2]   Jakšić, V., Pillet, C.-A.: On a model for quantum friction III: Ergodic properties of the spin-boson system. Commun. Math. Phys. **178**, 627 (1996).

[JP3]   Jakšić, V., Pillet, C.-A.: Spectral theory of thermal relaxation. J. Math. Phys. **38**, 1757 (1997).

[JP4]   Jakšić, V., Pillet, C.-A.: Mathematical theory of non-equilibrium quantum statistical mechanics. J. Stat. Phys. **108**, 787 (2002).

[JP5]   Jakšić, V., Pillet, C.-A.: Non-equilibrium steady states for finite quantum systems coupled to thermal reservoirs. Commun. Math. Phys. **226**, 131 (2002).

[JP6]   Jakšić, V., Pillet, C-A.: On entropy production in quantum statistical mechanics. Commun. Math. Phys. **217**, 285 (2001).

[JP7]   Jakšić, V., Pillet, C.-A.: A note on the entropy production formula. Contemp. Math. **327**, 175 (2003).

[JPR1]   Jakšić, V., Pillet, C.-A., Rey-Bellet, L.: Fluctuation of entropy production in classical statistical mechanics. In preparation.

[JPR2]   Jakšić, V., Pillet, C.-A., Rey-Bellet, L.: In preparation.

[KR]   Kadison R.V., Ringrose J.R.: *Fundamentals of the Theory of Operator Algebras II: Advanced Theory*, Graduate Studies in Mathematics **16**, AMS, Providence (1997).

[LeSp]   Lebowitz, J., Spohn, H.: Irreversible thermodynamics for quantum systems weakly coupled to thermal reservoirs. Adv. Chem. Phys. **39**, 109 (1978).

[Li]   Lindblad, G.: Completely positive maps and entropy inequalities. Commun. Math. Phys. **40**, 147 (1975).

[LMS]   Lieb, E.H., Schulz, T., Mathis, D.: Two soluble models of an anti-ferromagnetic chain. Ann. Phys. **28**, 407, (1961).

[Ma]   Matsui, T.: On the algebra of fluctuation in quantum spin chains. Ann. Henri Poincaré **4**, 63 (2003).

[Me1]   Merkli, M.: Positive commutators in non-equilibrium quantum statistical mechanics. Commun. Math. Phys. **223**, 327 (2001).

[Me2]   Merkli, M.: Stability of equilibria with a condensate. Commun. Math. Phys., in press.

[Me3]   Merkli, M.: The ideal quantum gas. Volume I of this series.

[Mes]   Messiah, A.: *Quantum Mechanics. Volume II.* Wiley, New York.

[Og]   Ogata, Y.: The stability of the non-equilibrium steady states. Commun. Math. Phys. **245**, 577 (2004).

[OP]   Ohya, M., Petz, D.: *Quantum Entropy and its Use.* Springer-Verlag, Berlin (1993).

[Pa]   Pais, A.: "*Subtle is the Lord...*", *The Science and Life of Albert Einstein*. Oxford University Press, Oxford (1982).

[Pi]   Pillet, C.-A.: Quantum dynamical systems. Volume I of this series.

[PoSt]   Powers, R. T., Stormer, E.: Free states of the canonical anticommutation relations. Commun. Math. Phys. **16**, 1 (1969).

[RC]   Rondoni, L., Cohen, E.G.D.: Gibbs entropy and irreversible thermodynamics. Nonlinearity **13**, 1905 (2000).

[Re]   Rey-Bellet, L.: Open classical systems. Volume II of this series.

[Ri]   Rideau, G.: On some representations of the anticommutation relations. Commun. Math. Phys. **9**, 229 (1968).

[Ro1]   Robinson, D.W.: Return to equilibrium. Commun. Math. Phys. **31**, 171 (1973).

[Ro2]   Robinson, D.W.: C^*-algebras in quantum statistical mechanics. In C^*-*algebras and their Applications to Statistical Mechanics and Quantum Field Theory*, (D. Kastler editor). North-Holand, Amsterdam (1976).

[Ru1]   Ruelle, D.: Natural nonequilibrium states in quantum statistical mechanics. J. Stat. Phys. **98**, 57 (2000).

66   Walter Aschbacher et al.

[Ru2]   Ruelle, D.: Entropy production in quantum spin systems. Commun. Math. Phys. **224**, 3 (2001).

[Ru3]   Ruelle, D.: Topics in quantum statistical mechanics and operator algebras. Preprint, mp-arc 01-257 (2001).

[Ru4]   Ruelle, D.: Smooth dynamics and new theoretical ideas in nonequilibrium statistical mechanics. J. Stat. Phys. **95**, 393 (1999).

[Ru5]   Ruelle, D.: Extending the definition of entropy to nonequilibrium steady states. Proc. Nat. Acad. Sci. USA **100**, 3054 (2003).

[Ru6]   Ruelle, D.: A remark on the equivalence of isokinetic and isoenergetic thermostats in the thermodynamic limit. J. Stat. Phys. **100**, 757 (2000).

[Ru7]   Ruelle, D.: Conversations on nonequilibrium physics with an extraterrestrial. Physics Today **57**, 48 (2004).

[RS]   Reed, M., Simon, B.: *Methods of Modern Mathematical Physics, I. Functional Analysis*, London, Academic Press (1980).

[Sp]   Spohn, H.: An algebraic condition for the approach to equilibrium of an open N-level system, Lett. Math. Phys. **2**, 33 (1977).

[Ta]   Takesaki, M.: *Theory of Operator Algebras I*. Springer, New-York (1979).

[TM]   Tasaki, S., Matsui, T.: Fluctuation theorem, nonequilibrium steady states and Mac-Lennan-Zubarev ensembles of a class of large quantum systems. Fundamental Aspects of Quantum Physics (Tokyo, 2001). QP–PQ: Quantum Probab. White Noise Anal., **17**, 100. World Sci., River Edge NJ, (2003).

[VH1]   van Hove, L.: Quantum-mechanical perturbations giving rise to a statistical transport equation. Physica **21**, 517.

[VH2]   van Hove, L.: The approach to equilibrium in quantum statistics. Physica **23**, 441.

[VH3]   van Hove, L.: Master equation and approach to equilibrium for quantum systems. In *Fundamental problems in statistical mechanics*, compiled by E.G.D. Cohen, North-Holand, Amsterdam 1962.

[WW]   Weisskopf, V., Wigner, E.: Berechnung der natürlichen Linienbreite auf Grund der Diracschen Lichttheorie. Zeitschrift für Physik **63**, 54 (1930).

Fermi Golden Rule and Open Quantum Systems

Jan Dereziński and Rafał Früboes

Department of Mathematical Methods in Physics, Warsaw University,
Hoża 74, 00-682, Warsaw, Poland
email: jan.derezinski@fuw.edu.pl, rafal.fruboes@fuw.edu.pl

1 Introduction

These lecture notes are an expanded version of the lectures given by the first author in the summer school "Open Quantum Systems" held in Grenoble, June 16—July 4, 2003. We are grateful to Stéphane Attal, Alain Joye, and Claude-Alain Pillet for their hospitality and invitation to speak.

Acknowledgments. The research of both authors was partly supported by the EU Postdoctoral Training Program HPRN-CT-2002-0277 and the Polish grants SPUB127 and 2 P03A 027 25. A part of this work was done during a visit of the first author to University of Montreal and to the Schrödinger Institute in Vienna. We acknowledge useful conversations with H. Spohn, C. A. Pillet, W. A. Majewski, and especially with V. Jakšić.

1.1 Fermi Golden Rule and Level Shift Operator in an Abstract Setting

We will use the name "the Fermi Golden Rule" to describe the well-known second order perturbative formula for the shift of eigenvalues of a family of operators $\mathbb{L}_\lambda = \mathbb{L}_0 + \lambda \mathbb{Q}$. Historically, the Fermi Golden Rule can be traced back to the early years of Quantum Mechanics, and in particular to the famous paper by Dirac [Di]. Two

"Golden Rules" describing the second order calculations for scattering amplitudes can be found in the Fermi lecture notes [Fe] on pages 142 and 148.

In its traditional form the Fermi Golden Rule is applied to Hamiltonians of quantum systems – self-adjoint operators on a Hilbert space. A nonzero imaginary shift of an eigenvalue of $\mathbb{L}_0$ indicates that the eigenvalue is unstable and that it has turned into a resonance under the influence of the perturbation $\lambda\mathbb{Q}$.

In our lectures we shall use the term Fermi Golden Rule in a slightly more general context, not restricted to Hilbert spaces. More precisely, we shall be interested in the case when $\mathbb{L}_\lambda$ is a generator of a 1-parameter group of isometries on a Banach space. For example, $\mathbb{L}_\lambda$ could be an anti-self-adjoint operator on a Hilbert space or the generator of a group of $*$-automorphisms of a W^*-algebra. These two special cases will be of particular importance for us.

Note that the spectrum of the generator of a group of isometries is purely imaginary. The shift computed by the Fermi Golden Rule may have a negative real part and this indicates that the eigenvalue has turned into a resonance. Hence, our convention differs from the traditional one by the factor of i.

In these lecture notes, we shall discuss several mathematically rigorous versions of the Fermi Golden Rule. In all of them, the central role is played by a certain operator that we call the Level Shift Operator (LSO). This operator will encode the second order shift of eigenvalues of $\mathbb{L}_\lambda$ under the influence of the perturbation. To define the LSO for $\mathbb{L}_\lambda = \mathbb{L}_0 + \lambda\mathbb{Q}$, we need to specify the projection $\mathbb{P}$ commuting with $\mathbb{L}_0$ (typically, the projection onto the point spectrum of $\mathbb{L}_0$) and a perturbation $\mathbb{Q}$. For the most part, we shall assume that $\mathbb{P}\mathbb{Q}\mathbb{P} = 0$, which guarantees the absence of the first order shift of the eigenvalues. Given the datum $(\mathbb{P}, \mathbb{L}_0, \mathbb{Q})$, we shall define the LSO as a certain operator on the range of the projection $\mathbb{P}$.

We shall describe several rigorous applications of the LSO for $(\mathbb{P}, \mathbb{L}_0, \mathbb{Q})$. One of them is the "weak coupling limit", called also the "van Hove limit". (We will not, however, use the latter name, since it often appears in a different meaning in statistical physics, denoting a special form of the thermodynamical limit). The time-dependent form of the weak coupling limit says that the reduced and rescaled dynamics $e^{-t\mathbb{L}_0/\lambda^2}\mathbb{P}e^{t\mathbb{L}_\lambda/\lambda^2}\mathbb{P}$ converges to the semigroup generated by the LSO. The time dependent weak coupling limit in its abstract form was proven by Davies [Da1, Da2, Da3]. In our lectures we give a detailed exposition of his results.

We describe also the so-called "stationary weak coupling limit", based on the recent work [DF2]. The stationary weak coupling limit says that appropriately rescaled and reduced resolvent of $\mathbb{L}_\lambda$ converges to the resolvent of the LSO.

The LSO has a number of other important applications. It can be used to describe approximate location and multiplicities of eigenvalues and resonances of $\mathbb{L}_\lambda$ for small nonzero λ. It also gives an upper bound on the number of eigenvalues of $\mathbb{L}_\lambda$ for small nonzero λ.

1.2 Applications of the Fermi Golden Rule to Open Quantum Systems

In these lectures, by an open quantum system we shall mean a "small" quantum system $\mathcal{S}$ interacting with a large "environment" or "reservoir" $\mathcal{R}$. The small quantum system is described by a finite dimensional Hilbert space $\mathcal{K}$ and a Hamiltonian K. The reservoir is described by a W^*-dynamical system $(\mathfrak{M}, \tau)$ and a reference state $\omega_\mathcal{R}$ (for a discussion of reference states see the lecture [AJPP]). We shall assume that $\omega_\mathcal{R}$ is normal and $\tau_\mathcal{R}$-invariant.

If $\omega_\mathcal{R}$ is a $(\tau_\mathcal{R}, \beta)$-KMS state, then we say that that the reservoir at inverse temperature β and that the open quantum system is thermal. Another important special case is when $\mathcal{R}$ has additional structure, namely consists of n independent parts $\mathcal{R}_1, \cdots, \mathcal{R}_n$, which are interpreted as sub-reservoirs. If the reference state of the sub-reservoir $\mathcal{R}_j$ is β_j-KMS (for $j = 1, \cdots, n$), then we shall call the corresponding open quantum system multi-thermal.

In the literature one can find at least two distinct important applications of the Fermi Golden Rule to the study of open quantum systems.

In the first application one considers the weak coupling limit for the dynamics in the Heisenberg picture reduced to the small system. This limit turns out to be an irreversible Markovian dynamics—a completely positive semigroup preserving the identity acting on the observables of the small system $\mathcal{S}$ ($n \times n$ matrices). The generator of this semigroup is given by the LSO for the generator of the dynamics. We will denote it by M.

The weak coupling limit and the derivation of the resulting irreversible Markovian dynamics goes back to the work of Pauli, Wigner-Weisskopf and van Hove [WW, VH1, VH2, VH3] see also [KTH, Haa]. In the mathematical literature it was studied in the well known papers of Davies [Da1, Da2, Da3], see also [LeSp, AL]. Therefore, the operator M is sometimes called the Davies generator in the Heisenberg picture.

One can also look at the dynamics in the Schrödinger picture (on the space of density matrices). In the weak coupling limit one then obtains a completely positive semigroup preserving the trace. It is generated by the adjoint of M, denoted by M^*, which is sometimes called the Davies generator in the Schrödinger picture.

The second application of the Fermi Golden Rule to the study of open quantum systems is relatively recent. It has appeared in papers on the so-called return to equilibrium [JP1, DJ1, DJ2, BFS2, M]. The main goal of these papers is to show that certain W^*-dynamics describing open quantum systems has only one stationary normal state or no stationary normal states at all. This problem can be reformulated into a question about the point spectrum of the so-called Liouvillean—the generator of the natural unitary implementation of the dynamics. To study this problem, it is convenient to introduce the LSO for the Liouvillean. We shall denote it by $i\Gamma$. It is an operator acting on Hilbert-Schmidt operators for the system $\mathcal{S}$—again $n \times n$ matrices.

The use of $i\Gamma$ in the spectral theory hinges on analytic techniques (Mourre theory, complex deformations), which we shall not describe in our lectures. We shall take it for granted that under suitable technical conditions such applications are possible and we will focus on the algebraic properties of M, $i\Gamma$ and M^*. To the best of our knowledge, some of these properties have not been discussed previously in the literature.

In Theorem 6.7 we give a simple characterization of the kernel of the imaginary part the operator Γ. This characterization implies that Γ has no nontrivial real eigenvalues in a generic nonthermal case. In [DJ2], this result was proven in the context of Pauli-Fierz systems and was used to show the absence of normal stationary states in a generic multithermal case. In our lectures we generalize the result of [DJ2] to a more general setting.

The characterization of the kernel of the imaginary part of Γ in the thermal case is given in Theorem 6.8. It implies that generically this kernel consists only of multiples of the square root of the Gibbs density matrix for the small system. In [DJ2], this result was proven in the more restrictive context of Pauli-Fierz systems and was used to show the return to equilibrium in the generic thermal case. A similar result was obtained earlier by Spohn [Sp].

The operators M, $i\Gamma$ and M^* act on the same vector space (the space of $n \times n$ matrices) and have similar forms. Naively, one may expect that $i\Gamma$ interpolates in some sense between M and M^*. Although this expectation is correct, its full description involves some advanced algebraic tools (the so-called noncommutative L^p-spaces associated to a von Neumann algebra), and for reasons of space we will not discuss it in these lecture notes (see [DJ4, JP6]).

In the thermal case, the relation between the operators M, $i\Gamma$ and M^* is considerably simpler—they are mutually similar and in particular have the same spectrum. This result has been recently proven in [DJ3] and we will describe it in detail in our lectures.

The similarity of $i\Gamma$ and M in the thermal case is closely related to the Detailed Balance Condition for M. In the literature one can find a number of different definitions of the Detailed Balance Condition applicable to irreversible quantum dynamics. In these lecture notes we shall propose another one and we will compare it with the definition due to Alicki [A] and Frigerio-Gorini-Kossakowski-Verri [FGKV].

For reason of space we have omitted many important topics in our lectures—they are treated in the review [DJ4], which is a continuation of these lecture notes. Some additional information about the weak coupling limit and the Davies generator can be also found in the lecture notes [AJPP].

2 Fermi Golden Rule in an Abstract Setting

2.1 Notation

Let L be an operator on a Banach space $\mathcal{X}$. $\mathrm{sp}L$, $\mathrm{sp}_{\mathrm{ess}}L$, $\mathrm{sp}_{\mathrm{p}}L$ will denote the spectrum, the essential spectrum and the point spectrum (the set of eigenvalues) of the operator L. If e is an isolated point in $\mathrm{sp}L$, then $\mathbf{1}_e(L)$ will denote the spectral projection of L onto e given by the usual contour integral. Sometimes we can also define $\mathbf{1}_e(L)$ if e is not an isolated point in the spectrum. This is well known if L is a normal operator on a Hilbert space. The definition of $\mathbf{1}_e(L)$ for some other classes of operators is discussed in Appendix, see (69), (70).

Let us now assume that L is a self-adjoint operator on a Hilbert space. Let A, B be bounded operators. Suppose that $p \in \mathbb{R}$. We define

$$A(p \pm \mathrm{i}0 - L)^{-1}B := \lim_{\epsilon \searrow 0} A(p \pm \mathrm{i}\epsilon - L)^{-1}B, \qquad (1)$$

provided that the right hand side of (1) exists. We will say that $A(p \pm i0 - L)^{-1}B$ exists if the limit in (1) exists.

The principal value of $(p - L)^{-1}$

$$A\mathcal{P}(p - L)^{-1}B := \frac{1}{2}\left(A(p + i0 - L)^{-1}B + A(p - i0 - L)^{-1}B\right)$$

and the delta function of $p - L$

$$A\delta(p - L)B := \frac{i}{2\pi}\left(A(p + i0 - L)^{-1}B - A(p - i0 - L)^{-1}B\right)$$

are then well defined.

$\mathcal{B}(\mathcal{X})$ denotes the algebra of bounded operators on $\mathcal{X}$. If $\mathcal{X}$ is a Hilbert space, then $\mathcal{B}^1(\mathcal{X})$ denotes the space of trace class operators and $\mathcal{B}^2(\mathcal{X})$ the space of Hilbert-Schmidt operators on $\mathcal{X}$. By a density matrix on $\mathcal{X}$ we mean $\rho \in \mathcal{B}^1(\mathcal{X})$ such that $\rho \geq 0$ and $\mathrm{Tr}\rho = 1$. We say that ρ is nondegenerate if $\mathrm{Ker}\rho = \{0\}$.

For more background material useful in our lectures we refer the reader to Appendix.

2.2 Level Shift Operator

In this subsection we introduce the definition of the Level Shift Operator. First we describe the basic setup needed to make this definition.

Assumption 2.1 *We assume that $\mathcal{X}$ is a Banach space, $\mathbb{P}$ is projection of norm 1 on $\mathcal{X}$ and $e^{t\mathbb{L}_0}$ is a 1-parameter C_0- group of isometries commuting with $\mathbb{P}$.*

We set $\mathbb{E} := \mathbb{L}_0\big|_{\mathrm{Ran}\mathbb{P}}$ and $\widetilde{\mathbb{P}} := \mathbf{1} - \mathbb{P}$. Clearly, $\mathbb{E}$ is the generator of a 1-parameter group of isometries on $\mathrm{Ran}\mathbb{P}$, and $\mathbb{L}_0\big|_{\mathrm{Ran}\widetilde{\mathbb{P}}}$ generates a 1-parameter group of isometries on $\mathrm{Ran}\widetilde{\mathbb{P}}$.

Later on, we will often write $\mathbb{L}_0\widetilde{\mathbb{P}}$ instead of $\mathbb{L}_0\big|_{\mathrm{Ran}\widetilde{\mathbb{P}}}$. For instance, in (2) $((ie + \xi)\widetilde{\mathbb{P}} - \mathbb{L}_0\widetilde{\mathbb{P}})^{-1}$ will denote the inverse of $(ie + \xi)\mathbf{1} - \mathbb{L}_0$ restricted to $\mathrm{Ran}\widetilde{\mathbb{P}}$. This is a slight abuse of notation, which we will make often without a comment.

Most of the time we will also assume that

Assumption 2.2 $\mathbb{P}$ *is finite dimensional.*

Under Assumption 2.1 and 2.2, the operator $\mathbb{E}$ is diagonalizable and we can write its spectral decomposition:

$$\mathbb{E} = \sum_{ie \in \mathrm{sp}\mathbb{E}} ie\mathbf{1}_{ie}(\mathbb{E}).$$

Note that $\mathbf{1}_{ie}(\mathbb{E})$ are projections of norm one.

In the remaining assumptions we impose our conditions on the perturbation:

Assumption 2.3 *We suppose that* $\mathbb{Q}$ *is an operator with* $\mathrm{Dom}\mathbb{Q} \supset \mathrm{Dom}\mathbb{L}_0$ *and, for* $|\lambda| < \lambda_0$, $\mathbb{L}_\lambda := \mathbb{L}_0 + \lambda\mathbb{Q}$ *is the generator of a 1-parameter C_0-semigroup of contractions.*

Assumption 2.3 implies that $\widetilde{\mathbb{P}}\mathbb{Q}\mathbb{P}$ and $\mathbb{P}\mathbb{Q}\widetilde{\mathbb{P}}$ are well defined.

Assumption 2.4 $\mathbb{P}\mathbb{Q}\mathbb{P} = 0$.

The above assumption is needed to guarantee that the first nontrivial contribution for the shift of eigenvalues of $\mathbb{L}_\lambda$ is 2nd order in λ.

It is also useful to note that if Assumption 2.2 holds, then $\widetilde{\mathbb{P}}\mathbb{Q}\mathbb{P}$ and $\mathbb{P}\mathbb{Q}\widetilde{\mathbb{P}}$ are bounded. Note also that in the definition of LSO only the terms $\widetilde{\mathbb{P}}\mathbb{Q}\mathbb{P}$ and $\mathbb{P}\mathbb{Q}\widetilde{\mathbb{P}}$ will play a role and the term $\widetilde{\mathbb{P}}\mathbb{Q}\widetilde{\mathbb{P}}$ will be irrelevant.

Assumption 2.5 *We assume that for all* $\mathrm{ie} \in \mathrm{sp}\mathbb{E}$ *there exists*

$$\mathbf{1}_{\mathrm{ie}}(\mathbb{E})\mathbb{Q}((\mathrm{ie}+0)\widetilde{\mathbb{P}} - \mathbb{L}_0\widetilde{\mathbb{P}})^{-1}\mathbb{Q}\mathbf{1}_{\mathrm{ie}}(\mathbb{E})$$
$$:= \lim_{\xi\searrow 0} \mathbf{1}_{\mathrm{ie}}(\mathbb{E})\mathbb{Q}((\mathrm{ie}+\xi)\widetilde{\mathbb{P}} - \mathbb{L}_0\widetilde{\mathbb{P}})^{-1}\mathbb{Q}\mathbf{1}_{\mathrm{ie}}(\mathbb{E}) \tag{2}$$

Under Assumptions 2.1, 2.2, 2.3, 2.4 and 2.5 we set

$$M := \sum_{\mathrm{ie}\in\mathrm{sp}\mathbb{E}} \mathbf{1}_{\mathrm{ie}}(\mathbb{E})\mathbb{Q}((\mathrm{ie}+0)\widetilde{\mathbb{P}} - \mathbb{L}_0\widetilde{\mathbb{P}})^{-1}\mathbb{Q}\mathbf{1}_{\mathrm{ie}}(\mathbb{E}) \tag{3}$$

and call it the Level Shift Operator (LSO) associated to the triple $(\mathbb{P}, \mathbb{L}_0, \mathbb{Q})$.

It is instructive to give time-dependent formulas for the LSO:

$$M = \lim_{\xi\searrow 0} \sum_{\mathrm{ie}\in\mathrm{sp}\mathbb{E}} \mathbf{1}_{\mathrm{ie}}(\mathbb{E})\int_0^\infty \mathrm{e}^{-\xi s}\mathbb{Q}\mathbb{Q}(s)\mathbf{1}_{\mathrm{ie}}(\mathbb{E})\mathrm{d}s$$
$$= \lim_{\xi\searrow 0} \sum_{\mathrm{ie}\in\mathrm{sp}\mathbb{E}} \mathbf{1}_{\mathrm{ie}}(\mathbb{E})\int_0^\infty \mathrm{e}^{-\xi s}\mathbb{Q}(-s/2)\mathbb{Q}(s/2)\mathbf{1}_{\mathrm{ie}}(\mathbb{E})\mathrm{d}s,$$

where $\mathbb{Q}(t) := \mathrm{e}^{t\mathbb{L}_0}\mathbb{Q}\mathrm{e}^{-t\mathbb{L}_0}$.

2.3 LSO for C_0^*-Dynamics

In the previous subsection we assumed that $\mathbb{L}_\lambda$ is a generator of a C_0-semigroup. In one of our applications, however, we will deal with another type of semigroups, the so-called C_0^*-semigroups (see Appendix for definitions and a discussion). In this case, we will need to replace Assumptions 2.1 and 2.3 by their "dual versions", which we state below:

Assumption 2.1* *We assume that* $\mathcal{Y}$ *is a Banach space and* $\mathcal{X}$ *is its dual, that is* $\mathcal{X} = \mathcal{Y}^*$, $\mathbb{P}$ *is a w^* continuous projection of norm 1 on $\mathcal{X}$ and $\mathrm{e}^{t\mathbb{L}_0}$ is a 1-parameter C_0^*- group of isometries commuting with $\mathbb{P}$.*

Assumption 2.3* *We suppose that* $\mathbb{Q}$ *is an operator with* $\mathrm{Dom}\mathbb{Q} \supset \mathrm{Dom}\mathbb{L}_0$ *and, for* $|\lambda| < \lambda_0$, $\mathbb{L}_\lambda := \mathbb{L}_0 + \lambda\mathbb{Q}$ *is the generator of a 1-parameter C_0^*-semigroup of contractions.*

2.4 LSO for W^*-Dynamics

The formalism of the Level Shift Operator will be applied to open quantum systems in two distinct situations.

In the first application, the Banach space $\mathcal{X}$ is a W^*-algebra, $\mathbb{P}$ is a normal conditional expectation and $e^{t\mathbb{L}_0}$ is a W^*-dynamics.

Note that W^*-algebras are usually not reflexive and W^*-dynamics are usually not C_0-groups. However, W^*-algebras are dual Banach spaces and W^*-dynamics are C_0^*-groups.

The perturbation has the form $i[V, \cdot]$ with V being a self-adjoint element of the W^*-algebra. Therefore, $e^{t\mathbb{L}_\lambda}$ will be a W^*-dynamics for all real λ – again a C_0^*-group.

2.5 LSO in Hilbert Spaces

In our second application, $\mathcal{X}$ is a Hilbert space. Hilbert spaces are reflexive, therefore we do not need to distinguish between C_0 and C_0^*-groups.

All strongly continuous groups of isometries on a Hilbert space are unitary groups. Therefore, the operator $\mathbb{L}_0$ has to be anti-self-adjoint (that means $\mathbb{L}_0 = iL_0$, where L_0 is self-adjoint).

All projections of norm one on a Hilbert space are orthogonal. Therefore, the distinguished projection has to be orthogonal.

In our applications to open quantum systems $e^{t\mathbb{L}_\lambda}$ is a unitary dynamics. This means in particular that $\mathbb{Q}$ has the form $\mathbb{Q} = iQ$, where Q is hermitian.

In the case of a Hilbert space the LSO will be denoted $i\Gamma$. Thus we will isolate the imaginary unit "i", which is consistent with the usual conventions for operators in Hilbert spaces, and also with the convention that was adopted in [DJ2].

Remark 2.1. In [DJ2] we used a formalism similar to that of Subsection 2.2 in the context of a Hilbert space. Note, however, that the terminology that was adopted there is not completely consistent with the terminology used in these lectures. In [DJ2] we considered a Hilbert space $\mathcal{X}$, an orthogonal projection P, and self-adjoint operators L_0, Q. If Γ is the LSO for the triple (P, L_0, Q) according to [DJ2], then $i\Gamma$ is the LSO for (P, iL_0, iQ) according to the present definition.

Let us quote the following easy fact valid in the case of a Hilbert space.

Theorem 2.1. *Suppose that $\mathcal{X}$ is a Hilbert space, Assumptions 2.1, 2.2, 2.3 and 2.5 hold and Q is self-adjoint. Then $e^{it\Gamma}$ is contractive for $t > 0$.*

Proof. We use the notation $\mathbb{E} = iE$, $\mathbb{L}_0 = iL$, $\mathbb{Q} = iQ$. We have

$$\frac{1}{2i}(\Gamma - \Gamma^*) = -\sum_{e \in spE} \mathbf{1}_e(E) Q \delta(e - L_0) Q \mathbf{1}_e(E) \leq 0$$

Therefore, $i\Gamma$ is a dissipative operator and $e^{it\Gamma}$ is contractive for $t > 0$. □

Note that in Theorem 3.4 we will show that the LSO is the generator of a contractive semigroup also in a more general situation, when $\mathcal{X}$ is a Banach space. The proof of this fact will be however more complicated and will require some additional technical assumptions.

2.6 The Choice of the Projection $\mathbb{P}$

In typical application of the LSO, the operators $\mathbb{L}_0$ and $\mathbb{Q}$ are given and our goal is to study the operator

$$\mathbb{L}_\lambda := \mathbb{L}_0 + \lambda \mathbb{Q}. \tag{4}$$

More precisely, we want to know what happens with its eigenvalues when we switch on the perturbation.

Therefore, it is natural to choose the projection $\mathbb{P}$ as "the projection onto the point spectrum of $\mathbb{L}_0$", that is

$$\mathbb{P} = \sum_{e \in \mathbb{R}} \mathbf{1}_{ie}(\mathbb{L}_0), \tag{5}$$

provided that (5) is well defined.

More generally, if we were interested only about what happens around some eigenvalues $\{ie_1, \ldots, ie_n\} \subset \mathrm{sp}_p \mathbb{L}_0$, then we could use the LSO defined with the projection

$$\mathbb{P} = \sum_{j=1}^{n} \mathbf{1}_{ie_j}(\mathbb{L}_0). \tag{6}$$

Clearly, if $\mathcal{X}$ is a Hilbert space and $\mathbb{L}_0$ is anti-self-adjoint, then $\mathbf{1}_{ie}(\mathbb{L}_0)$ are well defined for all $e \in \mathbb{R}$. Moreover, both (5) and (6) are projections of norm one commuting with $\mathbb{L}_0$, and hence they satisfy Assumption 2.1.

There is no guarantee that the spectral projections $\mathbf{1}_{ie}(\mathbb{L}_0)$ are well defined in the more general case when $\mathbb{L}_0$ is the generator of a group of isometries on a Banach space. If they are well defined, then they have norm one, however, we seem to have no guarantee that their sums have norm one. In Appendix we discuss the problem of defining spectral projections onto eigenvalues in this more general case.

Note, however, that in the situation considered by us later, we will have no such problems. In fact, $\mathbb{P}$ will be always given by (5) and will always have norm one.

If $\mathbf{1}_{ie}(\mathbb{L}_0)$ is well defined for all $e \in \mathbb{R}$ and we take $\mathbb{P}$ defined by (5), then $\mathbb{P}$ will be determined by the operator $\mathbb{L}_0$ itself. We will speak about "the LSO for $\mathbb{L}_\lambda$", if we have this projection in mind.

2.7 Three Kinds of the Fermi Golden Rule

Suppose that Assumptions 2.1, 2.2, 2.3, 2.4 and 2.5, or 2.1*, 2.2, 2.3*, 2.4 and 2.5 are satisfied. Let $\mathbb{P}$ be given by (5) and M be the LSO for $(\mathbb{P}, \mathbb{L}_0, \mathbb{Q})$. Our main object of interest is the operator $\mathbb{L}_\lambda$.

The assumption 2.4 ($\mathbb{PQP} = 0$) guarantees that there are no first order effects of the perturbation. The operator M describes what happens with the eigenvalues of $\mathbb{L}_0$ under the influence of the perturbation $\lambda\mathbb{Q}$ at the second order of λ. Following the tradition of quantum physics, we will use the name "the Fermi Golden Rule" to describe the second order effects of the perturbation.

The Fermi Golden Rule can be made rigorous in many ways under various technical assumptions. We can distinguish at least three varieties of the rigorous Fermi Golden Rule:

– **Analytic Fermi Golden Rule:** $\mathbb{E} + \lambda^2 M$ *predicts the approximate location (up to $o(\lambda^2)$) and the multiplicity of the resonances and eigenvalues of $\mathbb{L}_\lambda$ in a neighborhood of* $\mathrm{sp_p}\mathbb{L}_0$ *for small λ.*

   The Analytic Fermi Golden Rule is valid under some analyticity assumptions on $\mathbb{L}_\lambda$. It is well known and follows essentially by the standard perturbation theory for isolated eigenvalues ([Ka,RS4], see also [DF1]). The perturbation arguments are applied not to $\mathbb{L}_\lambda$ directly, but to the analytically deformed $\mathbb{L}_\lambda$. More or less explicitly, this idea was applied to Liouvilleans describing open quantum systems [JP1, JP2, BFS1, BFS2]. One can also apply it to the W^*-dynamics of open quantum systems [JP4, JP5].

   The **stationary weak coupling (or van Hove) limit** of [DF2], described in Theorem 3.1 and 3.4, can be viewed as an infinitesimal version of the Analytic Fermi Golden Rule.

– **Spectral Fermi Golden Rule:** *The intersection of the spectrum of $\mathbb{E} + \lambda^2 M$ with the imaginary line predicts possible location of eigenvalues of $\mathbb{L}_\lambda$ for small nonzero λ. It also gives an upper bound on their multiplicity.*

   Note that if the Analytic Fermi Golden Rule is true, then so is the Spectral Fermi Golden Rule. However, to prove the Analytic Fermi Golden Rule we need strong analytic assumption, whereas the Spectral Fermi Golden Rule can be shown under much weaker conditions. Roughly speaking, these assumptions should allow us to apply the so-called positive commutator method.

   The Spectral Fermi Golden Rule is stated in Theorem 6.7 of [DJ2], which is proven in [DJ1]. Strictly speaking, the analysis of [DJ1] and [DJ2] is restricted to Pauli-Fierz operators, but it is easy to see that their arguments extend to much larger classes of operators.

   To illustrate the usefulness of the Spectral Fermi Golden Rule, suppose that $\mathcal{X}$ is a Hilbert space, $\mathbb{L}_\lambda = iL_\lambda$ with L_λ self-adjoint and $i\Gamma$ is the LSO. Then the Spectral Fermi Golden Rule implies the bound

$$\dim \mathrm{Ran} \mathbf{1}_\mathrm{p}(L_\lambda) \leq \dim \mathrm{Ker}\Gamma^{\mathrm{I}},$$

   where $\Gamma^{\mathrm{I}} := \frac{1}{2\mathrm{i}}(\Gamma - \Gamma^*)$. Bounds of this type were used in various papers related to the Return to Equilibrium [JP1, JP2, DJ2, BFS2, M].

– **Dynamical Fermi Golden Rule.** *The operator* $\mathrm{e}^{t(\mathbb{E}+\lambda^2 M)}$ *describes approximately the reduced dynamics* $\mathbb{P}\mathrm{e}^{t\mathbb{L}_\lambda}\mathbb{P}$ *for small λ.*

   The Dynamical Fermi Golden Rule was rigorously expressed in the form of **the weak coupling** by Davies [Da1, Da2, Da3, LeSp]. Davies showed that under some weak assumptions we have

$$\lim_{\lambda \to 0} e^{-t\mathbb{E}/\lambda^2} \mathbb{P} e^{t\mathbb{L}_\lambda/\lambda^2} \mathbb{P} = e^{tM}.$$

We describe his result in Theorems 3.2 and 3.4.

3 Weak Coupling Limit

3.1 Stationary and Time-Dependent Weak Coupling Limit

In this section we describe in an abstract setting the weak coupling limit. We will show that, under some conditions, the dynamics restricted to an appropriate subspace, rescaled and renormalized by the free dynamics, converges to the dynamics generated by the LSO.

We will give two versions of the weak coupling limit: the time dependent and the stationary one. The time-dependent version is well known and in its rigorous form is due to Davies [Da1, Da2, Da3]. Our exposition is based on [Da3].

The stationary weak coupling limit describes the same phenomenon on the level of the resolvent. Our exposition is based on recent work [DF2]. Formally, one can pass from the time-dependent to stationary weak coupling limit by the Laplace transformation. However, one can argue that the assumptions needed to prove the stationary weak coupling limit are sometimes easier to verify. In fact, they involve the existence of certain matrix elements of the resolvent (a kind of the "Limiting Absorption Principle") only at the spectrum of $\mathbb{E}$, a discrete subset of the imaginary line. This is often possible to show by positive commutator methods.

Throughout the section we suppose that most of the assumptions of Subsection 2.2 are satisfied. We will, however, list explicitly the assumptions that we need for each particular result.

The first theorem describes the stationary weak coupling limit.

Theorem 3.1. *Suppose that Assumptions 2.1, 2.2, 2.3 and 2.4, or 2.1*, 2.2, 2.3* and 2.4 are true. We also assume the following conditions:*

1) *For* $\mathrm{ie} \in \mathrm{sp}\mathbb{E}$, $\xi > 0$, *we have* $\mathrm{ie} + \xi \notin \mathrm{sp}\widetilde{\mathbb{P}}\mathbb{L}_\lambda\widetilde{\mathbb{P}}$.

2) *There exists an operator* M_{st} *on* $\mathrm{Ran}\mathbb{P}$ *such that, for any* $\xi > 0$,

$$M_{\mathrm{st}} := \sum_{\mathrm{ie} \in \mathrm{sp}\mathbb{E}} \lim_{\lambda \to 0} \mathbf{1}_{\mathrm{ie}}(\mathbb{E})\mathbb{Q}\left((\mathrm{ie} + \lambda^2\xi)\widetilde{\mathbb{P}} - \widetilde{\mathbb{P}}\mathbb{L}_\lambda\widetilde{\mathbb{P}}\right)^{-1}\mathbb{Q}\mathbf{1}_{\mathrm{ie}}(\mathbb{E}). \quad (7)$$

   (Note that a priori the right hand side of (7) may depend on ξ; we assume that it does not).

3) *For any* $\mathrm{ie}, \mathrm{ie}' \in \mathrm{sp}\mathbb{E}$, $e \neq e'$ *and* $\xi > 0$,

$$\lim_{\lambda \to 0} \lambda \mathbf{1}_{\mathrm{ie}}(\mathbb{E})\mathbb{Q}\left((\mathrm{ie} + \lambda^2\xi)\widetilde{\mathbb{P}} - \widetilde{\mathbb{P}}\mathbb{L}_\lambda\widetilde{\mathbb{P}}\right)^{-1}\mathbb{Q}\mathbf{1}_{\mathrm{ie}'}(\mathbb{E}) = 0,$$

$$\lim_{\lambda \to 0} \lambda \mathbf{1}_{\mathrm{ie}'}(\mathbb{E})\mathbb{Q}\left((\mathrm{ie} + \lambda^2\xi)\widetilde{\mathbb{P}} - \widetilde{\mathbb{P}}\mathbb{L}_\lambda\widetilde{\mathbb{P}}\right)^{-1}\mathbb{Q}\mathbf{1}_{\mathrm{ie}}(\mathbb{E}) = 0.$$

Then the following holds:

1. $e^{tM_{st}}$ *is a contractive semigroup.*
2. *For any* $\xi > 0$

$$\sum_{ie\in sp\mathbb{E}} \lim_{\lambda\to 0} \mathbf{1}_{ie}(\mathbb{E}) \left(\xi - \lambda^{-2}(\mathbb{L}_\lambda - ie)\right)^{-1} \mathbb{P} = (\xi\mathbb{P} - M_{st})^{-1}.$$

3. *For any* $f \in C_0([0,\infty[)$,

$$\lim_{\lambda\to 0} \int_0^\infty f(t)e^{-t\mathbb{E}/\lambda^2}\mathbb{P}e^{t\mathbb{L}_\lambda/\lambda^2}\mathbb{P}dt = \int_0^\infty f(t)e^{tM_{st}}dt. \tag{8}$$

Next we describe the time-dependent version of the weak coupling limit for C_0-groups.

Theorem 3.2. *Suppose that Assumptions 2.1, 2.3 and 2.4 are true. We make also the following assumptions:*

1) $\mathbb{P}\mathbb{Q}\widetilde{\mathbb{P}}$ *and* $\widetilde{\mathbb{P}}\mathbb{Q}\mathbb{P}$ *are bounded. (Note that this assumption guarantees that* $\widetilde{\mathbb{P}}\mathbb{L}_\lambda\widetilde{\mathbb{P}}$ *is the generator of a C_0-semigroup on* $\mathrm{Ran}\widetilde{\mathbb{P}}$*).*
2) *Set*

$$K_\lambda(t) := \int_0^{\lambda^{-2}t} e^{-s\mathbb{E}}\mathbb{P}\mathbb{Q}e^{s\widetilde{\mathbb{P}}\mathbb{L}_\lambda\widetilde{\mathbb{P}}}\mathbb{Q}\mathbb{P}ds. \tag{9}$$

We suppose that for all $t_0 > 0$, there exists c such that

$$\sup_{|\lambda|<\lambda_0} \sup_{0\le t\le t_0} \|K_\lambda(t)\| \le c.$$

3) *There exists a bounded operator K on $\mathrm{Ran}\mathbb{P}$ such that*

$$\lim_{\lambda\to 0} K_\lambda(t) = K$$

for all $0 < t < \infty$.

4) *There exists an operator M_{dyn} such that*

$$\mathrm{s}-\lim_{t\to\infty} t^{-1}\int_0^t e^{s\mathbb{E}}Ke^{-s\mathbb{E}}ds = M_{dyn}.$$

Then the following holds:

1. $e^{tM_{dyn}}$ *is a contractive semigroup.*
2. *For any* $y \in \mathrm{Ran}\mathcal{Y}$ *and* $t_0 > 0$,

$$\lim_{\lambda\to 0} \sup_{0\le t\le t_0} \|e^{-\mathbb{E}t/\lambda^2}\mathbb{P}e^{t\mathbb{L}_\lambda/\lambda^2}\mathbb{P}y - e^{tM_{dyn}}y\| = 0.$$

One of possible C_0^*-versions of the above theorem is given below.

Theorem 3.2* *Suppose that Assumptions 2.1*, 2.3* and 2.4 are true. We make also the following assumptions:*

0) $e^{t\mathbb{E}}$ *is a C_0-group. (We already know that it is a C_0^*-group).*

1) $\mathbb{P}Q\widetilde{\mathbb{P}}$ *and* $\widetilde{\mathbb{P}}Q\mathbb{P}$ *are w^* continuous. (Note that this assumption guarantees that $\widetilde{\mathbb{P}}\mathbb{L}_\lambda\widetilde{\mathbb{P}}$ is a generator of a C_0^*-semigroup on $\mathrm{Ran}\widetilde{\mathbb{P}}$).*

2) *In the sense of a w^* integral [BR1] we set*

$$K_\lambda(t) := \int_0^{\lambda^{-2}t} e^{-s\mathbb{E}}\mathbb{P}Qe^{s\widetilde{\mathbb{P}}\mathbb{L}_\lambda\widetilde{\mathbb{P}}}Q\mathbb{P}\mathrm{d}s. \tag{10}$$

We suppose that for all $t_0 > 0$, there exists c such that

$$\sup_{|\lambda| < \lambda_0} \sup_{0 \le t \le t_0} \|K_\lambda(t)\| \le c.$$

3) *there exists a w^* continuous operator K on $\mathrm{Ran}\mathbb{P}$ such that*

$$\lim_{\lambda \to 0} K_\lambda(t) = K$$

for all $0 < t < \infty$.

4) *There exists an operator M_{dyn} such that*

$$\mathrm{s-}\lim_{t\to\infty} t^{-1}\int_0^t e^{s\mathbb{E}}Ke^{-s\mathbb{E}} = M_{\mathrm{dyn}}.$$

Then the same conclusions as in Theorem 3.2 hold.

Theorem 3.2 is due to Davies (we put together Theorem 5.18 and 5.11 from [Da3]). Note that, following Davies, in Theorems 3.2 and 3.2* we do not make Assumption 2.2 about the finite dimension of $\mathrm{Ran}\mathbb{P}$. Instead, we make the assumption 4) about spectral averaging. If we impose Assumption 2.2, then we can drop 4) and make some other minor simplifications, as is described below:

Theorem 3.3. *Suppose that Assumptions 2.1, 2.2, 2.3 and 2.4 or 2.1*, 2.2, 2.3* and 2.4 are true. Set*

$$K_\lambda(t) := \int_0^{\lambda^{-2}t} e^{-s\mathbb{E}}\mathbb{P}Qe^{s\widetilde{\mathbb{P}}\mathbb{L}_\lambda\widetilde{\mathbb{P}}}Q\mathbb{P}\mathrm{d}s.$$

We make also the following assumptions:

1) *We suppose that for all $t_0 > 0$, there exists c such that*

$$\sup_{|\lambda| < \lambda_0} \sup_{0 \le t \le t_0} \|K_\lambda(t)\| \le c.$$

2) *There exists an operator K on $\mathrm{Ran}\mathbb{P}$ such that*

$$\lim_{\lambda \to 0} K_\lambda(t) = K$$

for all $0 < t < \infty$. We set

$$M_{\mathrm{dyn}} := \sum_{i e \in \mathrm{sp}\mathbb{E}} \mathbf{1}_{ie}(\mathbb{E})K\mathbf{1}_{ie}(\mathbb{E})$$

Then the following holds:

1. $e^{tM_{\mathrm{dyn}}}$ *is a contractive semigroup.*
2. *For any* $t_0 > 0$,

$$\lim_{\lambda \to 0} \sup_{0 \le t \le t_0} \|e^{-\mathbb{E}t/\lambda^2} \mathbb{P} e^{t\mathbb{L}_\lambda/\lambda^2} \mathbb{P} - e^{tM_{\mathrm{dyn}}}\| = 0. \tag{11}$$

Note that if there exists an operator M_{st} satisfying (8), and an operator M_{dyn} satisfying (11), then they clearly coincide. In our last theorem of this section we will describe a connection between M_{st}, M_{dyn} and the LSO.

Theorem 3.4. *Suppose that Assumptions 2.1, 2.2, 2.3 and 2.4, or 2.1*, 2.2, 2.3* and 2.4 are true. Suppose also that the following conditions hold:*

1) $\int_0^\infty \sup_{|\lambda| \le \lambda_0} \|\mathbb{P}\mathbb{Q} e^{s\widetilde{\mathbb{P}}\mathbb{L}_\lambda \widetilde{\mathbb{P}}} \mathbb{Q}\mathbb{P}\| ds < \infty.$

2) *For any* $s > 0$, $\lim_{\lambda \to 0} \mathbb{P}\mathbb{Q} e^{s\widetilde{\mathbb{P}}\mathbb{L}_\lambda \widetilde{\mathbb{P}}} \mathbb{Q}\mathbb{P} = \mathbb{P}\mathbb{Q} e^{s\widetilde{\mathbb{P}}\mathbb{L}_0} \mathbb{Q}\mathbb{P}.$

Then

1. *Assumption 2.5 holds, and hence the LSO for* $(\mathbb{P}, \mathbb{L}_0, \mathbb{Q})$, *defined in (3) and denoted M, exists.*
2. e^{tM} *is a contractive semigroup.*
3. *The assumptions of Theorem 3.1 hold and* $M = M_{\mathrm{st}}$, *consequently, for any* $\xi > 0$

$$\lim_{\lambda \to 0} \sum_{ie \in \mathrm{sp}\mathbb{E}} \mathbf{1}_{ie}(\mathbb{E}) \left(\xi - \lambda^{-2}(\mathbb{L}_\lambda - ie)\right)^{-1} \mathbb{P} = (\xi\mathbb{P} - M)^{-1}.$$

4. *The assumptions of Theorem 3.3 hold and* $M = M_{\mathrm{dyn}}$, *consequently*

$$\lim_{\lambda \to 0} \sup_{0 \le t \le t_0} \|e^{-\mathbb{E}t/\lambda^2} \mathbb{P} e^{t\mathbb{L}_\lambda/\lambda^2} \mathbb{P} - e^{tM}\| = 0.$$

3.2 Proof of the Stationary Weak Coupling Limit

Proof of Theorem 3.1. We follow [DF2]. Let $ie \in \mathrm{sp}\mathbb{E}$. Set

$$G_\lambda(\xi, ie) := \xi\mathbb{P} + \lambda^{-2}(ie\mathbb{P} - \mathbb{E})$$

$$-\mathbb{P}\mathbb{Q}\left((\lambda^2\xi + ie)\widetilde{\mathbb{P}} - \widetilde{\mathbb{P}}\mathbb{L}_\lambda\widetilde{\mathbb{P}}\right)^{-1}\mathbb{Q}\mathbb{P}.$$

By the so-called Feshbach formula (see e.g. [DJ1, BFS1]), for $\xi > 0$ we have

$$G_\lambda(\xi, ie)^{-1} = \mathbb{P}\left(\xi + \lambda^{-2}(ie - \mathbb{L}_\lambda)\right)^{-1}\mathbb{P}$$

This and the dissipativity of $\mathbb{L}_\lambda$ implies the bound

$$\|G_\lambda(\xi, ie)^{-1}\| \leq \xi^{-1}. \tag{12}$$

Write for shortness G instead of $G_\lambda(\xi, ie)$. For $ie' \in \mathrm{sp}\mathbb{E}$, set

$$\mathbb{P}_{e'} := \mathbf{1}_{ie'}(\mathbb{E}),$$

$$\mathbb{P}_{\bar{e}'} := \mathbb{P} - \mathbf{1}_{ie'}(\mathbb{E}).$$

Decompose $G = G_{\mathrm{diag}} + G_{\mathrm{off}}$ into its diagonal and off-diagonal part:

$$G_{\mathrm{diag}} := \sum_{ie' \in \mathrm{sp}\mathbb{E}} \mathbb{P}_{e'} G \mathbb{P}_{e'},$$

$$G_{\mathrm{off}} := \sum_{ie' \in \mathrm{sp}\mathbb{E}} \mathbb{P}_{e'} G \mathbb{P}_{\bar{e}'} = \sum_{ie' \in \mathrm{sp}\mathbb{E}} \mathbb{P}_{\bar{e}'} G \mathbb{P}_{e'}.$$

First we would like to show that for $\xi > 0$ and small enough λ, G_{diag} is invertible. By an application of the Neumann series, $\mathbb{P}_{\bar{e}} G_{\mathrm{diag}}$ is invertible on $\mathrm{Ran}\mathbb{P}_{\bar{e}}$, and we have the bound

$$\|\mathbb{P}_{\bar{e}} G_{\mathrm{diag}}^{-1}\| \leq c\lambda^2. \tag{13}$$

It is more complicated to prove that $\mathbb{P}_e G_{\mathrm{diag}}$ is inverible on $\mathrm{Ran}\mathbb{P}_e$.

We fix $\xi > 0$. We know that G is invertible and $\|G^{-1}\| \leq \xi^{-1}$. Hence we can write

$$G_{\mathrm{diag}} G^{-1} = 1 - G_{\mathrm{off}} G^{-1}.$$

Therefore

$$\mathbb{P}_e G_{\mathrm{diag}} G^{-1} = \mathbb{P}_e - \mathbb{P}_e G_{\mathrm{off}} \mathbb{P}_{\bar{e}} G^{-1},$$

$$\mathbb{P}_{\bar{e}} G_{\mathrm{diag}} G^{-1} = \mathbb{P}_{\bar{e}} - \mathbb{P}_{\bar{e}} G_{\mathrm{off}} G^{-1}. \tag{14}$$

The latter identity can be for small enough λ transformed into

$$\mathbb{P}_{\bar{e}} G^{-1} = G_{\mathrm{diag}}^{-1} \mathbb{P}_{\bar{e}} - G_{\mathrm{diag}}^{-1} \mathbb{P}_{\bar{e}} G_{\mathrm{off}} G^{-1}. \tag{15}$$

We insert (15) into the first identity of (14) to obtain

$$\mathbb{P}_e G_{\mathrm{diag}} G^{-1} = \mathbb{P}_e - \mathbb{P}_e G_{\mathrm{off}} \mathbb{P}_{\bar{e}} G_{\mathrm{diag}}^{-1} + \mathbb{P}_e G_{\mathrm{off}} \mathbb{P}_{\bar{e}} G_{\mathrm{diag}}^{-1} G_{\mathrm{off}} G^{-1}. \tag{16}$$

We multiply (16) from the right by $\mathbb{P}_e$ to get

$$\mathbb{P}_e G_{\mathrm{diag}} \mathbb{P}_e G^{-1} \mathbb{P}_e = \mathbb{P}_e + \mathbb{P}_e G_{\mathrm{off}} \mathbb{P}_{\bar{e}} G_{\mathrm{diag}}^{-1} G_{\mathrm{off}} G^{-1} \mathbb{P}_e. \tag{17}$$

Now, using

$$\lim_{\lambda \to 0} \lambda \|G_{\mathrm{off}}\| = 0, \tag{18}$$

(12) and (13) we obtain

$$\lim_{\lambda \to 0} \mathbb{P}_e G_{\mathrm{off}} \mathbb{P}_{\bar{e}} G_{\mathrm{diag}}^{-1} G_{\mathrm{off}} G^{-1} \mathbb{P}_e = 0.$$

Thus, for small enough λ,

$$\mathbb{P}_e G_{\text{diag}} B_1 = \mathbb{P}_e,$$

where

$$B_1 := \mathbb{P}_e G^{-1} \mathbb{P}_e \left(\mathbb{P}_e + \mathbb{P}_e G_{\text{off}} \mathbb{P}_{\overline{e}} G_{\text{diag}}^{-1} G_{\text{off}} G^{-1} \mathbb{P}_e \right)^{-1}.$$

Similarly, for small enough λ, we find B_2 such that

$$B_2 \mathbb{P}_e G_{\text{diag}} = \mathbb{P}_e.$$

This implies that $\mathbb{P}_e G_{\text{diag}}$ is invertible on $\text{Ran}\mathbb{P}_e$.

Next, we can write

$$G^{-1} = G_{\text{diag}}^{-1} - G_{\text{diag}}^{-1} G_{\text{off}} G_{\text{diag}}^{-1} + G_{\text{diag}}^{-1} G_{\text{off}} G_{\text{diag}}^{-1} G_{\text{off}} G^{-1}.$$

Hence,

$$\mathbb{P}_e G^{-1} = \mathbb{P}_e G_{\text{diag}}^{-1} \left(\mathbf{1} - G_{\text{off}} \mathbb{P}_{\overline{e}} G_{\text{diag}}^{-1} + G_{\text{off}} \mathbb{P}_{\overline{e}} G_{\text{diag}}^{-1} G_{\text{off}} G^{-1} \right). \qquad (19)$$

Therefore, for a fixed ξ, by (12), (13) and (18) we see that as $\lambda \to 0$ we have

$$-G_{\text{off}} \mathbb{P}_{\overline{e}} G_{\text{diag}}^{-1} + G_{\text{off}} \mathbb{P}_{\overline{e}} G_{\text{diag}}^{-1} G_{\text{off}} G^{-1} \to 0.$$

Therefore, for small enough λ, we can invert the expression in the bracket of (19). Consequently,

$$\mathbb{P}_e(G_{\text{diag}}^{-1} - G^{-1}) = \mathbb{P}_e G^{-1} \left(1 - G_{\text{off}} \mathbb{P}_{\overline{e}} G_{\text{diag}}^{-1} + G_{\text{off}} \mathbb{P}_{\overline{e}} G_{\text{diag}}^{-1} G_{\text{off}} G^{-1} \right)^{-1}$$
$$\times \left(G_{\text{off}} \mathbb{P}_{\overline{e}} G_{\text{diag}}^{-1} - G_{\text{off}} \mathbb{P}_{\overline{e}} G_{\text{diag}}^{-1} G_{\text{off}} G^{-1} \right).$$
$$(20)$$

Therefore, for a fixed ξ, by (12), (13) and (18) we see that, as $\lambda \to 0$, we have

$$\mathbb{P}_e(G_{\text{diag}}^{-1} - G^{-1}) \to 0. \qquad (21)$$

Hence, (12) and (21) imply that $\mathbb{P}_e G_{\text{diag}}^{-1}$ is uniformly bounded as $\lambda \to 0$. We know that

$$\mathbb{P}_e G_{\text{diag}} \to \mathbb{P}_e \xi - \mathbb{P}_e M_{\text{st}}. \qquad (22)$$

Therefore, $\xi \mathbb{P}_e - \mathbb{P}_e M_{\text{st}}$ is invertible on $\text{Ran}\mathbb{P}_e$ and

$$\mathbb{P}_e G_{\text{diag}}^{-1} \to (\mathbb{P}_e \xi - \mathbb{P}_e M_{\text{st}})^{-1}.$$

Using again (21), we see that

$$\mathbb{P}_e G^{-1} \to (\mathbb{P}_e \xi - \mathbb{P}_e M_{\text{st}})^{-1}. \qquad (23)$$

Summing up (23) over e, we obtain

$$\sum_{ie \in \text{sp}\mathbb{E}} \mathbb{P}_e G_\lambda(\xi, ie)^{-1} \to (\xi \mathbb{P} - M_{\text{st}})^{-1}, \qquad (24)$$

which ends the proof of 2.

Let us now prove 1. We have

$$\sum_{ie\in\mathrm{spE}} \mathbb{P}_e G_\lambda(\xi,ie)^{-1} = \sum_{ie\in\mathrm{spE}} \int_0^\infty e^{-t(\xi+\lambda^{-2}ie)}\mathbb{P}_e e^{t\mathbb{L}_\lambda/\lambda^2}\mathbb{P}dt$$

$$= \int_0^\infty e^{-t\xi} e^{-t\mathbb{E}/\lambda^2}\mathbb{P}e^{t\mathbb{L}_\lambda/\lambda^2}\mathbb{P}dt \tag{25}$$

Clearly, $\|e^{-t\mathbb{E}/\lambda^2}\mathbb{P}e^{t\mathbb{L}_\lambda/\lambda^2}\mathbb{P}\| \leq 1$. Therefore,

$$\left\| \sum_{ie\in\mathrm{spE}} \mathbb{P}_e G_\lambda(\xi,ie)^{-1} \right\| \leq \xi^{-1}.$$

Hence, by (24),

$$\|(\xi\mathbb{P} - M_{\mathrm{st}})^{-1}\| \leq \xi^{-1},$$

which proves 1.

Let $f \in C_0([0,\infty[)$ and $\delta > 0$. By the Stone-Weierstrass Theorem, we can find a finite linear combination of functions of the form $e^{-t\xi}$ for $\xi > 0$, denoted g, such that $\|e^{t\delta}f - g\|_\infty < \epsilon$. Set

$$A_\lambda(t) := e^{-t\mathbb{E}/\lambda^2}\mathbb{P}e^{t\mathbb{L}_\lambda/\lambda^2}\mathbb{P}, \quad A_0(t) := e^{tM_{\mathrm{dyn}}}.$$

Note that $\|A_\lambda(t)\| \leq 1$ and $\|A_0(t)\| \leq 1$. Now

$$\| \int f(t)(A_\lambda(t) - A_0(t))dt\| \quad \leq \| \int e^{-\delta t}g(t)(A_\lambda(t) - A_0(t))dt\|$$

$$+\| \int (f(t) - e^{-\delta t}g(t))A_\lambda(t)dt\| +\| \int (f(t) - e^{-\delta t}g(t))A_0(t)dt\|.$$

By 2. and by the Laplace transformation, the first term on the right hand side goes to 0 as $\lambda \to 0$. The last two terms are estimated by $\epsilon \int_0^\infty e^{-\delta t}dt$, which can be made arbitrarily small by choosing ϵ small. This proves 3. $\square$

3.3 Spectral Averaging

Before we present the time-dependent version of the weak coupling limit, we discuss the spectral averaging of operators, following [Da3].

In this subsection, $\mathcal{Y}$ is an arbitrary Banach space and $e^{t\mathbb{E}}$ is a 1-parameter C_0-group of isometries on $\mathcal{Y}$. For $K \in \mathcal{B}(\mathcal{Y})$ we define

$$K^\natural := s-\lim_{t\to\infty} t^{-1} \int_0^t e^{s\mathbb{E}}Ke^{-s\mathbb{E}}ds, \tag{26}$$

provided that the right hand side exists.

Theorem 3.5. *Suppose that $K^\natural$ exists. Then, for any $t_0 > 0$, $y \in \mathcal{Y}$,*

$$\lim_{\lambda \to 0} \sup_{0 \le t \le t_0} \|e^{-t\mathbb{E}/\lambda} e^{t(\mathbb{E}+\lambda K)/\lambda} y - e^{tK^\natural} y\| = 0.$$

Proof. Consider the space $C([0, t_0], \mathcal{Y})$ with the supremum norm. Set $K(t) = e^{t\mathbb{E}/\lambda} K e^{-t\mathbb{E}/\lambda}$. For $f \in C([0, t_0], \mathcal{Y})$, define

$$B_\lambda f(t) := \int_0^t K(s/\lambda) f(s) \mathrm{d}s,$$

$$B_0 f(t) := K^\natural \int_0^t f(s) \mathrm{d}s.$$

Clearly, B_0 and B_λ are linear operators on $C([0, t_0], \mathcal{Y})$ satisfying

$$\|B_\lambda\| \le t_0 \|K\|. \tag{27}$$

Moreover

$$\lim_{\lambda \to 0} B_\lambda f = B_0 f. \tag{28}$$

To prove (28), by (27) it suffices to assume that $f \in C^1([0, t_0], \mathcal{Y})$. Now

$$B_\lambda f(t) = \left(\int_0^t K(s/\lambda) \mathrm{d}s \right) f(t) - \int_0^t \left(\int_0^s \mathrm{d}s_1 K(s_1/\lambda) \right) f'(s) \mathrm{d}s$$

$$\to tK^\natural f(t) - \int_0^t sK^\natural f'(s) \mathrm{d}s = B_0 f(t).$$

We easily get

$$\|B_\lambda^n\| \le \frac{t_0^n}{n!} \|K\|^n, \quad \|B_0^n\| \le \frac{t_0^n}{n!} \|K\|^n. \tag{29}$$

Let $y \in \mathcal{Y}$. Set $y_\lambda(t) := e^{-t\mathbb{E}/\lambda} e^{t(\mathbb{E}+\lambda K)/\lambda} y$. Note that

$$y_\lambda(t) = y + B_\lambda y_\lambda(t), \quad y_0(t) = y + B_0 y_0(t).$$

Treating y as an element of $C([0, t_0], \mathcal{Y})$ – the constant function equal to y we can write

$$(1 - B_\lambda)^{-1} y = \sum_{n=0}^{\infty} B_\lambda^n y, \quad (1 - B_0)^{-1} y = \sum_{n=0}^{\infty} B_0^n y,$$

where both Neumann series are absolutely convergent. Therefore, in the sense of the convergence in in $C([0, t_0], \mathcal{Y})$, we get

$$y_\lambda = \sum_{n=0}^{\infty} B_\lambda^n y \to \sum_{n=0}^{\infty} B_0^n y = y_0.$$

$\square$

Theorem 3.6. *Let $\mathcal{Y}$ be finite dimesional. Then $K^\natural$ exists for any $K \in \mathcal{B}(\mathcal{Y})$ and*

$$K^\natural = \sum_{ie \in \mathrm{sp}\mathbb{E}} \mathbb{1}_{ie}(\mathbb{E}) K \mathbb{1}_{ie}(\mathbb{E}) = \lim_{t \to \infty} t^{-1} \int_0^t e^{s\mathbb{E}} K e^{-s\mathbb{E}} ds,$$

$$\lim_{\lambda \to 0} \sup_{0 \leq t \leq t_0} \|e^{-t\mathbb{E}/\lambda} e^{t(\mathbb{E}+\lambda K)/\lambda} - e^{tK^\natural}\| = 0.$$

Proof. In finite dimension we can replace the strong limit by the norm limit. Moreover,

$$t^{-1} \int_0^t e^{s\mathbb{E}} K e^{-s\mathbb{E}} ds = \sum_{ie_1, ie_2 \in \mathrm{sp}\mathbb{E}} \mathbb{1}_{ie_1}(\mathbb{E}) K \mathbb{1}_{ie_2}(\mathbb{E}) \frac{e^{it(e_1 - e_2)} - 1}{i(e_1 - e_2)t}.$$

$\square$

Remark 3.1. The following results generalize some aspects of Theorem 3.6 to the case when $\mathbb{P}$ is not necessarily finite dimensional. They are proven in [Da3]. We will not need these results.

1) If $K^\natural$ exists, then it commutes with $e^{t\mathbb{E}}$.
2) If K is a compact operator and $\mathcal{Y}$ is a Hilbert space, then $K^\natural$ exists and we can replace the strong limit in (26) by the norm limit.
3) If $\mathbb{E}$ has a total set of eigenvectors, then $K^\natural$ exists as well.

3.4 Second Order Asymptotics of Evolution with the First Order Term.

In this subsection we consider a somewhat more general situation than in Subsection 3.1. We make the Assumptions 2.1, 2.3 and 2.4, or 2.1*, 2.3* and 2.4 but we do not assume that $\mathbb{P}$ is finite dimensional, nor that $\mathbb{PQP} = 0$. Thus we allow for a term of first order in λ in the asymptotics of the reduced dynamics. We again follow [Da3].

We assume also that $\mathbb{PQ}\widetilde{\mathbb{P}}$ and $\widetilde{\mathbb{P}}\mathbb{QP}$ are bounded or w* continuous and that $\mathbb{E} + \lambda\mathbb{PQP}$ generates a C_0- or C_0^*-group of isometries on $\mathrm{Ran}\mathbb{P}$.

Using the boundedness of off-diagonal elements $\mathbb{PQ}\widetilde{\mathbb{P}}$ and $\widetilde{\mathbb{P}}\mathbb{QP}$, we see that $\widetilde{\mathbb{P}}L_\lambda\widetilde{\mathbb{P}}$ is the generator of a continuous semigroup.

In this subsection, the definition of $K_\lambda(t)$ slightly changes as compared with (9):

$$K_\lambda(t) := \int_0^{\lambda^{-2}t} e^{-s(\mathbb{E}+\lambda\mathbb{PQP})} \mathbb{PQ} e^{s\widetilde{\mathbb{P}}L_\lambda\widetilde{\mathbb{P}}} \mathbb{QP} ds.$$

Theorem 3.7. *Suppose that the following assumptions are true:*
1) *For all $t_0 > 0$, there exists c such that*

$$\sup_{|\lambda| < \lambda_0} \sup_{0 \leq t \leq t_0} \|K_\lambda(t)\| \leq c.$$

2) *There exists a bounded (w* continuous in the C_0^* case) operator K on* $\mathrm{Ran}\mathbb{P}$
   such that

$$\lim_{\lambda \to 0} K_\lambda(t) = K$$

for all $0 < t < \infty$.
Then for $y \in \mathrm{Ran}\mathbb{P}$

$$\lim_{\lambda \to 0} \sup_{0 \leq t \leq t_1} \left\| \mathbb{P}e^{t\mathbb{L}_\lambda/\lambda^2}\mathbb{P}y - e^{t(\mathbb{E}+\lambda\mathbb{P}\mathbb{Q}\mathbb{P}+\lambda^2 K)/\lambda^2}y \right\| = 0.$$

Proof. Set $\mathcal{Y} := \mathrm{Ran}\mathbb{P}$. Consider the space $C([0,t_0],\mathcal{Y})$. For $f \in C([0,t_0],\mathcal{Y})$
define

$$H_\lambda f(t) := \int_0^t e^{(\mathbb{E}+\mathbb{P}\mathbb{Q}\mathbb{P})(t-s)/\lambda^2}K_\lambda(t-s)f(s)\mathrm{d}s,$$

$$G_\lambda f(t) := \int_0^t e^{(\mathbb{E}+\mathbb{P}\mathbb{Q}\mathbb{P})(t-s)/\lambda^2}Kf(s)\mathrm{d}s.$$

Note that H_λ and G_λ are linear operators on $C([0,t_0],\mathcal{Y})$ satisfying

$$\|H_\lambda^n\| \leq c^n t_0^n/n!, \quad \|G_\lambda^n\| \leq c^n t_0^n/n!,$$

Thus $1 - H_\lambda$ and $1 - G_\lambda$ are invertible. In fact, they can be defined by the Neumann
series:

$$(1 - H_\lambda)^{-1} = \sum_{j=0} H_\lambda^n, \quad (1 - G_\lambda)^{-1} = \sum_{j=0} G_\lambda^n.$$

Next we note that

$$\|H_\lambda^n - G_\lambda^n\| \leq \|H_\lambda - G_\lambda\|c^{n-1}t_0^{n-1}/(n-1)!, \tag{30}$$

because

$$\|H_\lambda^n - G_\lambda^n\| \leq \sum_{j=0}^{n-1} \|H_\lambda^j\|\|G_\lambda^{n-j-1}\|\|H_\lambda - G_\lambda\|$$

$$\leq \sum_{j=0}^{n-1} \frac{c^{n-1}t_0^{n-1}}{k!(n-k-1)!}\|H_\lambda - G_\lambda\| = (2ct_0)^{n-1}\|H_\lambda - G_\lambda\|/(n-1)!.$$

Therefore,

$$\|(1 - H_\lambda)^{-1} - (1 - G_\lambda)^{-1}\| \leq c\|H_\lambda - G_\lambda\|, \tag{31}$$

Next,

$$(H_\lambda - G_\lambda)f(t) = \int_0^t e^{(\mathbb{E}+\lambda\mathbb{P}\mathbb{Q}\mathbb{P})(t-s)/\lambda^2}(K_\lambda(t-s) - K)f(s)\mathrm{d}s.$$

and hence

$$\|H_\lambda - G_\lambda\| \leq \int_0^{t_0} \|K_\lambda(s) - K\|\mathrm{d}s \to 0.$$

Thus

$$\|(1 - H_\lambda)^{-1} - (1 - G_\lambda)^{-1}\| \to 0. \tag{32}$$

Let $y \in \mathcal{Y}$. Define the following elements of the space $C([0, t_0], \mathcal{Y})$:

$$k_\lambda(t) := e^{(\mathbb{E} + \lambda \mathbb{P} \mathbb{Q} \mathbb{P}) t / \lambda^2} y,$$

$$h_\lambda(t) := \mathbb{P} e^{\mathbb{L}_\lambda t / \lambda^2} \mathbb{P} y,$$

$$g_\lambda(t) := e^{(\mathbb{E} + \lambda \mathbb{P} \mathbb{Q} \mathbb{P} + \lambda^2 K) t / \lambda^2} y.$$

Now

$$h_\lambda = k_\lambda + H_\lambda h_\lambda,$$

$$g_\lambda = k_\lambda + G_\lambda g_\lambda.$$

Thus

$$h_\lambda - g_\lambda = (1 - H_\lambda)^{-1} k_\lambda - (1 - G_\lambda)^{-1} k_\lambda \to 0.$$

$\square$

3.5 Proof of Time Dependent Weak Coupling Limit

Proof of Theorem 3.2 and 3.2*. In addition to the assumptions of Theorem 3.7 we suppose that $\mathbb{P} \mathbb{Q} \mathbb{P} = 0$ and $K^\natural$ exists.

Theorem 3.6 implies that

$$\lim_{\lambda \to 0} \sup_{0 \le t \le t_0} \|e^{-\mathbb{E} t / \lambda^2} e^{t(\mathbb{E} + \lambda^2 K) / \lambda^2} y - e^{t K^\natural} y\| = 0.$$

Theorem 3.7 yields

$$\lim_{\lambda \to 0} \sup_{0 \le t \le t_0} \|\mathbb{P} e^{t \mathbb{L}_\lambda / \lambda^2} \mathbb{P} y - e^{t(\mathbb{E} + \lambda^2 K) / \lambda^2} y\| = 0.$$

Using that $e^{t \mathbb{E}}$ is isometric we obtain

$$\lim_{\lambda \to 0} \sup_{0 \le t \le t_0} \|e^{-\mathbb{E} t / \lambda^2} \mathbb{P} e^{t \mathbb{L}_\lambda / \lambda^2} \mathbb{P} y - e^{t K^\natural} y\| = 0. \tag{33}$$

It is clear from (33) that $e^{t K^\natural}$ is contractive. $\square$

Proof of Theorem 3.3 Because of the finite dimension all operators on $\mathrm{Ran}\mathbb{P}$ are w^* continuous and the strong and norm convergence coincide. Besides, we can apply Theorem 3.6 about the existence of $K^\natural$. $\square$

3.6 Proof of the Coincidence of M_{st} and M_{dyn} with the LSO

Proof of Theorem 3.4. Set

$$f(s) := \sup_{|\lambda| \le \lambda_0} \|\mathbb{P}\mathbb{Q}e^{s\widetilde{\mathbb{P}}\mathbb{L}_\lambda\widetilde{\mathbb{P}}}\mathbb{Q}\mathbb{P}\|.$$

We know that $f(t)$ is integrable.

For any $e \in \mathbb{R}$ and $\xi \ge 0$ we can dominate the integrand in the integral

$$F_\lambda(\mathrm{i}e, \xi) := \int_0^\infty \mathbb{P}\mathbb{Q}e^{s\widetilde{\mathbb{P}}\mathbb{L}_\lambda\widetilde{\mathbb{P}}}\mathbb{Q}\mathbb{P}e^{-(\mathrm{i}e+\lambda^2\xi)s}\mathrm{d}s$$

$$= \mathbb{P}\mathbb{Q}\left(\widetilde{\mathbb{P}}(\mathrm{i}e + \lambda^2\xi) - \widetilde{\mathbb{P}}\mathbb{L}_\lambda\widetilde{\mathbb{P}}\right)^{-1}\mathbb{Q}\mathbb{P} \qquad (34)$$

by $f(s)$. Hence, using the dominated convergence theorem we see that $F_\lambda(\mathrm{i}e, \xi)$ is continuous at $\lambda = 0$ and $\xi \ge 0$. But

$$\sum_{\mathrm{i}e \in \mathrm{sp}\mathbb{E}} \mathbf{1}_{\mathrm{i}e}(\mathbb{E})F_0(\mathrm{i}e, 0)\mathbf{1}_{\mathrm{i}e}(\mathbb{E})$$

$$= \sum_{\mathrm{i}e \in \mathrm{sp}\mathbb{E}} \lim_{\lambda \to 0} \mathbf{1}_{\mathrm{i}e}(\mathbb{E})\mathbb{Q}\left(\widetilde{\mathbb{P}}(\mathrm{i}e + \lambda^2\xi) - \widetilde{\mathbb{P}}\mathbb{L}_\lambda\widetilde{\mathbb{P}}\right)^{-1}\mathbb{Q}\mathbb{P}\mathbf{1}_{\mathrm{i}e}(\mathbb{E}) = M_{\mathrm{st}}.$$

Recall (9), the definition of $K_\lambda(t)$:

$$K_\lambda(t) := \int_0^{\lambda^{-2}t} e^{-s\mathbb{E}}\mathbb{P}\mathbb{Q}e^{s\widetilde{\mathbb{P}}\mathbb{L}_\lambda\widetilde{\mathbb{P}}}\mathbb{Q}\mathbb{P}\mathrm{d}s.$$

Its integrand can also be dominated by $f(s)$. Hence, using again the dominated convergence theorem, we see that, for $\lambda \to 0$, $K_\lambda(t)$ is convergent to

$$K = \int_0^\infty e^{-s\mathbb{E}}\mathbb{P}\mathbb{Q}e^{s\widetilde{\mathbb{P}}\mathbb{L}_0}\mathbb{Q}\mathbb{P}\mathrm{d}s.$$

Therefore,

$$K^\natural = \sum_{\mathrm{i}e \in \mathrm{sp}\mathbb{E}} \mathbf{1}_{\mathrm{i}e}(\mathbb{E})\int_0^\infty \mathbb{Q}e^{s\mathbb{L}_0\widetilde{\mathbb{P}}}\mathbb{Q}\mathbf{1}_{\mathrm{i}e}(\mathbb{E})e^{-\mathrm{i}es}\mathrm{d}s$$

$$= \sum_{\mathrm{i}e \in \mathrm{sp}\mathbb{E}} \mathbf{1}_{\mathrm{i}e}(\mathbb{E})F_0(\mathrm{i}e, 0)\mathbf{1}_{\mathrm{i}e}(\mathbb{E}).$$

□

4 Completely Positive Semigroups

In this section we recall basic information about completely positive maps and semigroups, which are often used to describe irreversible dynamics of quantum systems. For simplicity, most of the time we restrict ourselves to the finite dimensional case.

4.1 Completely Positive Maps

The following facts are well known and can be e.g. found in [BR2], Notes and Remarks to Section 5.3.1.

Let $\mathcal{K}_1, \mathcal{K}_2$ be Hilbert spaces. We say that a linear map $\Xi : \mathcal{B}(\mathcal{K}_1) \to \mathcal{B}(\mathcal{K}_2)$ is positive iff $A \geq 0$ implies $\Xi(A) \geq 0$. We say that it is completely positive (c.p. for short) iff for any n, $\Xi \otimes 1_{\mathcal{B}(\mathbb{C}^n)}$ is positive as a map $\mathcal{B}(\mathcal{K}_1 \otimes \mathbb{C}^n) \to \mathcal{B}(\mathcal{K}_2 \otimes \mathbb{C}^n)$.

We will say that a positive map Ξ is Markov if $\Xi(1) = 1$.

Recall that $\mathcal{B}^1(\mathcal{K}_i)$ denotes the space of trace class operators on $\mathcal{K}_i$. We can define positive and completely positive maps from $\mathcal{B}^1(\mathcal{K}_2)$ to $\mathcal{B}^1(\mathcal{K}_1)$ repeating verbatim the definition for the algebra of bounded operators. We will say that the map is Markov if it preserves the trace.

We can also speak of positive and completely positive maps on $\mathcal{B}^2(\mathcal{K})$.

We will sometimes say that maps on the algebra $\mathcal{B}(\mathcal{K})$ are "in the Heisenberg picture", maps on $\mathcal{B}^1(\mathcal{K})$ are "in the Schrödinger picture" and maps on $\mathcal{B}^2(\mathcal{K})$ are "in the standard picture" (see the notion of the standard representation later on and in [DJP]).

From now on, for simplicity, in this section we will assume that the spaces $\mathcal{K}_i$ are finite dimensional. Thus $\mathcal{B}(\mathcal{K}_i)$ and $\mathcal{B}^2(\mathcal{K}_i)$ and $\mathcal{B}^1(\mathcal{K}_i)$ coincide with one another as vector spaces. If Ξ is a map from matrices on $\mathcal{K}_1$ to matrices on $\mathcal{K}_2$, it is often useful to distinguish whether it is understood as a map from $\mathcal{B}(\mathcal{K}_1)$ to $\mathcal{B}(\mathcal{K}_2)$ (we then say that it is in the Heisenberg picture), as a map from $\mathcal{B}^2(\mathcal{K}_1)$ to $\mathcal{B}^2(\mathcal{K}_2)$ (we then say that it is in the standard picture) or as a map from $\mathcal{B}^1(\mathcal{K}_1)$ to $\mathcal{B}^1(\mathcal{K}_2)$ (we then say that it is in the Schrödinger picture).

Note that $\mathcal{B}^1(\mathcal{K}_i)$ and $\mathcal{B}(\mathcal{K}_i)$ are dual to one another. (This is one of the places where we use one of properties of finite dimensional spaces. In general, $\mathcal{B}(\mathcal{K}_i)$ is only dual to $\mathcal{B}^1(\mathcal{K}_i)$ and not the other way around.) The (sesquilinear) duality between $\mathcal{B}^1(\mathcal{K}_i)$ and $\mathcal{B}(\mathcal{K}_i)$ is given by

$$\mathrm{Tr}\rho^* A, \quad \rho \in \mathcal{B}^1(\mathcal{K}_i), \quad A \in \mathcal{B}(\mathcal{K}_i).$$

If Ξ is a map "in the Heisenberg picture", then its adjoint Ξ^*, is a map "in the Schrödinger picture" (and vice versa). Clearly, Ξ is a Markov transformation in the Heisenberg picture iff Ξ^* is Markov in the Schrödinger picture.

Note that (in a finite dimension) the definition of Ξ^* does not depend on whether we consider Ξ in the Heisenberg, standard or Schrödinger picture.

4.2 Stinespring Representation of a Completely Positive Map

By the Stinespring theorem [St], $\Xi : \mathcal{B}(\mathcal{K}_1) \to \mathcal{B}(\mathcal{K}_2)$ is completely positive iff there exists an auxilliary finite dimensional Hilbert space $\mathcal{H}$ and $W \in \mathcal{B}(\mathcal{K}_2, \mathcal{K}_1 \otimes \mathcal{H})$ such that

$$\Xi(B) = W^* B \otimes 1_{\mathcal{H}} W, \quad B \in \mathcal{B}(\mathcal{K}_1). \tag{35}$$

In practice it can be useful to transform (35) into a slightly different form. Let us fix an orthonormal basis $(e_1, \ldots, e_n)$ in $\mathcal{H}$. Then the operator W is completely determined by giving a family of operators $W_1, \ldots, W_n \in \mathcal{B}(\mathcal{K}_2, \mathcal{K}_1)$ such that

$$W\Psi_2 = \sum_{j=1}^{n}(W_j\Psi_2) \otimes e_j, \quad \Psi_2 \in \mathcal{K}_2.$$

Then

$$\Xi(B) = \sum_{j=1}^{n} W_j^* B W_j. \tag{36}$$

There exists a third way of writing (35), which is sometimes useful. Let $\overline{\mathcal{H}}$ be the space conjugate to $\mathcal{H}$ and let $\mathcal{H} \ni \Phi \mapsto \overline{\Phi} \in \overline{\mathcal{H}}$ be the corresponding conjugation (see e.g. [DJ2]). We define $W^\star \in \mathcal{B}(\mathcal{K}_1, \mathcal{K}_2 \otimes \overline{\mathcal{H}})$ by

$$(W^\star \Psi_1 | \Psi_2 \otimes \overline{\Phi})_{\mathcal{K}_2 \otimes \overline{\mathcal{H}}} = (\Psi_1 \otimes \Phi | W\Psi_2)_{\mathcal{K}_1 \otimes \mathcal{H}}, \tag{37}$$

(see [DJ2]). (Note that we use two different kinds of stars: $*$ for the hermitian conjugation and $\star$ for (37)). Let $\mathrm{Tr}_{\overline{\mathcal{H}}}$ denote the partial trace over $\overline{\mathcal{H}}$. Then

$$\Xi(B) = \mathrm{Tr}_{\overline{\mathcal{H}}} W^\star B W^{\star *}. \tag{38}$$

If Ξ is given by (35), then Ξ^* can be written in the following three forms:

$$\Xi^*(C) = \mathrm{Tr}_{\mathcal{H}} W C W^*$$

$$= \sum_{j=1}^{n} W_j C W_j^*$$

$$= W^{\star *} C \otimes 1_{\overline{\mathcal{H}}} W^\star,$$

where $C \in \mathcal{B}^1(\mathcal{K}_2)$.

4.3 Completely Positive Semigroups

Let $\mathcal{K}$ be a finite dimensional Hilbert space and $t \mapsto \Lambda(t)$ a continuous 1-parameter semigroup of operators on $\mathcal{B}(\mathcal{K})$. Let M be its generator, so that $\Lambda(t) = e^{tM}$.

We say that $\Lambda(t)$ is a completely positive semigroup iff $\Lambda(t)$ is completely positive for any $t \geq 0$. $\Lambda(t)$ is called a Markov semigroup iff $\Lambda(t)$ is Markov for any $t \geq 0$.

$\Lambda(t)$ is a completely positive semigroup iff there exists an operator Δ on $\mathcal{K}$ and a completely positive map Ξ on $\mathcal{B}(\mathcal{K})$ such that

$$M(B) = \Delta B + B\Delta^* + \Xi(B), \quad B \in \mathcal{B}(\mathcal{K}). \tag{39}$$

Operators of the form (39) are sometimes called Lindblad or Lindblad-Kossakowski generators [GKS,L] .

Let $[\cdot, \cdot]_+$ denote the anticommutator. $\Lambda(t)$ is Markov iff

$$M(B) = \mathrm{i}[\Theta, B] - \frac{1}{2}[\Xi(1), B]_+ + \Xi(B),$$

where $\Theta := \frac{1}{2}(\Delta + \Delta^*)$.

If Ξ is given by (35), then

$$M(B) = \mathrm{i}[\Theta, B] + \frac{1}{2}\left(W^*(WB - B\otimes 1W) + (BW^* - W^*B\otimes 1)W)\right)$$
$$= \mathrm{i}[\Theta, B] + \frac{1}{2}\sum_{j=1}^{n}(W_j^*[W_j, B] + [B, W_j^*]W_j),$$

(40)

and

$$M^*(B) = \mathrm{i}[\Theta, B] - \frac{1}{2}[W^*W, B]_+ + \mathrm{Tr}_{\mathcal{H}}WBW^*$$
$$= \mathrm{i}[\Theta, B] + \sum_{j=1}^{n}\left(-\frac{1}{2}[W_j^*W_j, B]_+ + W_j^{**}B\otimes 1W_j^*\right).$$

Suppose that e^{tM} is a positive Markov semigroup in the Heisenberg picture. We say that a density matrix ρ on $\mathcal{K}$ is stationary with respect to this semigroup iff $\mathrm{e}^{tM^*}(\rho) = \rho$. Every positive Markov semigroup in a finite dimension has a stationary density matrix.

Markov completely positive semigroups (both in the Heisenberg and Schrödinger picture) are often used in quantum physics. In the literature, they are called by many names such as quantum dynamical or quantum Markov semigroups.

4.4 Standard Detailed Balance Condition

In the literature one can find a number of various properties that are called the Detailed Balance Condition (DBC). In the quantum context, probably the best known is the definition due to Alicki [A] and Frigerio-Gorini-Kossakowski-Verri [FGKV], which we describe in the next subsection and call the DBC in the sense of AFGKV.

In this subsection we introduce a slightly different property that we think is the most satisfactory generalization of the DBC from the clasical to the quantum case. It is a modification of the DBC in the sense of AFGKV. To distinguish it from other kinds of the DBC, we will call it the standard Detailed Balance Condition. The name is justified by the close relationship of this condition to the standard representation. We have not seen the standard DBC in the literature, but we know that it belongs to the folklore of the subject. In particular, it was considered in the past by R. Alicki and A. Majewski (private communication).

In the literature one can also find other properties called the Detailed Balance Condition [Ma1, Ma2, MaSt]. Most of them involve the notion of the time reversal, which is not used in the case of the standard DBC or the DBC in the sense of AFGKV.

Let us assume that ρ is a nondegenerate density matrix on $\mathcal{K}$. (That means, $\rho > 0$, $\mathrm{Tr}\rho = 1$, and ρ^{-1} exists). On the space of operators on $\mathcal{K}$ we introduce the scalar product given by ρ:

$$(A|B)_\rho := \mathrm{Tr}\rho^{1/2}A^*\rho^{1/2}B.$$

(41)

This space equipped with the scalar product (41) will be denoted by $\mathcal{B}^2_\rho(\mathcal{K})$. Let $*\rho$ denote the hermitian conjugation with respect to this scalar product. Thus if M is a map on $\mathcal{B}(\mathcal{K})$, then $M^{*\rho}$ is defined by

$$(M^{*\rho}(A)|B)_\rho = (A|M(B))_\rho.$$

Explicitly,

$$M^{*\rho}(A) = \rho^{-1/2} M^*(\rho^{1/2} A \rho^{1/2}) \rho^{-1/2}.$$

Definition 4.1. *Let M be the generator of a Markov c.p. semigroup on $\mathcal{B}(\mathcal{K})$. We will say that M satisfies the standard Detailed Balance Condition with respect to ρ if there exists a self-adjoint operator Θ on $\mathcal{K}$ such that*

$$\frac{1}{2\mathrm{i}}(M - M^{*\rho}) = [\Theta, \cdot]. \tag{42}$$

Theorem 4.1. *Let M be the generator of a Markov c.p. semigroup on $\mathcal{B}(\mathcal{K})$.*
1) *Let M satisfy the standard DBC with respect to ρ. Then*

$$M(A) = \mathrm{i}[\Theta, A] + M_{\mathrm{d}}(A),$$
$$M^*(A) = -\mathrm{i}[\Theta, A] + \rho^{1/2} M_{\mathrm{d}}(\rho^{-1/2} A \rho^{-1/2}) \rho^{1/2}. \tag{43}$$

where M_{d} is a generator of another Markov c.p. semigroup satisfying $M_{\mathrm{d}} = M_{\mathrm{d}}^{*\rho}$ and Θ is a self-adjoint operator on $\mathcal{K}$. Moreover, $[\Theta, \rho] = 0$, $M^*(\rho) = M_{\mathrm{d}}^*(\rho) = 0$.
2) *Let M be given by (40). If there exists a unitary operator $U : \mathcal{H} \to \overline{\mathcal{H}}$ such that*

$$[\Theta, \rho] = 0, \quad [W^*W, \rho] = 0,$$
$$W^* = \rho^{-1/2} \otimes U \, W \rho^{1/2},$$

then M satisfies the standard DBC wrt ρ.

Proof. 1) By (42),

$$[\Theta, \cdot] = -[\Theta, \cdot]^{*\rho} = -\rho^{-1/2}[\Theta, \rho^{1/2} \cdot \rho^{1/2}] \rho^{-1/2}.$$

Using $[\Theta, 1] = 0$, we obtain $[\Theta, \rho] = 0$.

Setting $M_{\mathrm{d}} := \frac{1}{2}(M + M^{*\rho})$ we obtain the decomposition (43). Clearly, $0 = M(\mathbf{1}) = M_{\mathrm{d}}(\mathbf{1})$. Hence M_{d} is Markov. Next $0 = M_{\mathrm{d}}(\mathbf{1}) = M_{\mathrm{d}}^{*\rho}(\mathbf{1})$ gives $M_{\mathrm{d}}(\rho) = 0$.

To see 2) we note that if

$$M_{\mathrm{d}} = \frac{1}{2}[W^*W, B]_+ - W^* B \otimes 1 \, W,$$

then

$$M_{\mathrm{d}}^{*\rho}(B) = \rho^{-1/2} \left(\tfrac{1}{2}[W^*W, \rho^{1/2} B \rho^{1/2}]_+ - W^{**} \rho^{1/2} B \rho^{1/2} \otimes 1 \, W^* \right) \rho^{-1/2}$$
$$= \tfrac{1}{2}[W^*W, B]_+ - (\rho^{1/2} \otimes 1 \, W^* \rho^{1/2})^* B \otimes 1 \, \rho^{1/2} \otimes 1 \, W^* \rho^{-1/2}.$$

$\square$

M_{d} is called the dissipative part of the generator M.

4.5 Detailed Balance Condition in the Sense of Alicki-Frigerio-Gorini-Kossakowski-Verri

In this subsection we recall the definition of Detailed Balance Condition, which can be found in [A, FGKV].

Let us introduce the scalar product

$$(A|B)_{(\rho)} := \mathrm{Tr}\rho A^* B.$$

Let $\mathcal{B}^2_{(\rho)}(\mathcal{K})$ denote the space of operators on $\mathcal{K}$ equipped with this scalar product. Let $M^{*(\rho)}$ denote the conjugate of M with respect to this scalar product. Explicitly:

$$M^{*(\rho)}(A) = \rho^{-1} M^*(\rho A).$$

Definition 4.2. *We will say that M satisfies the Detailed Balance Condition with respect to ρ in the sense of AFGKV if there exists a self-adjoint operator Θ such that*

$$\frac{1}{2\mathrm{i}} (M - M^{*(\rho)}) = [\Theta, \cdot].$$

Note that for DBC in the sense of AFGKV, the analog of Theorem 4.1 1) holds, where we replace the scalar product $(\cdot|\cdot)_\rho$ with $(\cdot|\cdot)_{(\rho)}$.

In practical applications, c.p. semigroups usually originate from the weak coupling limit of reduced dynamics, as we describe further on in our lectures. In this case the standard DBC is equivalent to DBC in the sense of AFGKV, which follows from the following theorem:

Theorem 4.2. *Suppose that M satisfies*

$$\rho^{1/4} M(\rho^{-1/4} A \rho^{1/4}) \rho^{-1/4} = M(A).$$

Then M satisfies the DBC in the sense of (42) iff it satisfies DBC in the sense of AFGKV. Moreover, the decompositions $M = \mathrm{i}[\Theta, \cdot] + M_\mathrm{d}$ obtained in both cases concide.

Proof. It is enough to note that the map

$$\mathcal{B}^2_\rho(\mathcal{K}) \ni A \mapsto \rho^{-1/4} A \rho^{1/4} \in \mathcal{B}^2_{(\rho)}(\mathcal{K})$$

is unitary. $\square$

5 Small Quantum System Interacting with Reservoir

In this section we describe the class of W^*-dynamical systems that we consider in our notes. They are meant to describe a small quantum system $\mathcal{S}$ interacting with a large reservoir $\mathcal{R}$. Pauli-Fierz systems, considered in [DJ2], are typical examples of such systems.

In Subsect. 5.1 we recall basic elements of the theory of W^*-algebras (see [BR1, BR2, DJP] for more information). In Subsect. 5.2 we introduce the class of W^*-dynamical systems describing $\mathcal{S} + \mathcal{R}$ in purely algebraic (representation-independent) terms. In Subsect. 5.3 and 5.4 we explain the construction of two representations of our W^*-dynamical system: the semistandard and the standard representation. Both representations possess a distinguished unitary implementation of the dynamics. Its generator will be called the semi-Liouvillean in the former case and the Liouvillean in the latter case.

The standard representation and the Liouvillean can be defined for an arbitrary W^*-algebra (see next subsection, [DJP] and references therein). The semistandard representation and the semi-Liouvillean are concepts whose importance is limited to a system of the form $\mathcal{S} + \mathcal{R}$ considered in these notes. Their names were coined in [DJ2]. The advantage of the semistandard representation over the standard one is its simplicity, and this is the reason why it appears often in the literature [Da1,LeSp]. The semistandard representation is in particular well adapted to the study of the reduced dynamics.

5.1 W^*-Algebras

In this subsection we recall the definitions of basic concepts related to the theory of W^*-algebras (see [BR1, BR2, DJP]).

A W^*-dynamical system $(\mathfrak{M}, \tau)$ is a pair consisting of a W^*-algebra $\mathfrak{M}$ and a 1-parameter (pointwise) σ-weakly continuous group of $*$-automorphisms of $\mathfrak{M}$, $\mathbb{R} \ni t \mapsto \tau^t$.

A standard representation of a W^*-algebra $\mathfrak{M}$ is a quadruple $(\pi, \mathcal{H}, J, \mathcal{H}^+)$ consisting of a representation π, its Hilbert space $\mathcal{H}$, an antilinear involution J and a self-dual cone $\mathcal{H}^+$ satisfying the following conditions:

1) $J\pi(\mathfrak{M})J = \pi(\mathfrak{M})'$;
2) $J\pi(A)J = \pi(A)^*$ for A in the center of $\mathfrak{M}$;
3) $J\Psi = \Psi$ for $\Psi \in \mathcal{H}^+$;
4) $\pi(A)J\pi(A)\mathcal{H}^+ \subset \mathcal{H}^+$ for $A \in \mathfrak{M}$.

J is called the modular conjugation and $\mathcal{H}^+$ the modular cone. Every W^*-algebra possesses a standard representation, unique up to the unitary equivalence.

Suppose that we are given a faithful state ω on $\mathfrak{M}$. In the corresponding GNS representation $\pi_\omega : \mathfrak{M} \to \mathcal{B}(\mathcal{H}_\omega)$, the state ω is given by a cyclic and separating vector Ω_ω. The Tomita-Takesaki theory yields the modular W^*-dynamics $t \mapsto \sigma_\omega^t$, the modular conjugation J_ω and the modular cone $\mathcal{H}_\omega^+ := \{AJ_\omega A\Omega_\omega : A \in \mathfrak{M}\}^{\mathrm{cl}}$, where cl denotes the closure. The state ω satisfies the -1-KMS condition for the dynamics σ_ω. The quadruple $(\pi_\omega, \mathcal{H}_\omega, J_\omega, \mathcal{H}_\omega^+)$ is a standard representation of $\mathfrak{M}$.

Until the end of this subsection, we suppose that a standard representation $(\pi, \mathcal{H}, J, \mathcal{H}^+)$ of $\mathfrak{M}$ is given.

Let ω be a state on $\mathfrak{M}$. Then there exists a unique vector in the modular cone $\Omega \in \mathcal{H}^+$ representing ω. Ω is cyclic iff Ω is separating iff ω is faithful.

Let $t \mapsto \tau^t$ be a W^*-dynamics on $\mathfrak{M}$. The Liouvillean L of τ is a self-adjoint operator on $\mathcal{H}$ uniquely defined by demanding that

$$\pi(\tau^t(A)) = e^{itL}\pi(A)e^{-itL}, \qquad e^{itL}\mathcal{H}^+ = \mathcal{H}^+, \quad t \in \mathbb{R}.$$

(L implements the dynamics in the representation π and preserves the modular cone). It has many useful properties that make it an efficient tool in the study of the ergodic properties of the dynamics τ. In particular, L has no point spectrum iff τ has no normal invariant states, and L has a 1-dimensional kernel iff τ has a single invariant normal state.

5.2 Algebraic Description

The Hilbert space of the system $\mathcal{S}$ is denoted by $\mathcal{K}$. Throughout the notes we will assume that $\dim \mathcal{K} < \infty$. Let the self-adjoint operator K be the Hamiltonian of the small system. The free dynamics of the small system is $\tau_S^t(B) := e^{itK}Be^{-itK}, B \in \mathcal{B}(\mathcal{K})$. Thus the small system is described by the W^*-dynamical system $(\mathcal{B}(\mathcal{K}), \tau_S)$.

The reservoir $\mathcal{R}$ is described by a W^*-dynamical system $(\mathfrak{M}_\mathcal{R}, \tau_\mathcal{R})$. We assume that it has a unique normal stationary state $\omega_\mathcal{R}$ (not necessarily a KMS state). The generator of $\tau_\mathcal{R}$ is denoted by $\delta_\mathcal{R}$ (that is $\tau_\mathcal{R}^t = e^{\delta_\mathcal{R} t}$).

The coupled system $\mathcal{S} + \mathcal{R}$ is described by the W^*-algebra $\mathfrak{M} := \mathcal{B}(\mathcal{K}) \otimes \mathfrak{M}_\mathcal{R}$. The free dynamics is given by the tensor product of the dynamics of its constituents:

$$\tau_0^t(A) := \left(\tau_S^t \otimes \tau_\mathcal{R}^t\right)(A), \quad A \in \mathfrak{M}.$$

We will denote by δ_0 the generator of τ_0.

Let V be a self-adjoint element of $\mathfrak{M}$. The full dynamics $t \mapsto \tau_\lambda^t := e^{t\delta_\lambda}$ is defined by

$$\delta_\lambda := \delta_0 + i\lambda[V, \cdot].$$

(One can consider also a more general situation, where V is only affilliated to $\mathfrak{M}$— see [DJP] for details).

5.3 Semistandard Representation

Suppose that $\mathfrak{M}_\mathcal{R}$ is given in the standard form on the Hilbert space $\mathcal{H}_\mathcal{R}$. Let $1_\mathcal{R}$ stand for the identity on $\mathcal{H}_\mathcal{R}$. We denote by $\mathcal{H}_\mathcal{R}^+$, $J_\mathcal{R}$, and $L_\mathcal{R}$ the corresponding modular cone, modular conjugation, and standard Liouvillean. Let $\Omega_\mathcal{R}$ be the (unique) vector representative in $\mathcal{H}_\mathcal{R}^+$ of the state $\omega_\mathcal{R}$. Clearly, $\Omega_\mathcal{R}$ is an eigenvector of $L_\mathcal{R}$. $|\Omega_\mathcal{R}\rangle\langle\Omega_\mathcal{R}|$ denotes projection on $\Omega_\mathcal{R}$.

Let us represent $\mathcal{B}(\mathcal{K})$ on $\mathcal{K}$ and take the representation of $\mathfrak{M}$ in the Hilbert space $\mathcal{K} \otimes \mathcal{H}_\mathcal{R}$. We will call it the semistandard representation and denote by π^{semi} : $\mathfrak{M} \to \mathcal{B}(\mathcal{K} \otimes \mathcal{H}_\mathcal{R})$. (To justify its name, note that it is standard on its reservoir part, but not standard on the small system part). We will usually drop π^{semi} and treat $\mathfrak{M}$ as a subalgebra of $\mathcal{B}(\mathcal{K} \otimes \mathcal{H}_\mathcal{R})$.

Let us introduce the so-called free semi-Liouvillean

$$L_0^{\text{semi}} = K \otimes 1 + 1 \otimes L_\mathcal{R}. \tag{44}$$

The full semi-Liouvillean is defined as

$$L_\lambda^{\mathrm{semi}} = L_0^{\mathrm{semi}} + \lambda V.$$

It is the generator of the distinguished unitary implementation of the dynamics τ_λ:

$$\tau_\lambda^t(A) = \mathrm{e}^{\mathrm{i}tL_\lambda^{\mathrm{semi}}} A \mathrm{e}^{-\mathrm{i}tL_\lambda^{\mathrm{semi}}}, \qquad A \in \mathfrak{M}, \tag{45}$$

with

$$\delta_\lambda = \mathrm{i}[L_\lambda^{\mathrm{semi}}, \cdot].$$

5.4 Standard Representation

Let us recall how one constructs the standard representation for the algebra $\mathcal{B}(\mathcal{K})$. Recall that $\mathcal{B}^2(\mathcal{K})$ denotes the space of Hilbert-Schmidt operators on $\mathcal{K}$. Equipped with the inner product $(X|B) = \mathrm{Tr}(X^*B)$ it is a Hilbert space. Note that $\mathcal{B}(\mathcal{K})$ acts naturally on $\mathcal{B}^2(\mathcal{K})$ by the left multiplication. This defines a representation $\pi_s : \mathcal{B}(\mathcal{K}) \to \mathcal{B}(\mathcal{B}^2(\mathcal{K}))$. Let $J_s : \mathcal{B}^2(\mathcal{K}) \to \mathcal{B}^2(\mathcal{K})$ be defined by $J_s(X) = X^*$, and let $\mathcal{B}_+^2(\mathcal{K})$ be the set of all positive $X \in \mathcal{B}^2(\mathcal{K})$. The algebra $\pi_s(\mathcal{B}(\mathcal{K}))$ is in the standard form on the Hilbert space $\mathcal{B}^2(\mathcal{K})$, and its modular cone and modular conjugation are $\mathcal{B}_+^2(\mathcal{K})$ and J_s.

There exists a unique representation $\pi : \mathfrak{M} \to \mathcal{B}(\mathcal{B}^2(\mathcal{K}) \otimes \mathcal{H}_\mathcal{R})$ satisfying

$$\pi(B \otimes C) = \pi_s(B) \otimes C. \tag{46}$$

The von Neumann algebra $\pi(\mathfrak{M})$ is in standard form on the Hilbert space $\mathcal{B}^2(\mathcal{K}) \otimes \mathcal{H}_\mathcal{R}$. The modular conjugation is $J = J_s \otimes J_\mathcal{R}$. The modular cone can be obtained as

$$\mathcal{H}^+ := \{\pi(A)J\pi(A)\,(\rho \otimes \Omega_\mathcal{R}) \,:\, A \in \mathfrak{M}\}^{\mathrm{cl}},$$

where ρ is an arbitrary nondegenerate element of $\mathcal{B}_+^2(\mathcal{K})$.

The Liouvillean of the free dynamics (the free Liouvillean) equals

$$L_0 = [K, \cdot] \otimes 1 + 1 \otimes L_\mathcal{R}, \tag{47}$$

and the Liouvillean of the full dynamics (the full Liouvillean) equals

$$L_\lambda = L_0 + \lambda(\pi(V) - J\pi(V)J). \tag{48}$$

Sometimes we will assume that the reservoir is thermal. By this we mean that $\omega_\mathcal{R}$ is a β-KMS state for the dynamics $\tau_\mathcal{R}$. Set

$$\Psi_0 := \mathrm{e}^{-\beta K/2} \otimes \Omega_\mathcal{R}.$$

Then the state $(\Psi_0|\pi(\cdot)\Psi_0)/\|\Psi_0\|^2$ is a (τ_0, β)-KMS state.

The Araki perturbation theory yields that

$$\Psi_0 \in \mathrm{Dom}(\mathrm{e}^{-\beta(L_0 + \lambda\pi(V))/2}),$$

the vector

$$\Psi_\lambda := e^{-\beta(L_0 + \lambda\pi(V))/2}\Psi_0 \tag{49}$$

belongs to $\mathcal{H}^+ \cap \mathrm{Ker}L_\lambda$, and that $(\Psi_\lambda|\pi(\cdot)\Psi_\lambda)/\|\Psi_\lambda\|^2$ is a (τ_λ, β)-KMS state (see [BR2, DJP]). In particular, zero is always an eigenvalue of L_λ. Thus, in the thermal case, $(\mathfrak{M}, \tau_\lambda)$ has at least one stationary state.

6 Two Applications of the Fermi Golden Rule to Open Quantum Systems

In this section we keep all the notation and assumtions of the preceding section. We will describe two applications of the Fermi Golden Rule to the W^*-dynamical system $(\mathfrak{M}, \tau_\lambda)$ introduced in the previous section.

In the first application we compute the LSO for the generator of the dynamics δ_λ. We will call it the Davies generator and denote by M. In the literature, M appears in the context of the Dynamical Fermi Golden Rule. It is the generator of the semigroup obtained by the weak coupling limit to the reduced dynamics. This result can be used to partly justify the use of completely positive semigroups to describe dynamics of small quantum systems weakly interacting with environment [Da1, LeSp].

In the second application we consider the standard representation of the W^*-dynamical system in the Hilbert space $\mathcal{H}$ with the Liouvillean L. We will compute the LSO for iL_λ. We denote it by $i\Gamma$. In the literature, $i\Gamma$ appears in the context of the Spectral Golden Rule. It is used to study the point spectrum of the Liouvillean L_λ. The main goal of this study is a proof of the uniqueness of a stationary state in the thermal case and of the nonexistence of a stationary state in the non-thermal state under generic conditions [DJ1, DJ2, DJP]. (See also [JP1, JP2, BFS2] for related results).

In Subsection 6.3, we will describe the result of [DJ3], which gives a relationship between the two kinds of LSO's in the thermal case.

In Subsections 6.4–6.6 we compute both LSO's explicitly. In the case of the Davies generator, these formulas are essentially contained in the literature, in the case of the LSO for the Liouvillean, they are generalizations of the analoguous formulas from [DJ2]. Both LSO's can be expressed in a number of distinct forms, each having a different advantage. In particular, as a result of our computations, we describe a simple characterization of the kernel of imaginary part of Γ, which can be used in the proof of the return to equilibrium. This characterization is a generalization of a result from [DJ2].

6.1 LSO for the Reduced Dynamics

It is easy to see that there exists a unique bounded linear map $\mathbb{P}$ on $\mathfrak{M}$ such that for $B \otimes C \in \mathfrak{M} \subset \mathcal{B}(\mathcal{K} \otimes \mathcal{H}_\mathcal{R})$

$$\mathbb{P}(B \otimes C) = \omega_\mathcal{R}(C)B \otimes 1_\mathcal{R}.$$

$\mathbb{P} \in \mathcal{B}(\mathfrak{M})$ is a projection of norm 1. (It is an example of a *conditional expectation*). We identify $\mathcal{B}(\mathcal{K})$ with $\text{Ran}\mathbb{P}$ by

$$\mathcal{B}(\mathcal{K}) \ni B \mapsto B \otimes \mathbf{1}_{\mathcal{R}} \in \text{Ran}\mathbb{P}. \tag{50}$$

Note that $\delta_0\big|_{\text{Ran}\mathbb{P}}$ can be identified with $i[K, \cdot]$.

We assume that $\omega_{\mathcal{R}}(V) = 0$. That implies $\mathbb{P}[V, \cdot]\mathbb{P} = 0$.

Note that Assumptions 2.1*, 2.2, 2.3* and 2.4 are satisfied for the Banach space $\mathfrak{M}$, the projection $\mathbb{P}$, the C_0^*-group of isometries $e^{t\delta_0}$, and the perturbation $i[V, \cdot]$.

Remark 6.1. One can ask whether the above defined projection $\mathbb{P}$ is given by the formula (5). Note that $\mathfrak{M}$ is not a reflexive Banach space, so it is even not clear if this formula makes sense.

Assume that $\delta_{\mathcal{R}}$ has no eigenvectors apart from scalar operators. Then the set of eigenvalues of δ_0 equals $\{i(k - k') : k, k' \in \text{sp}K\}$. One can also show that for any $e \in \mathbb{R}$, δ_0 is globally ergodic at $ie \in i\mathbb{R}$ (see Appendix) and the corresponding eigenprojection is given by

$$\mathbf{1}_{ie}(\delta_0)(B \otimes C) = \sum_{k \in \text{sp}K} \omega_{\mathcal{R}}(C) \left(\mathbf{1}_k(K)B\mathbf{1}_{k-e}(K)\right) \otimes \mathbf{1}_{\mathcal{R}}.$$

Therefore, in this case the answer to our question is positive and

$$\mathbb{P} = \sum_{e \in \mathbb{R}} \mathbf{1}_{ie}(\delta_0),$$

as suggested in Subsection 2.6.

We make the following assumption:

Assumption 6.1 *Assumption 2.5 holds for* $(\mathbb{P}, \delta_0, i[V, \cdot])$. *This means that there exists*

$$M := - \sum_{e \in \text{sp}([K, \cdot])} \mathbf{1}_e([K, \cdot])[V, \cdot](ie + 0 - \delta_0)^{-1}[V, \cdot]\mathbf{1}_e([K, \cdot]). \tag{51}$$

M is the LSO for $(\mathbb{P}, \delta_0, i[V, \cdot])$. It will be called the Davies generator (in the Heisenberg picture).

To describe the physical interpretation of M, suppose that we are interested only in the evolution of the observables corresponding to system $\mathcal{S}$ (taking, however, into account the influence of $\mathcal{R}$). We also suppose that initially the reservoir is given by the state $\omega_{\mathcal{R}}$. Let X be a density matrix on the Hilbert space $\mathcal{K}$, such that the initial state of the system is described by the density matrix $X \otimes |\Omega_{\mathcal{R}})(\Omega_{\mathcal{R}}|$. Let $B \in \mathcal{B}(\mathcal{K})$ be an observable for the system $\mathcal{S}$, such that the measurement at the final time t is given by the operator $B \otimes \mathbf{1}_{\mathcal{R}}$. Then the expectation value of the measurement is given by

$$\mathrm{Tr}_{\mathcal{K}}\Big(X\otimes|\Omega)(\Omega|\,\tau_\lambda^t(B\otimes\mathbf{1}_{\mathcal{R}})\Big). \tag{52}$$

Obviously, (52) tensored with $\mathbf{1}_{\mathcal{R}}$ equals

$$\mathrm{Tr}_{\mathcal{K}}\left(X\mathbb{P}\tau_\lambda^t\mathbb{P}(B\otimes\mathbf{1}_{\mathcal{R}})\right).$$

Now under quite general conditions [Da1, Da2, Da3] we have

$$\lim_{\lambda\to 0}\mathrm{e}^{-\mathrm{i}t[K,\cdot]/\lambda^2}\mathbb{P}\tau_\lambda^{t/\lambda^2}\mathbb{P}=\mathrm{e}^{tM}. \tag{53}$$

Thus M describes the reduced dynamics renormalized by $[K,\cdot]/\lambda^2$ in the limit of the weak coupling, where we rescale the time by λ^2.

Let us note the following fact:

Theorem 6.1. *Suppose Assumption 6.1 holds. Then M is the generator of a Markov c.p. semigroup and for any $z\in\mathbb{C}$,*

$$M(B)=\mathrm{e}^{zK}M(\mathrm{e}^{-zK}B\mathrm{e}^{zK})\mathrm{e}^{-zK}. \tag{54}$$

Proof. We know that LSO M commutes with $\mathbb{E}=\mathrm{i}[K,\cdot]$. This is equivalent to $\mathrm{e}^{z\mathbb{E}}M\mathrm{e}^{-z\mathbb{E}}=M$, which means (54).

The fact that M is a Lindblad-Kossakowski generator and annihilates 1 will follow immediately from explicit formulas given in Subsection 6.4.

If we can prove 53, then an alternative proof is possible: we immediately see that the left hand side of (53) is a Markov c.p. map for any t and λ, hence so is e^{tM}. $\square$

6.2 LSO for the Liouvillean

Consider the the Hilbert space $\mathcal{B}^2(\mathcal{K})\otimes\mathcal{H}_{\mathcal{R}}$ and the orthogonal projection

$$P:=\mathbf{1}_{\mathcal{B}^2(\mathcal{K})}\otimes|\Omega_{\mathcal{R}})(\Omega_{\mathcal{R}}|.$$

We have $PL_0=L_0P=[K,\cdot]P$. We identify $\mathcal{B}^2(\mathcal{K})$ with $\mathrm{Ran}P$ by

$$\mathcal{B}^2(\mathcal{K})\ni B\mapsto B\otimes\Omega_{\mathcal{R}}\in\mathrm{Ran}P. \tag{55}$$

We again assume that $\omega_{\mathcal{R}}(V)=0$. This implies $P\pi(V)P=PJ\pi(V)JP=0$.

Note that Assumptions 2.1, 2.2, 2.3 and 2.4 are satisfied for the Hilbert space $\mathcal{B}^2(\mathcal{K})\otimes\mathcal{H}_{\mathcal{R}}$, the projection P, the strongly continuous unitary group $\mathrm{e}^{\mathrm{i}tL_0}$, and the perturbation $\mathrm{i}(\pi(Q)-J\pi(Q)J)$.

Remark 6.2. Assume that $L_{\mathcal{R}}$ has no eigenvectors apart from $\Omega_{\mathcal{R}}$. Then the set of eigenvalues of δ_0 equals $\{\mathrm{i}(k-k')\ :\ k,k'\in\mathrm{sp}K\}$ and

$$1_e(L_0)B \otimes \Psi = (\Omega_\mathcal{R}|\Psi) \sum_{k \in \mathrm{sp}K} (1_k(K)B1_{k-e}(K)) \otimes \Omega_\mathcal{R}.$$

Therefore,

$$P = \sum_{e \in \mathbb{R}} 1_{ie}(iL_0)$$

is the spectral projection on the point spectrum of iL_0, as suggested in Subsection 2.6.

Assumption 6.2 *Assumption 2.5 for* $(P, iL_0, i(\pi(V) - J\pi(V)J))$ *is satisfied. This means that there exists*

$$i\Gamma := - \sum_{e \in \mathrm{sp}([K,\cdot])} 1_e([K,\cdot])(\pi(V) - J\pi(V)J)$$

$$\times (ie + 0 - iL_0)^{-1}(\pi(V) - J\pi(V)J)1_e([K,\cdot]).$$

$i\Gamma$ is the LSO for $(P, iL_0, i(\pi(V) - J\pi(V)J))$. We will call it the LSO for the Liouvillean. The operator Γ appeared in [DJ1], where it was used to give an upper bound on the point spectrum of L_λ for small nonzero λ.

Theorem 6.2. *Suppose that Assumption 6.2 holds. Then* $i\Gamma$ *is the generator of a contractive c.p. semigroup and for any* $z \in \mathbb{C}$,

$$\Gamma(B) = e^{zK}\Gamma(e^{-zK}Be^{zK})e^{-zK}. \tag{56}$$

Proof. The proof of (56) is the same as that of (54). $e^{ti\Gamma}$ is contractive by Theorem 2.1. The proof of its complete positivity will be given later on (after (60)). □

6.3 Relationship Between the Davies Generator and the LSO for the Liouvillean in Thermal Case

Obviously, as vector spaces, $\mathcal{B}(\mathcal{K})$ and $\mathcal{B}^2(\mathcal{K})$ coincide. We are interested in the relation between $i\Gamma$ and generator M. We will see that in the thermal case the two operators are similar to one another.

The following theorem was proven in [DJ3]:

Theorem 6.3. *Suppose that* $\omega_\mathcal{R}$ *is a* $(\tau_\mathcal{R}, \beta)$-*KMS state. Assumption 6.1 holds if and only if Assumption 6.2 holds. If these assumptions hold, then for* $B \in \mathcal{B}(\mathcal{K})$, *we have*

$$M(B) = i\Gamma(Be^{-\beta K/2})e^{\beta K/2}$$

$$= e^{\beta K/4}i\Gamma(e^{-\beta K/4}Be^{-\beta K/4})e^{\beta K/4}. \tag{57}$$

Remark 6.3. Let $\rho := e^{-\beta K}$ and $\gamma_\rho : \mathcal{B}(\mathcal{K}) \to \mathcal{B}^2(\mathcal{K})$ be the linear invertible map defined by

$$\gamma_\rho(B) := B\rho^{1/2}. \tag{58}$$

Then the first identity of Theorem 6.3 can be written as $M = i\gamma_\rho^{-1} \circ \Gamma \circ \gamma_\rho$. Therefore, both $i\Gamma$ and M have the same spectrum.

Theorem 6.3 follows from the explicit formulas for M and $i\Gamma$ given in Subsections 6.4–6.6. It is, however, instructive to give an alternative, time dependent proof of Identity (57), which avoids calculating both LSO's. Strictly speaking, the identity will be proven for the "the dynamical Level Shift Operators" M_{dyn} and $i\Gamma_{\mathrm{dyn}}$ which, however, according to the Dynamical Fermi Golden Rule, under broad conditions, coincide with the usual Level Shift Operators M and $i\Gamma$.

Theorem 6.4. *Suppose that $\omega_\mathcal{R}$ is a $(\tau_\mathcal{R}, \beta)$-KMS state. Then the following statements are equivalent:*

1) *there exists an operator M_{dyn} satisfying*

$$\lim_{\lambda \to 0} e^{-it[K,\cdot]/\lambda^2} \mathbb{P}\tau_\lambda^{t/\lambda^2} \mathbb{P} = e^{tM_{\mathrm{dyn}}}.$$

2) *there exists an operator Γ_{dyn} satisfying*

$$\lim_{\lambda \to 0} e^{-it[K,\cdot]/\lambda^2} P e^{-itL_\lambda/\lambda^2} P = e^{it\Gamma_{\mathrm{dyn}}}.$$

Moreover,

$$M_{\mathrm{dyn}} = \gamma_\rho^{-1} \circ i\Gamma_{\mathrm{dyn}} \circ \gamma_\rho.$$

Proof. The Araki perturbation theory (see [DJP] and references therein) yields that the vector Ψ_λ, defined by (49), satisfies $\Psi_\lambda = \Psi_0 + O(\lambda)$ and $L_\lambda \Psi_\lambda = 0$. For $X, B \in \mathcal{B}(\mathcal{K}) = \mathcal{B}^2(\mathcal{K})$, using the identifications (50) and (55), we have

$$\mathrm{Tr}_\mathcal{K}\left(X^*\mathbb{P}\tau_0^{-t}\tau_\lambda^t(B \otimes 1_\mathcal{R})\right)$$

$$= \left(Xe^{\beta K/2} \otimes \Omega_\mathcal{R} \,\middle|\, (e^{-itL_0}e^{itL_\lambda} B \otimes 1_\mathcal{R}\, e^{-itL_\lambda}e^{itL_0})\, e^{-\beta K/2} \otimes \Omega_\mathcal{R}\right)$$

$$\overset{O(\lambda)}{=} \left(Xe^{\beta K/2} \otimes \Omega_\mathcal{R} \,\middle|\, e^{-itL_0}e^{itL_\lambda}\, B \otimes 1_\mathcal{R}\, e^{-itL_\lambda}\Psi_\lambda\right)$$

$$\overset{O(\lambda)}{=} \left(Xe^{\beta K/2} \otimes \Omega_\mathcal{R} \,\middle|\, e^{-itL_0}e^{itL_\lambda}\, B \otimes 1_\mathcal{R}\, e^{-\beta K/2} \otimes \Omega_\mathcal{R}\right)$$

$$= \left(X \,\middle|\, (Pe^{-itL_0}e^{itL_\lambda}\left(Be^{-\beta K/2} \otimes \Omega_\mathcal{R}\right))e^{\beta K/2}\right)$$

uniformly for $t \geq 0$. Hence, since $\dim \mathcal{K} < \infty$,

$$e^{-it[K,\cdot]/\lambda^2}\mathbb{P}\tau_\lambda^t(B \otimes 1_\mathcal{R}) = \left(e^{-it[K,\cdot]/\lambda^2} Pe^{itL_\lambda}\left(Be^{-\beta K/2} \otimes \Omega_\mathcal{R}\right)\right)e^{\beta K/2} + O(\lambda)$$

uniformly for $t \geq 0$. We conclude that for a given t the limit

$$\lim_{\lambda \to 0} e^{-it[K,\cdot]/\lambda^2} P e^{itL_\lambda/\lambda^2} P =: T^t$$

exists iff the limit

$$\lim_{\lambda \to 0} e^{-it[K,\cdot]/\lambda^2} \mathbb{P} \tau_\lambda^{t/\lambda^2} \mathbb{P} =: \mathbb{T}^t$$

exists. Moreover, if the limits exist, then

$$\mathbb{T}^t = \gamma_\rho^{-1} \circ T^t \circ \gamma_\rho.$$

In particular, $\mathbb{T}^t$ is a semigroup iff T^t is a semigroup and their generators (M_{dyn} and $i\Gamma_{\mathrm{dyn}}$ respectively) satisfy (57). □

It is perhaps interesting that Theorem 6.4 can be immediately generalized to some non-thermal cases.

Theorem 6.5. *Suppose that instead of assuming that $\omega_\mathcal{R}$ is KMS, we make the following stability assumption: We suppose that ρ is a nondegenerate density matrix on $\mathcal{K}$, and for $|\lambda| \leq \lambda_0$ there exists a normalized vector $\Psi_\lambda \in \mathcal{H}$ such that $\Psi_\lambda = \rho^{1/2} \otimes \Omega_\mathcal{R} + o(\lambda^0)$ and $L_\lambda \Psi_\lambda = 0$. Then all the statements of Theorem 6.4 remain true, with ρ replacing $e^{-\beta K}$.*

Let us return to the thermal case. It is well known [A, FGKV] that in this case the Davies generator satisfies the Detailed Balance Condition. We will see that this fact is essentially equivalent to Relation (57).

Theorem 6.6. *Suppose that $\omega_\mathcal{R}$ is a $(\tau_\mathcal{R}, \beta)$-KMS state and Assumption 6.1 holds. Then the Davies generator M satisfies DBC for $e^{-\beta K}$ both in the standard sense and in the sense of AFGKV.*

Proof. Recall that the operator γ_ρ defined in (58) is unitary from $\mathcal{B}^2_{(\rho)}(\mathcal{K})$ to $\mathcal{B}^2(\mathcal{K})$. Recall also that in the thermal case

$$M = \gamma_\rho^{-1} \circ i\Gamma \circ \gamma_\rho.$$

Hence,

$$M^{*(\rho)} = -\gamma_\rho^{-1} \circ i\Gamma^* \circ \gamma_\rho.$$

Thus,

$$\begin{aligned} \tfrac{1}{2i}(M - M^{*(\rho)}) &= \gamma_\rho^{-1} \circ \tfrac{1}{2}(\Gamma + \Gamma^*) \circ \gamma_\rho \\ &= \gamma_\rho^{-1} \circ [\Delta^{\mathrm{R}}, \cdot] \circ \gamma_\rho = [\Delta^{\mathrm{R}}, \cdot], \end{aligned}$$

(where Δ^{R} will be defined in the next subsection). This proves DBC in the sense of AFGKV.

By Theorem 6.1 and the fact that ρ is proportional to $e^{-\beta K}$, for any $z \in \mathbb{C}$ we have

$$M(B) = \rho^z M(\rho^{-z} B \rho^z) \rho^{-z}.$$

Therefore, by Theorem 4.2, the DBC in the sense of AFGKV is equivalent to the standard DBC. □

6.4 Explicit Formula for the Davies Generator

In this subsection we suppose that Assumption 6.1 is true and we describe an explicit formula for the Davies generator M.

We introduce the following notation for the set of allowed transition frequencies and the set of allowed transition frequencies from $k \in \mathrm{sp}K$:

$$\mathcal{F} := \{k_1 - k_2 \ : \ k_1, k_2 \in \mathrm{sp}K\} = \mathrm{sp}[K, \cdot], \quad \mathcal{F}_k := \{k - k_1 \ : \ k_1 \in \mathrm{sp}K\}.$$

Let $|\Omega)$ denote the map

$$\mathbb{C} \ni z \mapsto |\Omega)z := z\Omega \in \mathcal{H}_\mathcal{R}.$$

Then $1_\mathcal{K} \otimes |\Omega) \in \mathcal{B}(\mathcal{K}, \mathcal{K} \otimes \mathcal{H}_\mathcal{R})$. Set

$$v := V \, 1_\mathcal{K} \otimes |\Omega)$$

Note that v belongs to $\mathcal{B}(\mathcal{K}, \mathcal{K} \otimes \mathcal{H}_\mathcal{R})$. We also define

$$v^{k_1, k_2} := 1_{k_1}(K) \otimes 1_\mathcal{R} \, v \, 1_{k_2}(K);$$

$$\tilde{v}^p := \sum_{k \in \mathrm{sp}K} v^{k, k-p};$$

$$\Delta = \sum_{k \in \mathrm{sp}K} \sum_{p \in \mathcal{F}_k} (v^*)^{k, k-p} 1 \otimes (p + \mathrm{i}0 - L_\mathcal{R})^{-1} v^{k-p, k}$$

$$= \sum_{p \in \mathcal{F}} (\tilde{v}^p)^* 1 \otimes (p + \mathrm{i}0 - L_\mathcal{R})^{-1} \tilde{v}^p.$$

The real and the imaginary part of Δ are given by

$$\Delta^\mathrm{R} := \tfrac{1}{2}(\Delta + \Delta^*) = \sum_{k \in \mathrm{sp}K} \sum_{p \in \mathcal{F}_k} (v^*)^{k, k-p} 1 \otimes \mathcal{P}(p - L_\mathcal{R})^{-1} v^{k-p, k}$$

$$= \sum_{p \in \mathcal{F}} (\tilde{v}^p)^* 1 \otimes \mathcal{P}(p - L_\mathcal{R})^{-1} \tilde{v}^p;$$

$$\Delta^\mathrm{I} := \tfrac{1}{2\mathrm{i}}(\Delta - \Delta^*) = \pi \sum_{k \in \mathrm{sp}K} \sum_{p \in \mathcal{F}_k} (v^*)^{k, k-p} 1 \otimes \delta(p - L_\mathcal{R}) v^{k-p, k}$$

$$= \pi \sum_{p \in \mathcal{F}} (\tilde{v}^p)^* 1 \otimes \delta(p - L_\mathcal{R}) \tilde{v}^p;$$

Note that $\Delta^\mathrm{I} \geq 0$. Below we give four explicit formulas for the Davies generator in the Heisenberg picture:

$$M(B) = \mathrm{i}(\Delta B - B\Delta^*)$$

$$+2\pi \sum_{p\in\mathcal{F}} (\tilde{v}^p)^* \, B \otimes \delta(p - L_{\mathcal{R}}) \tilde{v}^p$$

$$= \mathrm{i} \sum_{p\in\mathcal{F}} (\tilde{v}^p)^* \mathbf{1} \otimes (p - \mathrm{i}0 - L_{\mathcal{R}})^{-1} (\tilde{v}^p B - B \otimes \mathbf{1}_{\mathcal{R}} \tilde{v}^p)$$

$$-\mathrm{i} \sum_{p\in\mathcal{F}} (B(\tilde{v}^p)^* - (\tilde{v}^p)^* B \otimes \mathbf{1}_{\mathcal{R}}) \, \mathbf{1} \otimes (p + \mathrm{i}0 - L_{\mathcal{R}})^{-1} \tilde{v}^p$$

$$= \mathrm{i}[\Delta^{\mathrm{R}}, B]$$

$$+\pi \sum_{p\in\mathcal{F}} (\tilde{v}^p)^* \mathbf{1} \otimes \delta(p - L_{\mathcal{R}}) \, (B \otimes \mathbf{1}_{\mathcal{R}} \tilde{v}^p - \tilde{v}^p B)$$

$$+\pi \sum_{p\in\mathcal{F}} ((\tilde{v}^p)^* B \otimes \mathbf{1}_{\mathcal{R}} - B(\tilde{v}^p)^*) \, \mathbf{1} \otimes \delta(p - L_{\mathcal{R}}) \tilde{v}^p$$

$$= \mathrm{i} \sum_{k\in\mathrm{sp}K} \sum_{p\in\mathcal{F}_k} \int_0^\infty \mathbf{1}_k(K)(\Omega|V\mathbf{1}_{k-p}(K)\tau_0^s(V)\Omega)\mathbf{1}_k(K)B\,\mathrm{d}s$$

$$-\mathrm{i} \sum_{k\in\mathrm{sp}K} \sum_{p\in\mathcal{F}_k} \int_{-\infty}^0 B\mathbf{1}_k(K)(\Omega|V\mathbf{1}_{k-p}(K)\tau_0^s(V)\Omega)\mathbf{1}_k(K)\,\mathrm{d}s$$

$$+2\pi \sum_{\substack{k,k'\in\mathrm{sp}K \\ p\in\mathcal{F}_k\cap\mathcal{F}_{k'}}} \int_{-\infty}^\infty \mathbf{1}_k(K)(\Omega|V\mathbf{1}_{k-p}(K)B\mathbf{1}_{k'-p}(K)\tau_0^s(V)\Omega)\mathbf{1}_{k'}(K)\,\mathrm{d}s.$$

The first expression on the right has the standard form of a Lindblad-Kossakowski generator (39). The second expression can be used in a characterization of the kernel of M. In particular, it implies immediately that $\mathbf{1}_{\mathcal{K}} \in \mathrm{Ker}M$. The third expression shows the splitting of M into a reversible part and an irreversible part. The fourth expression uses uses time-dependent quantities and is analoguous to formulas appearing often in the physics literature.

6.5 Explicit Formulas for LSO for the Liouvillean

In this subsection we suppose that Assumption 6.2 is true and we describe an explicit formula for $\mathrm{i}\Gamma$, the LSO for the Liouvillean.

Recall that π denotes the standard representation of $\mathfrak{M}$ and $L_{\mathcal{R}}$ is the Liouvillean of the free reservoir dynamics $\tau_{\mathcal{R}}$. Let $L_{\mathcal{R}}^0$ denote the Liouvillean of the modular dynamics for the state $\omega_{\mathcal{R}}$. The fact that $\omega_{\mathcal{R}}$ is stationary for $\tau_{\mathcal{R}}^t$ implies that the two Liouvilleans commute:

$$\mathrm{e}^{\mathrm{i}tL_{\mathcal{R}}}\mathrm{e}^{\mathrm{i}sL_{\mathcal{R}}^0} = \mathrm{e}^{\mathrm{i}sL_{\mathcal{R}}^0}\mathrm{e}^{\mathrm{i}tL_{\mathcal{R}}}, \quad t,s \in \mathbb{R}.$$

The following identities follow from the modular theory and will be useful in our explicit formulas for Γ:

Proposition 6.1. *The following identities are true for $B \in \mathcal{B}^2(\mathcal{K})$:*

$$\pi(V)\, B {\otimes} \Omega_{\mathcal{R}} = vB,$$

$$J\pi(V)J\, B {\otimes} \Omega_{\mathcal{R}} = B {\otimes} e^{L^0_{\mathcal{R}}/2} v.$$

Moreover, if $B_1, B_2 \in \mathcal{B}^2(\mathcal{K})$ and $\Phi \in \mathcal{H}_{\mathcal{R}}$, then

$$(B_1 \otimes \Phi | v B_2) = (e^{L^0_{\mathcal{R}}/2} v B_1 | B_2 \otimes J_{\mathcal{R}} \Phi). \tag{59}$$

Proof. To prove the second identity we note that

$$J\, B {\otimes} \Omega_{\mathcal{R}} = B^* {\otimes} \Omega_{\mathcal{R}},$$

$$J\pi(V) B^* {\otimes} \Omega_{\mathcal{R}} = e^{L^0_{\mathcal{R}}/2} B {\otimes} \pi(V) \Omega_{\mathcal{R}}.$$

To see (59), we note that it is enough to assume that $\Phi = A' \Omega_{\mathcal{R}}$, where $A' \in \pi(\mathfrak{M}_{\mathcal{R}})'$ and $\pi(\mathfrak{M}_{\mathcal{R}})'$ denotes the commutant of $\pi(\mathfrak{M}_{\mathcal{R}})$. Then

$$(B_1 \otimes \Phi | v B_2) = (B_1 \otimes A' \Omega_{\mathcal{R}} | \pi(V) B_2 \otimes \Omega_{\mathcal{R}})$$

$$= (\pi(V) B_1 \otimes \Omega_{\mathcal{R}} | B_2 \otimes A'^* \Omega_{\mathcal{R}})$$

$$= (v B_1 | B_2 \otimes e^{L^0_{\mathcal{R}}/2} J_{\mathcal{R}} A' \Omega_{\mathcal{R}}).$$

$\square$

Note that if we compare (59) with the definition of the $\star$-operation (37), and if we make the identification $\overline{\Phi} = J_{\mathcal{R}} \Phi$, then we see that (59) can be rewritten as

$$v^\star = e^{L^0_{\mathcal{R}}/2} v.$$

The LSO for the Liouvillean equals

$$i\Gamma(B) = i\Delta B - iB\Delta^*$$
$$+ 2\pi \sum_{p \in \mathcal{F}} (\tilde{v}^p)^* B {\otimes} \delta(p - L_{\mathcal{R}}) e^{L^0_{\mathcal{R}}/2} \tilde{v}^p. \tag{60}$$

Note that the term on the second line of (60) is completely positive. Therefore, (60) is in the Lindblad-Kossakowski form. Hence $e^{it\Gamma}$ is a c.p. semigroup. This completes the proof of Theorem 6.2.

Let us split Γ into its real and imaginary part:

$$\Gamma^{\mathrm{R}} := \frac{1}{2}(\Gamma + \Gamma^*), \quad \Gamma^{\mathrm{I}} := \frac{1}{2i}(\Gamma - \Gamma^*).$$

(Γ^* is defined using the natural scalar product in $\mathcal{B}^2(\mathcal{K})$). Then the real part is given by

$$\Gamma^{\mathrm{R}}(B) = [\Delta^{\mathrm{R}}, B]. \tag{61}$$

The imaginary part equals

$$\Gamma^{\mathrm{I}} = \pi \sum_{p \in \mathcal{F}} (\tilde{v}^p)^* 1 \otimes \delta(p - L_{\mathcal{R}}) \left(B \otimes e^{L_{\mathcal{R}}^0/2} \tilde{v}^p - \tilde{v}^p B \right)$$
$$+ \pi \sum_{p \in \mathcal{F}} \left((\tilde{v}^p)^* B \otimes e^{L_{\mathcal{R}}^0/2} - B(\tilde{v}^p)^* \right) 1 \otimes \delta(p - L_{\mathcal{R}}) \tilde{v}^p. \tag{62}$$

Another useful formula for Γ^{I} represents it as a quadratic form:

$$\mathrm{Tr} B_1 \Gamma^{\mathrm{I}}(B_2)$$
$$= \pi \sum_{p \in \mathcal{F}} \mathrm{Tr}(\tilde{v}^p B_1 - B_1 \otimes e^{L_{\mathcal{R}}^0/2} \tilde{v}^p)^* 1 \otimes \delta(p - L_{\mathcal{R}})(\tilde{v}^p B_2 - B_2 \otimes e^{L_{\mathcal{R}}^0/2} \tilde{v}^p). \tag{63}$$

To see (63) we note the following identities:

$$(\tilde{v}^p)^* 1 \otimes \delta(p - L_{\mathcal{R}}) \tilde{v}^p = \mathrm{Tr}_{\mathcal{H}_{\mathcal{R}}} 1 \otimes \delta(p - L_{\mathcal{R}}) e^{L_{\mathcal{R}}^0} \tilde{v}^p (\tilde{v}^p)^*,$$

$$(\tilde{v}^p)^* B \otimes \delta(p - L_{\mathcal{R}}) e^{L_{\mathcal{R}}^0/2} \tilde{v}^p = \mathrm{Tr}_{\mathcal{H}_{\mathcal{R}}} 1 \otimes \delta(p - L_{\mathcal{R}}) e^{L_{\mathcal{R}}^0/2} \tilde{v}^p B(\tilde{v}^p)^*,$$

which follow from (59).

The study of the kernel of Γ^{I} is important in applications based on the Spectral Fermi Golden Rule. The identity (63) is very convenient for this purpose. It was first discovered in the context of Pauli-Fierz systems in [DJ2].

In the thermal case (63) can be transformed into

$$\mathrm{Tr} B_1 \Gamma^{\mathrm{I}}(B_2) = \pi \sum_{p \in \mathcal{F}} \mathrm{Tr} e^{-\beta K} (\tilde{v}^p B_1 e^{\beta K/2} - B_1 e^{\beta K/2} \otimes 1_{\mathcal{R}} \tilde{v}^p)^*$$
$$\times 1 \otimes \delta(p - L_{\mathcal{R}})(\tilde{v}^p B_2 e^{\beta K/2} - B_2 e^{\beta K/2} \otimes 1_{\mathcal{R}} \tilde{v}^p). \tag{64}$$

6.6 Identities Using the Fibered Representation

Using the decomposition of the Hilbert space $\mathcal{H}_{\mathcal{R}}$ into the fibered integral given by the spectral decomposition of $L_{\mathcal{R}}$, we can rewrite (63) in an even more convenient form. To describe the fibered form of (63), we will not strive at the greatest generality. We will make the following assumptions (which are modelled after the version of the Jakšić-Pillet gluing condition considered in [DJ2]):

Assumption 6.3 *There exists a Hilbert space $\mathcal{G}$ and a linear isometry $U : \mathcal{G} \otimes L^2(\mathbb{R}) \to \mathcal{H}_{\mathcal{R}}$ such that* $\mathrm{Ran}\, v, \mathrm{Ran}\, e^{\beta L_{\mathcal{R}}^0/2} v \subset \mathcal{K} \otimes \mathrm{Ran} U$ *and $U^* L_{\mathcal{R}} U$ is the operator of the multiplication by the variable in $\mathbb{R}$.*

We will identify $\mathrm{Ran} U$ with $L^2(\mathbb{R}) \otimes \mathcal{G}$. Note that $\Psi \in L^2(\mathbb{R}) \otimes \mathcal{G}$ can be identified with an almost everywhere defined function $\mathbb{R} \ni p \mapsto \Psi(p) \in \mathcal{G}$ such that

$$(L_{\mathcal{R}} \Psi)(p) = p \Psi(p),$$

(see e.g. [DJ2]). We can (at least formally) write $L_{\mathcal{R}}^0$ as the direct integral:

$$(L_\mathcal{R}^0 \Psi)(p) = L_\mathcal{R}^0(p)\Psi(p),$$

where $L_\mathcal{R}^0(p)$ are operators on $\mathcal{G}$.

Likewise, $v \in \mathcal{B}(\mathcal{K}, \mathcal{K}\otimes\mathcal{H}_\mathcal{R})$ can be interpreted as an almost everywhere defined function $\mathbb{R} \ni p \mapsto v(p) \in \mathcal{B}(\mathcal{K}, \mathcal{K} \otimes \mathcal{G})$ such that

$$(L_\mathcal{R} v\Phi)(p) = pv(p)\Phi, \quad \Phi \in \mathcal{K}.$$

Assumption 6.4 $\mathbb{R} \ni p \mapsto v(p)$, $L_\mathcal{R}^0(p)$ *are continuous at $p \in \mathcal{F}$, so that we can define unambiguously $v(p)$, $L_\mathcal{R}^0(p)$ for those values of p.*

Under the above two assumptions we can define

$$w^p := \tilde{v}^p(p) \quad p \in \mathcal{F}.$$

Then we can rewrite the formula (63) as

$$
\begin{aligned}
&\mathrm{Tr}B_1\Gamma^\mathrm{I}(B_2) \\
&= \pi \sum_{p\in\mathcal{F}} \mathrm{Tr}(w^p B_1 - B_1\otimes e^{L_\mathcal{R}^0(p)/2}w^p)^*(w^p B_2 - B_2\otimes e^{L_\mathcal{R}^0(p)/2}w^p).
\end{aligned}
\tag{65}
$$

(65) implies immediately

Theorem 6.7. *The kernel of Γ^I consists of $B \in \mathcal{B}^2(\mathcal{K})$ such that*

$$w^p B = B\otimes e^{L_\mathcal{R}^0(p)/2} w^p, \quad p \in \mathcal{F}.$$

Note that Theorem 6.7 implies that generically $\mathrm{Ker}\Gamma^\mathrm{I} = \{0\}$. Therefore, for a generic open quantum system, if the Spectral Fermi Golden Rule can be applied, then the Liouvillean L_λ has no point spectrum for small nonzero λ. Therefore, for the same λ, the W^*-dynamical system $(\mathfrak{M}, \tau_\lambda)$ has no invariant normal states.

Identities (63), (65) and Theorem 6.7 are generalizations of similar statements from [DJ2]. In [DJ2] the reader will find their rigorous application to Pauli-Fierz systems.

If $\omega_\mathcal{R}$ is a $(\tau_\mathcal{R}, \beta)$-KMS state, we can transform (65) as follows:

$$
\begin{aligned}
\mathrm{Tr}B_1\Gamma^\mathrm{I}(B_2) = \pi \sum_{p\in\mathcal{F}} \mathrm{Tr}\, e^{-\beta K}(w^p\, B_1 e^{\beta K/2} - B_1 e^{\beta K/2}\otimes 1_\mathcal{R}\, w^p)^* \\
\times(w^p\, B_2 e^{\beta K/2} - B_2 e^{\beta K/2}\otimes 1_\mathcal{R}\, w^p).
\end{aligned}
\tag{66}
$$

Following [DJ2], define

$$\mathcal{N} := \{C \ : \ w^p\, C = C\otimes 1_\mathcal{R}\, w^p, \ p \in \mathcal{F}\}. \tag{67}$$

Repeating the arguments of [DJ2] we get

Theorem 6.8.   1) $\mathcal{N}$ *is a $*$-algebra invariant wrt $e^{itK} \cdot e^{-itK}$ and containing $\mathbb{C}1$.*

2) *The kernel of Γ^{I} consists of* $e^{-\beta K/2} C$ *with* $C \in \mathcal{N}$.

Theorem 6.8 implies that in a thermal case, generically, $\mathrm{Ker}\,\Gamma^{\mathrm{I}} = \{0\}$. Therefore, if the Spectral Fermi Golden Rule can be applied, for a generic open quantum system, for small nonzero λ, the Liouvillean L_λ has no point spectrum except for a nondegenerate eigenvalue at zero. Therefore, for the same λ, the W^*-dynamical system $(\mathfrak{M}, \tau_\lambda)$ has a unique stationary normal state.

Again, Identity (66) and Theorem 6.8 are generalizations of similar statements from [DJ2], where they were used to study the return to equilibrium for thermal Pauli-Fierz systems.

7 Fermi Golden Rule for a Composite Reservoir

In this section we describe a small quantum system interacting with several reservoirs. We will assume that the reservoirs $\mathcal{R}_1, \ldots, \mathcal{R}_n$ do not interact directly—they interact with one another only through the small system $\mathcal{S}$. We will compute both kinds of the LSO for the composite system. We will see that it is equal to the sum of the LSO's corresponding to the interaction of $\mathcal{S}$ with a single reservoir $\mathcal{R}_i$.

Our presentation is divided into 3 subsections. The first uses the framework of Section 2, the second—that of Section 5 and the third—that of Section 6.

7.1 LSO for a Sum of Perturbations

Let $\mathcal{X}$ be a Banach space. Let $\mathbb{P}^1, \ldots, \mathbb{P}^n$ be projections of norm 1 on $\mathcal{X}$ such that $\mathbb{P}^i \mathbb{P}^j = \mathbb{P}^i \mathbb{P}^j$. Let $\mathbb{L}_0$ be the generator of a group of isometries such that $\mathbb{L}_0 \mathbb{P}^i = \mathbb{P}^i \mathbb{L}_0$, $i = 1, \ldots, n$. Let $\mathbb{Q}^i$ be operators such that $\mathrm{Ran}\mathbb{P}^i \subset \mathrm{Dom}\mathbb{Q}^i$ and $\mathbb{Q}^i \mathbb{P}^j = \mathbb{P}^j \mathbb{Q}^i$, $i \neq j$. Set

$$\mathbb{Q} := \sum_{j=1}^{n} \mathbb{Q}^j, \quad \mathbb{P} := \prod_{j=1}^{n} \mathbb{P}^j, \quad \mathcal{X}_j := \mathrm{Ran} \prod_{i \neq j} \mathbb{P}^i.$$

Clearly, $\mathcal{X}_j$ is left invariant by $\mathbb{L}_0$, $\mathbb{P}^j$, $\mathbb{Q}^j$. Therefore, these operators can be restricted to $\mathcal{X}_j$. We set

$$\mathbb{L}_{0,j} := \mathbb{L}_0 \big|_{\mathcal{X}_j}, \quad \mathbb{P}_j := \mathbb{P}^j \big|_{\mathcal{X}_j} = \mathbb{P}\big|_{\mathcal{X}_j}, \quad \mathbb{Q}_j := \mathbb{Q}^j \big|_{\mathcal{X}_j}.$$

Clearly,

$$\mathrm{Ran}\mathbb{P} = \mathrm{Ran}\mathbb{P}_j \quad \mathbb{L}_0 \big|_{\mathrm{Ran}\mathbb{P}} = \mathbb{L}_{0,j} \big|_{\mathrm{Ran}\mathbb{P}_j}.$$

We set $\mathbb{E} := \mathbb{L}_0 \big|_{\mathrm{Ran}\mathbb{P}}$.

Theorem 7.1. *Suppose that* $\mathbb{P}^j \mathbb{Q}^j \mathbb{P}^j = 0$, $j = 1, \ldots, n$. *Then:*
1) $\mathbb{P}\mathbb{Q}\mathbb{P} = 0$, $\mathbb{P}_j \mathbb{Q}_j \mathbb{P}_j = 0$, $j = 1, \ldots, n$.

2) *Suppose in addition that the LSO's for* $(\mathbb{P}_i, \mathbb{L}_{0,i}, \mathbb{Q}_i)$*, denoted* M_i*, exist. Then the LSO for* $(\mathbb{P}, \mathbb{L}_0, \mathbb{Q})$*, denoted* M*, exists as well and*

$$M = \sum_{i=1}^{n} M_i.$$

Proof. Set $\mathbb{J}_j := \prod_{i \neq j} \mathbb{P}^i$.

1) It is obvious that $\mathbb{P}^i \mathbb{Q}^i \mathbb{P}^i = 0$ implies $\mathbb{P}_i \mathbb{Q}_i \mathbb{P}_i = 0$.

2) We have

$$M = \sum_{i,j=1}^{n} \sum_{ie \in sp\mathbb{E}} \mathbf{1}_{ie}(\mathbb{E}) \mathbb{Q}^i (ie + 0 - \mathbb{L}_0)^{-1} \mathbb{Q}^j \mathbf{1}_{ie}(\mathbb{E}),$$

$$M_j = \sum_{ie \in sp\mathbb{E}} \mathbf{1}_{ie}(\mathbb{E}) \mathbb{Q}_j (ie + 0 - \mathbb{L}_{0,j})^{-1} \mathbb{Q}_j \mathbf{1}_{ie}(\mathbb{E}).$$

For $i \neq j$,

$$\mathbb{P} \mathbb{Q}^i (ie + 0 - \mathbb{L}_0)^{-1} \mathbb{Q}^j \mathbb{P} = \mathbb{P} \mathbb{Q}^i \mathbb{J}_j (ie + 0 - \mathbb{L}_0)^{-1} \mathbb{Q}^j \mathbb{P} = 0,$$

since $\mathbb{P} \mathbb{Q}^i \mathbb{J}_j = \mathbb{P} \mathbb{P}^i \mathbb{Q}^i \mathbb{P}^i \mathbb{J}_j = 0$. Clearly,

$$\mathbb{P} \mathbb{Q}^i (ie + 0 - \mathbb{L}_0)^{-1} \mathbb{Q}^i \mathbb{P} = \mathbb{P} \mathbb{Q}_i (ie + 0 - \mathbb{L}_{0,i})^{-1} \mathbb{Q}_i \mathbb{P}.$$

$\square$

7.2 Multiple Reservoirs

Suppose that $(\mathfrak{M}_{\mathcal{R}_1}, \tau_{\mathcal{R}_1}), \ldots, (\mathfrak{M}_{\mathcal{R}_n}, \tau_{\mathcal{R}_n})$ are W^*-dynamical systems with $\tau^t_{\mathcal{R}_i} = e^{t\delta_{\mathcal{R}_i}}$. Let $\mathbf{1}_{\mathcal{R}_i}$ denote the identity on $\mathcal{H}_{\mathcal{R}_i}$. Suppose that $\mathfrak{M}_{\mathcal{R}_i}$ have a standard representation in Hilbert spaces $\mathcal{H}_{\mathcal{R}_i}$ with the modular conjugations $J_{\mathcal{R}_i}$. Let $L_{\mathcal{R}_i}$ be the Liouvillean of the dynamics $\tau_{\mathcal{R}_i}$.

Let $(\mathcal{B}(\mathcal{K}), \tau_s)$ describe the small quantum system, with $\tau^t_s := e^{it[K,\cdot]}$, as in Section 5. Define the free systems $(\mathfrak{M}_i, \tau_{0,i})$ where

$$\mathfrak{M}_i := \mathcal{B}(\mathcal{K}) \otimes \mathfrak{M}_{\mathcal{R}_i},$$

$$\mathcal{H}_i := \mathcal{B}^2(\mathcal{K}) \otimes \mathcal{H}_{\mathcal{R}_i},$$

$$J_i := J_s \otimes J_{\mathcal{R}_i},$$

$$\tau^t_{0,i} := \tau^t_s \otimes \tau^t_{\mathcal{R}_i} = e^{t\delta_{0,\lambda}},$$

$$\delta_{0,i} = i[K, \cdot] + \delta_{\mathcal{R}_i},$$

$$L_{0,i} = [K, \cdot] + L_{\mathcal{R}_i}.$$

Let π_i be the standard representation of $\mathfrak{M}_i$ in $\mathcal{H}_i$ and J_i the corresponding conjugations.

Let $V_i \in \mathfrak{M}_i$ and define the perturbed systems $(\mathfrak{M}_i, \tau_{\lambda,i})$ where $\tau_{\lambda,i}^t := e^{t\delta_{\lambda,i}}$ and

$$\delta_{\lambda,i} := \delta_{0,i} + i\lambda[V_i, \cdot],$$

$$L_{\lambda,i} = L_{0,i} + \lambda(\pi_i(V_i) - J_i\pi_i(V_i)J_i).$$

Likewise, consider the composite reservoir $\mathcal{R}$ described by the W^*-dynamical system $(\mathfrak{M}_\mathcal{R}, \tau_\mathcal{R})$, where

$$\mathfrak{M}_\mathcal{R} := \mathfrak{M}_{\mathcal{R}_1} \otimes \cdots \otimes \mathfrak{M}_{\mathcal{R}_n},$$

$$\mathcal{H}_\mathcal{R} := \mathcal{H}_{\mathcal{R}_1} \otimes \cdots \otimes \mathcal{H}_{\mathcal{R}_n},$$

$$J_\mathcal{R} := J_{\mathcal{R}_1} \otimes \cdots \otimes J_{\mathcal{R}_n},$$

$$\tau_\mathcal{R}^t := \tau_{\mathcal{R}_1}^t \otimes \cdots \otimes \tau_{\mathcal{R}_n}^t = e^{t\delta_\mathcal{R}},$$

$$\delta_\mathcal{R} := \delta_{\mathcal{R}_1} + \cdots + \delta_{\mathcal{R}_n},$$

$$L_\mathcal{R} = L_{\mathcal{R}_1} + \cdots + L_{\mathcal{R}_n}.$$

Define the free composite system $(\mathfrak{M}, \tau_0)$ where

$$\mathfrak{M} := \mathcal{B}(\mathcal{K}) \otimes \mathfrak{M}_\mathcal{R},$$

$$\mathcal{H} := \mathcal{B}^2(\mathcal{K}) \otimes \mathcal{H}_\mathcal{R},$$

$$J = J_s \otimes J_\mathcal{R},$$

$$\tau_0^t := \tau_s^t \otimes \tau_\mathcal{R}^t = e^{t\delta_0},$$

$$\delta_0 = i[K, \cdot] + \delta_\mathcal{R},$$

$$L_0 = [K, \cdot] + L_\mathcal{R}.$$

Let π be the standard representation of $\mathfrak{M}$ in $\mathcal{H}$.

Set $V = V_1 + \cdots + V_n$. The perturbed composite system describing the small system $\mathcal{S}$ interacting with the composite reservoir $\mathcal{R}$ is $(\mathfrak{M}, \tau_\lambda)$, where $\tau_\lambda^t := e^{t\delta_\lambda}$,

$$\delta_\lambda := \delta_0 + i\lambda[V, \cdot],$$

$$L_\lambda := L_0 + \lambda(\pi(V) - J\pi(V)J).$$

7.3 LSO for the Reduced Dynamics in the Case of a Composite Reservoir

Suppose that the reservoir dynamics $\tau_{\mathcal{R}_i}$ have stationary states $\omega_{\mathcal{R}_i}$. We introduce a projection of norm one in $\mathfrak{M}$, denoted $\mathbb{P}^i$, such that

$$\mathbb{P}^i(B \otimes A_1 \otimes, \cdots \otimes A_i \otimes \cdots \otimes A_n) = \omega_{\mathcal{R}_i}(A_i)B \otimes A_1 \otimes \cdots \otimes 1_{\mathcal{R}_i} \otimes \cdots \otimes A_n.$$

Set $\mathbb{P} := \prod_{i=1}^{n} \mathbb{P}^i$. The projection $\mathbb{P}^i$ restricted to $\mathfrak{M}_i$ (which can be viewed as a subalgebra of $\mathfrak{M}$) is denoted by $\mathbb{P}_i$. Explicitly,

$$\mathbb{P}_i(B \otimes A_i) = \omega_{\mathcal{R}_i}(A_i)B \otimes 1_{\mathcal{R}_i}.$$

Assume that $\omega_{\mathcal{R}_i}(V_i) = 0$ for $i = 1, \ldots, n$.

Note that we can apply the formalism of Subsection 7.1, where the Banach space is $\mathcal{X}$ is $\mathfrak{M}$, the projections $\mathbb{P}^i$ are $\mathbb{P}^i$, the generator of an isometric dynamics $\mathbb{L}_0$ is δ_0 and the perturbations $\mathbb{Q}^i$ are $i[V_i, \cdot]$. Clearly, $\mathcal{X}_i$ can be identified with $\mathfrak{M}_i$ and Ran$\mathbb{P}$ with $\mathcal{B}(\mathcal{K})$.

We obtain the LSO for $(\mathbb{P}, \delta_0, i[V, \cdot])$, denoted M, and the LSO's for $(\mathbb{P}_i, \delta_{0,i}, i[V_i, \cdot])$, denoted M_i. By Theorem 7.1, we have

$$M = \sum_{i=1}^{n} M_i,$$

7.4 LSO for the Liovillean in the Case of a Composite Reservoir

Let $\Omega_{\mathcal{R}_i}$ be the standard vector representative of $\omega_{\mathcal{R}_i}$. We define the orthogonal projection in $\mathcal{B}(\mathcal{H})$

$$P^i := 1_{\mathcal{B}^2(\mathcal{K})} \otimes 1_{\mathcal{R}_1} \otimes \cdots \otimes |\Omega_{\mathcal{R}_i})(\Omega_{\mathcal{R}_i}| \otimes \cdots \otimes 1_{\mathcal{R}_n}.$$

The projection P^i restricted to $\mathcal{H}_i$ is denoted by P_i and equals

$$P_i = 1_{\mathcal{B}^2(\mathcal{K})} \otimes |\Omega_{\mathcal{R}_i})(\Omega_{\mathcal{R}_i}|.$$

Set $P = \prod_{i=1}^{n} P^i$.

We can apply the formalism of Subsection 7.1, where the Banach space is $\mathcal{X}$ is $\mathcal{H}$, the projections $\mathbb{P}^i$ are P^i, the generator of an isometric dynamics $\mathbb{L}_0$ is iL_0 and the perturbations $\mathbb{Q}^i$ are $i(V_i - J_i V_i J_i)$. Clearly, $\mathcal{X}_i$ can be identified with $\mathcal{H}_i$ and Ran$P$ with $\mathcal{B}^2(\mathcal{K})$ (which as a vector space coincides with $\mathcal{B}(\mathcal{K})$).

We obtain the LSO for $(P, iL_0, i(V - JVJ))$, denoted $i\Gamma$, and the LSO for $(P_i, iL_{0,i}, i(V_i - J_i V_i J_i))$, denoted $i\Gamma_i$. By Theorem 7.1, we have

$$i\Gamma = \sum_{i=1}^{n} i\Gamma_i.$$

The following theorem follows from obvious properties of negative operators:

Theorem 7.2. *Suppose that for some $i \neq j$, $\dim \mathrm{Ker}\Gamma_i^{\mathrm{I}} = \dim \mathrm{Ker}\Gamma_j^{\mathrm{I}} = 1$ and $\mathrm{Ker}\Gamma_i^{\mathrm{I}} \neq \mathrm{Ker}\Gamma_j^{\mathrm{I}}$. Then $\mathrm{Ker}\Gamma = \{0\}$.*

Corollary 7.1. *Suppose that for some $i \neq j$, the states $\omega_{\mathcal{R}_i}$ and $\omega_{\mathcal{R}_j}$ are $(\tau_{\mathcal{R}_i}, \beta_i)$ and $(\tau_{\mathcal{R}_j}, \beta_j)$-KMS. Let $\mathcal{N}_i$ and $\mathcal{N}_j$ be the corresponding $*$-algebras defined as in (67). Suppose that $\beta_i \neq \beta_j$ and $\mathcal{N}_i' = \mathcal{N}_j' = \mathbb{C}1$. Then $\mathrm{Ker}\Gamma = \{0\}$.*

If we can apply the Spectral Fermi Golden Rule, then under the assumptions of 7.1, for sufficiently small nonzero λ, L_λ has no point spectrum. Consequently, for the same λ, the system $(\mathfrak{M}_\lambda, \tau_\lambda)$, has no invariant normal states.

A Appendix – One-Parameter Semigroups

In this section we would like to discuss some concepts related to one-parameter semigroups of operators in Banach spaces, which are used in our lectures. Even though the material that we present is quite standard, we could not find a reference that presents all of it in a convenient way. Most of it can be found in [BR1]. Less pedantic readers may skip this appendix altogether.

Let $\mathcal{X}$ be a Banach space. Recall that $[0,\infty[\ni t \mapsto U(t) \in B(\mathcal{X})$ is called a 1-parameter semigroup iff $U(0) = 1$ and $U(t_1)U(t_2) = U(t_1 + t_2)$. If $[0,\infty[$ is replaced with $\mathbb{R}$, then we speak about a one-parameter group instead of a one-parameter semigroup.

We say that $U(t)$ is a strongly continuous semigroup (or a C_0-semigroup) iff for any $\Phi \in \mathcal{X}$, $t \mapsto U(t)\Phi$ is continuous. Every C_0-semigroup possesses its generator, that is the operator A defined as follows:

$$\Phi \in \mathrm{Dom}A \ \Leftrightarrow\ \lim_{t\searrow 0} t^{-1}(U(t) - 1)\Phi =: A\Phi \ \text{ exists.}$$

The generator is always closed and densely defined and uniquely determines the semigroup. We write $U(t) = \mathrm{e}^{tA}$.

Recall also the following well known characterization of contractive semigroups:

Theorem A.1. *The following conditions are equivalent:*
1) e^{tA} *is contractive for all $t \geq 0$.*
2) A *is densely defined,* $\mathrm{sp}A \subset \{z \in \mathbb{C}\ :\ \mathrm{Re}z \leq 0\}$ *and* $\|(z - A)^{-1}\| \leq (\mathrm{Re}z)^{-1}$ *for* $\mathrm{Re}z > 0$.
3)   (i) A *is densely defined and for some z_+ with $\mathrm{Re}z_+ > 0$, $z_+ \notin \mathrm{sp}A$,*
    (ii) A *is dissipative, that is for any $\Phi \in \mathrm{Dom}A$ there exists $\xi \in \mathcal{X}^*$ with* $(\xi|\Phi) = \|\Phi\|$ *and* $(\xi|A\Phi) \leq 0$.
Moreover, if A is bounded, then we can omit (i) *in* 3).

There exists an obvious corollary of the above theorem for groups of isometries:

Theorem A.2. *The following conditions are equivalent:*
1) e^{tA} *is isometric for all $t \in \mathbb{R}$.*
2) A *is densely defined,* $\mathrm{sp}A \subset i\mathbb{R}$ *and* $\|(z - A)^{-1}\| \leq |\mathrm{Re}z|^{-1}$ *for* $\mathrm{Re}z \neq 0$.
3)   (i) A *is densely defined and for some $z_\pm$ with $\pm\mathrm{Re}z_\pm > 0$, $z_\pm \notin \mathrm{sp}A$,*
    (ii) A *is conservative, that is for any $\Phi \in \mathrm{Dom}A$ there exists $\xi \in \mathcal{X}^*$ with* $(\xi|\Phi) = \|\Phi\|$ *and* $\mathrm{Re}(\xi|A\Phi) = 0$.
Morover, if A is bounded, then we can omit (i) *in* (3).

Not all semigroups considered in our lectures are C_0-semigroups. An important role in our lectures (and in applications to statistical physics) is played by somewhat less known C_0^*-semigroups. In order to discuss them, first we need to say a few words about dual Banach spaces.

Let $\mathcal{X}^*$ denote the Banach space dual to $\mathcal{X}$ (the space of continuous linear functionals on $\mathcal{X}$). We will use the sesquilinear duality between $\mathcal{X}^*$ and $\mathcal{X}$: the form $(\xi|\Phi)$ will be antilinear in $\xi \in \mathcal{X}^*$ and linear in $\Phi \in \mathcal{X}$.

The so-called weak∗ (w∗) topology on $\mathcal{X}^*$ is defined by the seminorms $|(\cdot|\Phi)|$, where $\Phi \in \mathcal{X}$.

The space of w∗ continuous linear operators on $\mathcal{X}^*$ will be denoted by $\mathcal{B}_{w*}(\mathcal{X}^*)$. Note that $\mathcal{B}_{w*}(\mathcal{X}^*) \subset \mathcal{B}(\mathcal{X}^*)$. If $A \in \mathcal{B}(\mathcal{X})$, and A^* is its adjoint, then $A^* \in \mathcal{B}_{w*}(\mathcal{X}^*)$. Conversely, if $B \in \mathcal{B}_{w*}(\mathcal{X}^*)$, then there exists a unique $A \in \mathcal{B}(\mathcal{X})$, sometimes called the preadjoint of B, such that $B = A^*$. Likewise, if A is closed and densely defined on $\mathcal{X}$, then A^* is w∗ closed and w∗ densely defined on $\mathcal{X}^*$.

We say that $[0, \infty[\ni t \mapsto W(t) \in \mathcal{B}_{w*}(\mathcal{X}^*)$ is a w∗ continuous semigroup (or a C_0^*-semigroup) iff $t \ni W(t)\xi$ is w∗ continuous for any $\xi \in \mathcal{X}^*$. Note that if $U(t)$ is a C_0-semigroup, then $U(t)^*$ is a C_0^*-semigroup. Conversely, if $W(t)$ is a C_0^*-semigroup on $\mathcal{X}^*$, then there exists a unique C_0-semigroup $U(t)$ on $\mathcal{X}$ such that $W(t) = U(t)^*$.

Every C_0^*-semigroup $W(t)$ possesses its generator, that is the operator B defined as follows:

$$\xi \in \mathrm{Dom}B \iff \mathrm{w}*-\lim_{t \searrow 0} t^{-1}(W(t) - 1)\xi =: B\xi \ \text{ exists.}$$

The generator is always w∗-closed and w∗-densely defined and uniquely determines the semigroup. We write $W(t) = \mathrm{e}^{tB}$. We have

$$(\mathrm{e}^{tA})^* = \mathrm{e}^{tA^*}.$$

On a reflexive Banach space, e.g. on a Hilbert space, the concepts of a C_0- and C_0^*-semigroup coincide. Unfortunately, W^*-algebras are usually not reflexive. They are, however, dual Banach spaces: they are dual to the space of normal functionals. In the context of W^*-algebras the w∗-topology is usually called the σ-weak or ultraweak topology.

Groups of automorphisms of W^*-algebras are rarely C_0-groups. To see this note that if H is a self-adjoint operator on a Hilbert space $\mathcal{H}$, then

$$t \mapsto \mathrm{e}^{itH} \cdot \mathrm{e}^{-itH} \tag{68}$$

is always a C_0^*-group on $\mathcal{B}(\mathcal{H})$. It is a C_0-group (and even a norm continuous group) iff H is bounded, which is usually a very severe restriction.

In the context of W^*-algebras, C_0^* groups are usually called (pointwise) σ-weakly continuous groups. C_0^*-groups of ∗-automorphisms are often called W^*-dynamics.

So far, all the material that we recalled can be found e.g. in [BR1]. Now we would like to discuss how to define the spectral projection onto a (not necessarily isolated) eigenvalue of a generator of contractive semigroup. We will see that a fully satisfactory answer is available for purely imaginary eigenvalues in the case of a reflexive Banach spaces. For non-reflexive Banach spaces the situation is much more complicated. Our discussion is adapted from [Zs] and partly from [Da3].

Let A be the generator of a contractive C_0-semigroup on $\mathcal{X}$ and $e \in \mathbb{R}$. Following [Zs], we say that A is ergodic at ie iff

$$\mathbf{1}_{\mathrm{ie}}(A) := \lim_{\xi \searrow 0} \xi(\xi + \mathrm{ie} - A)^{-1} \tag{69}$$

exists.

Let B be the generator of a contractive C_0^*-semigroup on $\mathcal{X}^*$ and $e \in \mathbb{R}$. Following [Zs], we say that B is globally ergodic at ie iff

$$\mathbf{1}_{\mathrm{ie}}(B) := \mathrm{w}* - \lim_{\xi \searrow 0} \xi(\xi + \mathrm{ie} - \mathrm{B})^{-1} \tag{70}$$

exists and is w∗-continuous.

As we will see from the theorem below, (69) and (70) can be called spectral projections onto the eigenvalue ie.

Theorem A.3. *Let A, B and $e \in \mathbb{R}$ be as above.*
   1) *If A is ergodic at ie, then $\mathbf{1}_{\mathrm{ie}}(A)$ is a projection of norm 1 such that*

$$\mathrm{Ran}\,\mathbf{1}_{\mathrm{ie}}(A) = \mathrm{Ker}(A - \mathrm{ie}), \quad \mathrm{Ker}\,\mathbf{1}_{\mathrm{ie}}(A) = (\mathrm{Ran}\,(A - \mathrm{ie}))^{\mathrm{cl}}.$$

   2) *On a reflexive Banach space, we have always the ergodic property for all generators of contractive semigroups and all ie $\in i\mathbb{R}$.*
   3) *If B is globally ergodic at ie, then $\mathbf{1}_{\mathrm{ie}}(B)$ is a w∗-continuous projection of norm 1 such that*

$$\mathrm{Ran}\,\mathbf{1}_{\mathrm{ie}}(B) = \mathrm{Ker}(B - \mathrm{ie}), \quad \mathrm{Ker}\,\mathbf{1}_{\mathrm{ie}}(B) = (\mathrm{Ran}\,(A - \mathrm{ie}))^{w*\mathrm{cl}}.$$

   4) *A is ergodic at ie iff A^* is globally ergodic at $-$ie and*

$$\mathbf{1}_{\mathrm{ie}}(A)^* = \mathbf{1}_{-\mathrm{ie}}(A^*).$$

1) and 2) are proven in [Da3] Theorem 5.1 and Corollary 5.2. 3) and 4) can be proven by adapting the arguments of [Zs] Theorem 3.4 and Corollary 3.5.

As an ilustration of the above concepts consider the W^*-dynamics (68). Clearly, it is a group of isometries and the spectrum of its generator $i[H, \cdot]$ is contained in $i\mathbb{R}$. If H possesses only point spectrum, then $i[H, \cdot]$ is globally ergodic for any ie $\in i\mathbb{R}$. In fact, we have the following formula for

$$\mathbf{1}_{\mathrm{ie}}(i[H, \cdot])(C) = \sum_{x \in \mathbb{R}} \mathbf{1}_{x+e}(H)C\mathbf{1}_x(H).$$

Note that $i[H, \cdot]$ always possesses an eigenvalue 0 and the corresponding eigenvectors are all operators commuting with H. It is never globally ergodic at 0 if H has some continuous spectrum.

References

[AJPP] Aschbacher, W., Jakšić, V., Pautrat, Y., Pillet, C.-A.: Introduction to non-equilibrium quantum statistical mechanics. Grenoble lecture notes.

[A]   Alicki, R.: On the detailed balance condition for non-hamiltonian systems, Rep. Math. Phys. 10 (1976) 249

[AL]   Alicki, R., Lendi, K.: *Quantum dynamical semigroups and applications*, Lecture Notes in Physics no 286, Springer 1991

[BR1] Brattelli, O., Robinson D. W.: *Operator Algebras and Quantum Statistical Mechanics, Volume 1*. Springer-Verlag, Berlin, second edition 1987.

[BR2] Brattelli, O., Robinson D. W.: *Operator Algebras and Quantum Statistical Mechanics, Volume 2*. Springer-Verlag, Berlin, second edition 1996.

[BFS1] Bach, V., Fröhlich, J., Sigal, I.: Quantum electrodynamics of confined non-relativistic particles, Adv. Math. 137, 299 (1998).

[BFS2] Bach, V., Fröhlich, J., Sigal, I.: Return to equilibrium. J. Math. Phys. 41, 3985 (2000).

[Da1] Davies, E. B.: Markovian master equations. Commun. Math. Phys. 39, 91 (1974).

[Da2] Davies, E. B.: Markovian master equations II. Math. Ann. 219, 147 (1976).

[Da3] Davies, E. B.: One parameter semigroups, Academic Press 1980

[DJ1] Dereziński, J., Jakšić, V.: Spectral theory of Pauli-Fierz operators. J. Func. Anal. 180, 243 (2001).

[DJ2] Dereziński, J., Jakšić, V.: Return to equilibrium for Pauli-Fierz systems. Ann. Henri Poincaré 4, 739 (2003).

[DJ3] Dereziński, J., Jakšić, V.: On the nature of Fermi golden rule for open quantum systems, J. Statist. Phys. 116 (2004), 411 (2004).

[DJ4] Dereziński, J., Jakšić, V.: In preparation.

[Di]   Dirac, P.A.M.: The quantum theory of the emission and absorption of radiation. Proc. Royal Soc. London, Series A 114, 243 (1927).

[DJP] Dereziński, J., Jakšić, V., Pillet, C. A.:Perturbation theory of W^*-dynamics, Liouvilleans and KMS-states, to appear in Rev. Math. Phys

[FGKV] Frigerio, A., Gorini, V., Kossakowski A., Verri, M.: Quantum detailed balance and KMS condition, Comm. Math. Phys. 57 (1977) 97-110

[DF1] Dereziński, J., Früboes, R.: Level Shift Operator and second order perturbation theory. J. Math. Phys. 46, 033512 (2005).

[DF2] Dereziński, J., Früboes, R.: Stationary van Hove limit. J. Math. Phys. 46, 063511 (2005).

[Fe]   Fermi, E.: Nuclear Physics, University of Chicago Press, Chicago 1950

[GKS] Gorini, V., Kossakowski, A., Sudarshan, E.C.G. Journ. Math. Phys. 17 (1976) 821

[Haa] Haake, F.: *Statistical treatment of open systems by generalized master equation.* Springer Tracts in Modern Physics 66, Springer-Verlag, Berlin, 1973.

[JP1] Jakšić, V., Pillet, C.-A.: On a model for quantum friction III. Ergodic properties of the spin-boson system. Commun. Math. Phys. 178, 627 (1996).

[JP2] Jakšić, V., Pillet, C.-A.: Spectral theory of thermal relaxation. J. Math. Phys. 38, 1757 (1997).

[JP3] Jakšić, V., Pillet, C.-A.: From resonances to master equations. Ann. Inst. Henri Poincaré 67, 425 (1997).

[JP4] Jakšić, V., Pillet, C.-A.: Non-equilibrium steady states for finite quantum systems coupled to thermal reservoirs. Commun. Math. Phys. 226, 131 (2002).

[JP5] Jakšić, V., Pillet, C.-A.: Mathematical theory of non-equilibrium quantum statistical mechanics. J. Stat. Phys. **108**, 787 (2002).

[JP6] Jakšić, V., Pillet, C.-A.: In preparation.

[Ka]   Kato, T.: *Perturbation Theory for Linear Operators*, second edition,

[KTH] Kubo, R., Toda, M., Hashitsume, N.: *Statistical Physics II. Nonequilibrium Statistical Mechanics*. Springer-Verlag, Berlin, 1985.

[L]   Lindblad, G.: On the generators of quantum dynamical semigroups, Comm. Math. Phys. 48 (1976) 119-130

[M]   Merkli, M.: Positive commutators in non-equilibrium quantum statistical mechanics. Commun. Math. Phys. **223**, 327 (2001).

[LeSp] Lebowitz, J., Spohn, H.: Irreversible thermodynamics for quantum systems weakly coupled to thermal reservoirs. Adv. Chem. Phys. **39**, 109 (1978).

[Ma1] Majewski, W. A.: Dynamical semigroups in the algebraic formulation of statistical mechanics, Fortschritte der Physik 32 (1984) 89-133

[Ma2] Majewski, W. A.: Journ. Math. Phys. The detailed balance condition in quantum statistical mechanics 25 (1984) 614

[MaSt] Majewski, W. A., Streater, R. F.: Detailed balance and quantum dynamical maps Journ. Phys. A: Math. Gen. 31 (1998) 7981-7995

[VH1] Van Hove, L.: Quantum-mechanical perturbations giving rise to a statistical transport equation. Physica **21** (1955) 517.

[VH2] Van Hove, L.: The approach to equilibrium in quantum statitsics. Physica **23** (1957) 441.

[VH3] Van Hove, L.: Master equation and approach to equilibrium for quantum systems. In *Fundamental problems in statistical mechanics*, compiled by E.G.D. Cohen, North-Holland, Amsterdam 1962.

[RS4] Reed, M., Simon, B.: *Methods of Modern Mathematical Physics, IV. Analysis of Operators*, London, Academic Press 1978.

[Sp]   Spohn, H.: Entropy production for quantum dynamical semigroups, Journ. Math. Phys. 19 (1978) 1227

[St]   Stinespring, W. F.: Positive functions on C^*-algebras, Proc. Am. Math. Soc. 6 (1955) 211-216

[WW]   Weisskopf, V., Wigner, E.: Berechnung der natürlichen Linienbreite auf Grund der Dirakschen Lichttheorie, Zeitschrift für Physik 63 (130) 54

[Zs]   Zsidó, L.: Spectral and ergodic properties of analytic generators, Journ. Approximation Theory 20 (1977) 77-138

Decoherence as Irreversible Dynamical Process in Open Quantum Systems

Philippe Blanchard[1] and Robert Olkiewicz[2]

[1] Physics Faculty and BiBoS, University of Bielefeld,
   Universitätsstrasse 25, 33615 Bielefeld, Germany
   e-mail: physik.uni-bielefeld.de
[2] Institute of Theoretical Physics, University of Wrocław,
   pl. M. Borna 9, 50-204 Wrocław, Poland
   e-mail: rolek@ift.uni.wroc.pl

1 Physical and Mathematical Prologue

1.1 Physical Background

Quantum mechanics, whose basic laws were formulated in the twenties, still remains the most fundamental theory we know. It describes, among other things, behavior of electrons in atoms, molecules and solids with astonishing accuracy. Since its appearance one has seen its remarkable successes and ever increasing range of applicability. There seems to be no limit to the Schrödinger equation and to the power of quantum theory as an incredibly accurate computational tool for physicists and chemists. However, quantum mechanics, when applied to the objects surrounding us, results in contradictions to what is observed. The question why those objects always appear localized and obey the laws of classical physics is in the center of this issue. The most transparent example illustrating that problem is a hypothetical experiment proposed by Schrödinger in 1935 which can be described briefly as follows [53].

A cat is penned up in a steel chamber, along with the following device. In a Geiger counter there is a tiny amount of radioactive substance, so small that perhaps in the course of one hour one of the atoms decays, but also, with equal probability, perhaps none. If it happens, the counter tube discharges and through a relay releases a hammer which shatters a small flask of hydrocyanic acid. If one has left the entire system to itself for one hour, one would say that the cat still lives if meanwhile no atom has decayed. The first atomic decay would have poisoned it.

Hence after one hour the system should be described by a schizophrenic quantum state which is a linear superposition over live and dead states of the cat. It is not a surprise that such an experiment caused some confusion among physicists. The first resolution of that paradox, advocated by Bohr, was to outlaw the use of quantum theory for the objects which are classical. He insisted on considering the interaction between a quantum object and apparatus described in classical terms as an indivisible whole and, consequently, claimed that atomic objects have no specific properties of their own and only factual phenomena exist. This orthodox Copenhagen interpretation places the emergence of a fact at the point where it is first registered by a classical measuring instrument. Such an interpretation had obviously several flaws. Firstly, it would have forced quantum theory to depend on classical physics for its very existence, and, secondly, it would have also meant that neither quantum nor classical theory were universal. Thus it was not convincing for some physicists. They could not accept the fact that quantum mechanics requires us to regard any question concerning the status of the cat as meaningless until we establish an observational relationship with it. For example, in a letter to Schrödinger in 1950 Einstein complained:

Most of them simply do not see what sort of risky game they are playing with reality - reality as something independent of what is experimentally established. Their interpretation is, however, refuted most elegantly by your system of radioactive atom + amplifier + charge of gun powder + cat in a box, in which the ψ-function of the system contains the cat both alive and blown to bits. Nobody really doubts that the presence or absence of the cat is something independent of the act of observation.

Less radical but going into the same direction was Heisenberg's point of view on the subject. According to him, the physical world can be split into two parts, the observed object and the observing system which should be described in quantum and classical terms, respectively. Although such a recipe was useful, it was ambiguous in principle. Since there is no fundamental reason why the physics involved in measurements should differ from how other physical interactions are described, the very legitimacy of such a conceptual discontinuity and the precise location of its position, the so-called Heisenberg's cut, was the crux of the issue.

In contrast to that of Bohr and Heisenberg, the Dirac-von Neumann approach sought to solve the measurement riddle by invoking an additional axiom, known as the projection postulate. However, the wave function collapse as a dynamical transition from a pure state to a mixed state cannot be achieved by a unitary transformation. Hence, the hypothesis that at some stage during a measurement the Schrödinger evolution must be suspended and replaced by a different physical process whose detailed dynamics remains unspecified has some obvious flaws. Apart from the ad hoc way such a postulate is inserted to the theory, we are not told exactly at what point the collapse occurs or how long it takes [33, 45]. The reality of the reduction of the wave packet was discussed by Omnès who concluded: "Reduction is nothing but a convenient shortcut to avoid keeping track of all measurements by dealing explicitly with histories involving many outside devices, when one needs only to compute some probabilities. It has no physical content." [44].

1.2 Environmental Decoherence

A different point of view was taken in a program of environment induced decoherence which emphasizes the role of an inevitable dissipative coupling of macrosystems with the surrounding environment. The aim of that program was to describe consequences of such openness of quantum systems to their environments and to study emergence of the effective classicality of some of the quantum states and of the associated observables.

In recent years decoherence has received much attention and has been accepted as the mechanism responsible for the appearance of classicality in quantum measurements and the absence in the real world of Schrödinger-cat-like states [5, 29, 35, 47, 67]. It was also shown that decoherence is a universal short time phenomenon independent of the character of the system and its reservoir [15]. Decoherence is a dynamical effect taking place in a bulk of matter. It is most of the time extraordinary efficient and happens so quickly that it was extremely difficult to catch it [16, 34]. For that reason it is also the main obstacle for the experimental implementation of quantum state processing in quantum information theory. Nevertheless, the intuitive idea of environmental decoherence is rather clear: Quantum interference effects for macroscopic systems are practically unobservable because superpositions of their quantum states are rapidly destroyed by unavoidable interaction and entanglement with the surrounding environment. More precisely, it accepts the wave function description of such a system but contends that it is practically impossible to distinguish between vast majority of its pure states and the corresponding statistical mixtures. Moreover, the emergent localized wave packets at a macroscopic level obey classical dynamics because their spreading cannot occur since the coupling to the environment dominates the system's internal quantum dynamics. It was summed up by Zeh who concluded: "All quasi-classical phenomena, even those representing reversible mechanics, are based on de facto irreversible decoherence" [66]. Decoherence therefore provides an answer, at least for all practical purposes, to one of the most difficult problems by which quantum mechanics was plagued for many years by taking advantage of the unavoidable entanglement of the system with its environment. However, in spite of the continuing progress in the theoretical and experimental understanding of this effect, its range of validity and its full meaning still need to be revealed [46, 50].

1.3 Algebraic Framework

Everybody agrees that concepts of classical and quantum physics are opposite in many aspects. Therefore, in order to discuss the emergence of classical behavior of quantum systems we need a single theory which allows a coherent description of quantum physics, classical mechanics and electrodynamics in the same mathematical language. A convenient framework for such a unified description is the formalism of algebraic quantum theory [3]. A recent presentation of this approach is given in [31]. It is a representation theory of the basic kinematical symmetry group and the associated canonical commutation or anticommutation relations. Its usefulness stems from the fact that it is valid for microscopic, mesoscopic and macroscopic systems with both finitely and infinitely many degrees of freedom. They may be purely quantal, purely classical or mixed (quantal with some classical properties). In this approach description of any individual physical system can be given in terms of an abstract unital C^*-algebra $\mathcal{A}$ [32]. A C^*-algebra is a topological algebra with an extraordinary property that its topology (the so-called norm topology) is determined algebraically and so does not depend on any experimental context [49]. It is believed that all intrinsic properties of a physical system can be represented by self-adjoint elements of $\mathcal{A}$. A state of a system is represented by a positive linear functional ω on $\mathcal{A}$ such that $\omega(\mathbf{1}) = 1$, where $\mathbf{1}$ stands for the unit element in $\mathcal{A}$. The set of all states we shall denote by $\mathcal{S}$. A state ω is called faithful if $\omega(A^*A) = 0$ implies that $A = 0$.

Algebraic quantum mechanics provides also mathematical tools for entering the traditional framework based on Hilbert space theory. For any faithful state $\omega \in \mathcal{S}$ the so-called GNS representation (according to Gelfand, Naimark and Segal [13]) allows the construction of a Hilbert space $\mathcal{H}_\omega$ and a faithful representation $\pi_\omega : \mathcal{A} \to B(\mathcal{H}_\omega)$, the algebra of all bounded linear operators acting on $\mathcal{H}_\omega$. In general, C^*-algebras admit an uncountable number of unitary inequivalent GNS representations, most of which presumably have no physical interpretation. Hence, in order to enter the traditional framework one has to select a subset $\mathcal{S}_a$ of admissible states, which through the GNS construction would lead to physically meaningful structures.

The von Neumann algebra $\mathcal{M} = \pi_\omega(\mathcal{A})''$, the bicommutant of $\pi_\omega(\mathcal{A})$, contains all physical observables of the system as its Hermitian operators or, more generally, self-adjoint operators affiliated to $\mathcal{M}$. Generalizing the notion of a density matrix representing mixture of states in the traditional framework of quantum mechanics we say that density matrices (statistical states) of the system are represented by positive normal and normalized functionals on $\mathcal{M}$. The set of density matrices we denote by D. Hence $\phi \in D$ iff $\phi(A) \geq 0$ whenever $A \geq 0$, $\phi(1) = 1$ and ϕ is continuous in the σ-weak topology on $\mathcal{M}$. The linear space generated by D is a Banach space called the predual space of $\mathcal{M}$ and denoted by $\mathcal{M}_*$. The connection between a Hermitian operator A representing an observable and experimentally measured values of this observable, when the system is described by a density matrix ϕ, is obtained in following way. Suppose $\int \lambda dE(\lambda)$ is the spectral decomposition of A. The probability that the measured value is in an interval $[a, b]$ is given by $\phi(E[a, b])$, and so the expectation value of A in the state ϕ equals to $< A > = \int \lambda d\phi(E(\lambda))$. Let us observe that $d\phi(E(\lambda))$ is a probability measure on $\sigma(A)$, the spectrum of A. Genuine quantum systems are represented by factors, i.e. von Neumann algebras with a trivial center $Z(\mathcal{M}) = \mathbf{C} \cdot \mathbf{1}$, whereas classical systems are described by commutative algebras. The triviality of the center expresses the fact that for any two orthogonal pure states their superpositions are physically distinguishable from the corresponding statistical mixtures [9]. Since a classical observable by definition commutes with all other observables so it belongs to the center of algebra $\mathcal{M}$. Hence, the appearance of classical properties of a quantum system has to result in the emergence of an algebra with a nontrivial center, while transition from a factor to commutative algebra corresponds to the passage from purely quantum to classical description of the system. Since automorphic evolutions preserve the center of any algebra so this program may be accomplished only if we admit the loss of quantum coherence, i.e. that quantum systems are open and interact with their environment.

1.4 Quantum Dynamical Semigroups

Generally, an open quantum system S can be described as a subsystem of the joint system $S + E$, where E denotes its environment. Guided by discussion in the previous subsection we associate with the system $S + E$ a von Neumann algebra $\mathcal{N}$ acting in a Hilbert space $\mathcal{H}_{SE}$, and with S a von Neumann algebra $\mathcal{M} \subset \mathrm{B}(\mathcal{H}_S)$. Since $S \subset S + E$ so there exists an injective normal *-homomorphism $i : \mathcal{M} \to \mathcal{N}$ such that $i(1) = 1$. Hence the image $i(\mathcal{M})$ is a von Neumann subalgebra in $\mathcal{N}$. The joint system is assumed to be closed and so evolves in a reversible way. Physical examples indicate that a conservative dynamical system can be described, under quite general circumstances, by a triple $(\mathcal{N}, \psi, \alpha_t)$, $t \in \mathbf{R}$, where ψ is a faithful normal and semifinite weight on $\mathcal{N}$ and α_t is a σ-weakly continuous one parameter group of *-automorphisms of $\mathcal{N}$ such that $\psi \circ \alpha_t(A) = \psi(A)$ for all $t \in \mathbf{R}$ and all positive

$A \in \mathcal{N}$. Moreover, we assume that the weight ϕ on $\mathcal{M}$ given by $\phi = \psi \circ i$ is also semifinite. We recall that a weight on a von Neumann algebra generalizes the notion of a state in the same way as the replacement of a compact space with a probability measure by a locally compact space with a σ-finite measure, for a broad discussion of weights see, for example, [58].

Dynamics of the system S is the reduced evolution obtained by tracing out the environmental degrees of freedom. Suppose that $E : \mathcal{N} \to \mathcal{M}$ is a ψ-compatible conditional expectation (a normal projection of norm one which satisfies $E \circ i = $ id and $\psi \circ i \circ E = \psi$) of $\mathcal{N}$ onto $\mathcal{M}$. Such a projection always exists if $\sigma_t^{\psi}(i(\mathcal{M})) = i(\mathcal{M})$, where σ_t^{ψ} is the modular group of automorhisms corresponding to the weight ψ [58]. Then the evolution of the system S is given by $T_t = E \circ \alpha_t \circ i$, for $t \geq 0$. Generalizing a bit the result of Emch and Varilly [22] one can show the following.

Theorem 1. $t \to T_t$ is a σ-weakly continuous family of completely positive normal and norm contractive linear operators on $\mathcal{M}$ such that for any $t \geq 0$ the following properties hold:

(i) $T_t(\mathbf{1}) = \mathbf{1}$,
(ii) $\phi \circ T_t(A) = \phi(A)$ for all positive $A \in \mathcal{M}$,
(iii) $T_t \circ \sigma_s^{\phi} = \sigma_s^{\phi} \circ T_t$ for all $s \in \mathbf{R}$.

We now say that the reduced dynamics $(\mathcal{M}, \phi, T_t)$ is Markovian if $T_t \circ T_s = T_{t+s}$ for all $t, s \geq 0$. It is worth noting that operators T_t being defined as the composition of an automorphism with a conditional expectation satisfy in general a complicated integro-differential equation. However, for a large class of physical models one can derive, using certain limiting procedures, an approximated Markovian evolution $T_t = e^{tL}$, whose generator L is given by the so-called master equation [2, 17, 47]. In such a case we shall speak of quantum dynamical semigroups. Since operators T_t are normal so there exists a semigroup $T_* = (T_{t*})_{t \geq 0}$ on the predual space $\mathcal{M}_*$ also called a quantum dynamical semigroup. For a general discussion of these concepts see Rebolledo's contribution in vol. II of this proceedings.

In physical models one takes usually $\mathcal{H}_{SE} = \mathcal{H}_S \otimes \mathcal{H}_E$, $\mathcal{N} = \mathcal{M} \otimes \mathcal{M}_E$ and $i(A) = A \otimes 1$, $A \in \mathcal{M}$. The conditional expectation $E : \mathcal{N} \to \mathcal{M}$ depends on a reference state ω_E of the environment E and is given by $\phi(E(B)) = (\phi \otimes \omega_E)(B)$, $B \in \mathcal{N}$, for all states $\phi \in D$. The joint system evolves unitarily with the Hamiltonian H (a self-adjoint operator on $\mathcal{H}_{SE}$) consisting of three parts

$$H = H_S \otimes 1 + 1 \otimes H_E + H_I, \tag{1}$$

where H_S describes the free evolution of the system, H_E the free evolution of the environment, and H_I describes the interaction between S and E. The reduced dynamics of the system is then given by

$$T_t(A) = E(e^{itH}(A \otimes 1)e^{-itH}). \tag{2}$$

In the representation free framework of C^*-algebras we say that $T = (T_t)_{t \geq 0}$ is a quantum dynamical semigroup on $\mathcal{A}$ if $t \to T_t$ is a C_0-semigroup of completely positive and norm contractive linear operators on $\mathcal{A}$ such that $T_t(1) = 1$ for all $t \geq 0$. It is worth noting that in a number of cases the assumption of the strong continuity of the semigroup T is too restrictive. Thus we also consider a weaker type of continuity and shall say that T is ω-continuous, $\omega \in \mathcal{S}$, if for all $A \in \mathcal{A}$ the mapping $t \rightarrow \pi_\omega(T_t A)$ is continuous in the σ-weak topology on $\pi_\omega(\mathcal{A})$.

1.5 A Model of a Discrete Pointer Basis

Before turning to a general description of decoherence let us first consider a well known physical example of environmentally induced pointer states. An analysis from the point of view of physics of mutually exclusive quantum states, the so-called privileged basis, was given in [68]. Suppose that
- $\mathcal{M} = B(\mathcal{H}_S)$, $\mathcal{M}_E = B(\mathcal{H}_E)$, where both $\mathcal{H}_S$ and $\mathcal{H}_E$ are separable,
- $H_I = A \otimes B$,
- $A = \sum_{n=1}^{\infty} \lambda_n P_n$, $\lambda \in \mathbf{R}$, $\min_{n \neq m} |\lambda_n - \lambda_m| = \delta > 0$ and P_n are mutually orthogonal one-dimensional projections summing up to the identity operator,
- $B = B^*$ has an absolutely continuous spectrum,
- $H_S = \sum_{n=1}^{\infty} \gamma_n P_n$, $\gamma_n \in \mathbf{R}$,
- $[H_E, B] = 0$ (we say that two self-adjoint operators commute when their spectral measures commute),
- ω_E is an arbitrary statistical state of the environment represented by a density matrix ρ, i.e. a positive trace class operator with $\mathrm{Tr}\rho = 1$.

It is clear that A is self-adjoint and so is H_I. Because all three terms in equation (1) commute so H is essentially self-adjoint on an appropriate domain and generates a one parameter unitary group

$$e^{itH} = e^{itH_S \otimes 1} e^{it1 \otimes H_E} e^{itH_I}.$$

Let E be the conditional expectation from $B(\mathcal{H}_S) \otimes B(\mathcal{H}_E)$ onto $B(\mathcal{H}_S)$ with respect to the reference state ω_E. Then, for any $X \in B(\mathcal{H}_S)$,

$$T_t(X) = E[e^{itH}(X \otimes 1)e^{-itH}] = e^{itH_S} E[e^{itH_I}(X \otimes 1)e^{-itH_I}]e^{-itH_S}.$$

Because

$$e^{itH_I} = \sum_{n=1}^{\infty} P_n \otimes e^{it\lambda_n B}$$

so

$$T_t(X) = \sum_{n,m=1}^{\infty} \chi_{n,m}(t) e^{it(\gamma_n - \gamma_m)} P_n X P_m,$$

where

$$\chi_{n,m}(t) = \int e^{it(\lambda_n - \lambda_m)s} d\mathrm{Tr}(\rho F(s)),$$

and $dF(s)$ is the spectral measure of B. Since this measure is absolutely continuous so $d\mathrm{Tr}(\rho F(s))$ is a probability measure absolutely continuous with respect to the Lebesgue measure. Hence, by the Riemann-Lebesgue lemma, $\chi_{n,m} \in C_0(\mathbf{R})$. Because $\min_{n \neq m} |\lambda_n - \lambda_m| = \delta > 0$ so for any $\epsilon > 0$ there exists $t_0 > 0$ such that $|\chi_{n,m}(t)| < \epsilon$ for all $n \neq m$ and all $t > t_0$. The triple $(B(\mathcal{H}_S), \mathrm{Tr}, T_t)$, where Tr denotes the standard trace on $B(\mathcal{H}_S)$, satisfies all properties listed in Theorem 1 and hence represents the reduced dynamics of the system. Let us notice that in general the family of maps $(T_t)_{t \geq 0}$ is not a semigroup. A Markov approximation when applied to this case would result in a special form of the correlation function $\chi_{n,m}(t)$, namely $\chi_{n,m}(t) = e^{-|\lambda_n - \lambda_m|t}$. Let

$$\hat{P}(B(\mathcal{H}_S)) := \sum_{n=1}^{\infty} P_n B(\mathcal{H}_S) P_n \equiv l^{\infty}(\mathbf{N}).$$

We show now that all expectation values of any observable $X \in B(\mathcal{H}_S)$ such that $\hat{P}(X) = 0$ decrease to zero uniformly on bounded sets. Suppose that $X = (\mathrm{id} - \hat{P})X$, and $\|X\|_{\infty} \leq 1$. Then for any $\rho_S \in D$,

$$|\mathrm{Tr}\rho_S T_t(X)| \leq \|T_{t*}\rho_S - \hat{P}(T_{t*}\rho_S)\|_1 \|X\|_{\infty} \leq \|T_{t*}\rho_S - \hat{P}(T_{t*}\rho_S)\|_1,$$

where $\|\cdot\|_1$ is the trace norm and

$$T_{t*}\rho_S = \sum_{n,m=1}^{\infty} \chi_{n,m}(t) e^{it(\gamma_n - \gamma_m)} P_m \rho_S P_n.$$

Hence

$$|\mathrm{Tr}\rho_S T_t(X)| \leq \| \sum_{n \neq m} P_m \rho_S P_n \chi_{n,m}(t) e^{it(\gamma_n - \gamma_m)} \|_1.$$

Suppose first that ρ_S is a pure state associated with a unit vector from a dense subspace M of $\mathcal{H}_S$ given by finite linear combinations of vectors $|e_n>$ such that $P_n = |e_n><e_n|$, i.e. $\rho_S = |f><f|$, where $|f> = \sum_{k=1}^{N} z_k |e_k>$ and $\sum_{k=1}^{N} |z_k|^2 = 1$. Then

$$\rho_s = \sum_{k,l=1}^{N} \bar{z}_k z_l |e_l><e_k|$$

and so

$$|\mathrm{Tr}\rho_S T_t(X)| \leq \sum_{1=n<m}^{N} \| \bar{z}_n z_m \chi_{n,m}(t) e^{it(\gamma_n - \gamma_m)} |e_m><e_n| + \text{h.c.} \|_1$$

$$= \sum_{1=n<m}^{N} |z_n||z_m||\chi_{n,m}(t)| \mathrm{Tr}(P_n + P_m) \leq \sum_{1=n<m}^{N} (|z_n|^2 + |z_m|^2)|\chi_{n,m}(t)|,$$

where h.c. stands for the Hermitian conjugate. Suppose that $\epsilon > 0$ is given. Let us take $t_0 > 0$ such that for all $t > t_0$, $|\chi_{n,m}(t)| < \frac{\epsilon}{N-1}$ for all $n \neq m$. Then

$$|\mathrm{Tr}\rho_S T_t(X)| \leq \frac{\epsilon}{N-1} \cdot (N-1) \sum_{n=1}^{N} |z_n|^2 = \epsilon.$$

If ρ_S is an arbitrary pure state, i.e $\rho_S = |v><v|$ with a unit vector $v \in \mathcal{H}_S$, then for any unit vector $w \in \mathcal{H}_S$

$$\| |v><v| - |w><w| \|_1 \leq 2\sqrt{1 - |<v,w>|^2},$$

see for example [62], and so for any $\epsilon > 0$ there exists a unit vector $f \in M$ such that $\|\rho_S - |f><f| \|_1 < \epsilon$. Hence

$$\|(\mathrm{id} - \hat{P})T_{t*}\rho_S\|_1 \leq \|(\mathrm{id} - \hat{P})T_{t*}(\rho_S - |f><f|)\|_1$$

$$+\|(\mathrm{id} - \hat{P})T_{t*}(|f><f|)\|_1 < 3\epsilon$$

for all $t > t_0$. Finally, suppose that ρ_S is an arbitrary statistical state. Then $\rho_S = \sum_{n=1}^{\infty} a_n Q_n$, where Q_n are mutually orthogonal one-dimensional projections summing up to the identity operator, and $\sum_{n=1}^{\infty} a_n = 1$. Let $N \in \mathbf{N}$ be such that $\|\rho_S - \tilde{\rho}_S\|_1 < \epsilon$, where $\tilde{\rho}_S = \sum_{n=1}^{N} a_n Q_n$. By the previous step, for each $n \in \{1, 2, ..., N\}$ we can find $t_0^{(n)}$ such that $\|T_{t*}Q_n - \hat{P}(T_{t*}Q_n)\|_1 < \epsilon$ for all $t > t_0^{(n)}$. Let $t_0 = \max_{1 \leq n \leq N} t_0^{(n)}$. Then for all $t > t_0$

$$\|T_{t*}\tilde{\rho}_S - \hat{P}(T_{t*}\tilde{\rho}_S)\|_1 \leq \sum_{n=1}^{N} a_n \|T_{t*}Q_n - \hat{P}(T_{t*}Q_n)\|_1 < \epsilon,$$

and so

$$\|T_{t*}\rho_S - \hat{P}(T_{t*}\rho_S)\|_1 < 3\epsilon.$$

This implies that

$$\lim_{t \to \infty} \|T_{t*}\rho_S - \hat{P}(T_{t*}\rho_S)\|_1 = 0,$$

and hence

$$\lim_{t \to \infty} |\mathrm{Tr}\rho_S T_t(X)| = 0,$$

uniformly in X provided they belong to the unit ball. It means that for any experimental setup there exists a decoherence time t_d (usually very short) after which the expectation value of any observable X such that $\hat{P}(X) = 0$ is beyond the experimental resolution. On the other hand, if $\hat{P}(X) = X$, then $T_t X = X$ for all $t \geq 0$. Because any observable $X \in B(\mathcal{H}_S)$ can be written as $X = \hat{P}(X) + (\mathrm{id} - \hat{P})(X)$ so we conclude that

$$\lim_{t \to \infty} T_t(X) = \hat{P}(X)$$

in the σ-weak topology.

From the above example we can learn that there exists a decomposition of the von Neumann algebra $\mathcal{M} = B(\mathcal{H}_S)$ of the system onto two parts $\mathcal{M} = \mathcal{M}_1 \oplus \mathcal{M}_2$ such that $\mathcal{M}_1 = \hat{P}(B(\mathcal{H}_S))$ is a commutative von Neumann algebra with a trivial evolution on it, and $\mathcal{M}_2 = (\mathrm{id} - \hat{P})(B(\mathcal{H}_S))$ is a Banach subspace whose observables cannot be detected in practice. All projections $P_n = |e_n \rangle\langle e_n|$ belong to the algebra $\mathcal{M}_1$ and so do not evolve in time. However, any superpositions of the states $|e_n \rangle$ deteriorate to classical probability distributions over pure states P_n. Therefore such states have been termed classical quantum states [8]. In the next section we show that such a splitting is typical for the reduced dynamics of a quantum open system.

2 The Asymptotic Decomposition of T

2.1 Notation

Since we have a considerable amount of mathematics to discuss in this section it is perhaps useful to describe briefly the general strategy. Guided by the above example our main objective here is to establish a decomposition $\mathcal{M} = \mathcal{M}_1 \oplus \mathcal{M}_2$, which we shall call the isometric-sweeping decomposition, under the most possible general assumptions (Section 2.2). To this end we, firstly, implement the dynamics in the associated Hilbert space, where the unitary-completely nonunitary decomposition holds (Section 2.3), next we extend this decomposition to the predual space, and finally that result will be transformed by duality to the algebra $\mathcal{M}$ (Section 2.4). We start with introducing the following notation. Let $\mathcal{M}$ be a von Neumann algebra with a distinguished normal semifinite and faithful weight ϕ. The predual space of $\mathcal{M}$ we shall denote by $\mathcal{M}_*$ with the duality between $\mathcal{M}_*$ and $\mathcal{M}$ given by $\psi(x)$, $\psi \in \mathcal{M}_*$, $x \in \mathcal{M}$. For any subset $\mathcal{N} \subset \mathcal{M}$ we shall denote by $\mathcal{N}_h$ (respectively $\mathcal{N}_+$) the set of all Hermitian (respectively positive) operators from $\mathcal{N}$. The same will apply for sets in $\mathcal{M}$. We shall use the standard notation for objects associated with ϕ in the Tomita-Takesaki theory (because the weight ϕ is fixed we omit the subscript ϕ in notation) such as

$$N = \{x \in \mathcal{M} : \phi(x^*x) < \infty\},$$

$$M = \text{span}\{y^*x : x, y \in N\} = \text{span}\{x \in \mathcal{M}_+ : \phi(x) < \infty\},$$

Λ the canonical injection of N into its Hilbert space completion $\mathcal{H}$ with the scalar product given by $\langle \Lambda(x), \Lambda(y) \rangle = \phi(y^*x)$, $x, y \in N$, π the canonical representation of $\mathcal{M}$ in $\mathcal{H}$ being an isometry and a σ-weak homeomorphism of $\mathcal{M}$ onto $\pi(\mathcal{M})$, Δ the modular operator in $\mathcal{H}$ arising from the left Hilbert algebra $\mathcal{U} = \Lambda(N \cap N^*)$, J the corresponding isometric involution in $\mathcal{H}$, and $(\sigma_t)_{t\in\mathbf{R}}$ the group of modular automorphisms on $\mathcal{M}$ associated with ϕ. By S (respectively F) we shall denote the corresponding sharp (respectively flat) operators in $\mathcal{H}$, i.e. for any $\xi \in D_S$, $S(\xi) = \xi^\sharp$, and for any $\eta \in D_F$, $F(\eta) = \eta^\flat$. By the same symbol π we shall denote the canonical faithful representation of $\mathcal{U}$ in $B(\mathcal{H})$ given by $\pi(\xi)\eta = \xi\eta, \xi, \eta \in \mathcal{U}$. Thus, for any $x \in N \cap N^*$, $\pi(\Lambda(x)) = \pi(x)$. Definitions of all these concepts can be found for example in [56,57].

The operator norm in $\mathcal{M}$ we shall denote by $\|\cdot\|_\infty$, the norm in $\mathcal{M}_*$ by $\|\cdot\|_1$, and the norm in $\mathcal{H}$ induced by the scalar product by $\|\cdot\|_2$. The operator norm in

$B(\mathcal{H})$ will be denoted by $\|\cdot\|_{op}$. Thus $\|\pi(x)\|_{op} = \|x\|_\infty$ for all $x \in \mathcal{M}$. Because in the following we use operators acting in all spaces $\mathcal{M}$, $\mathcal{M}_*$, and $\mathcal{H}$ so in order to avoid confusion we shall mark operators acting in $\mathcal{M}_*$ with the subscript 1, and in $\mathcal{H}$ with the subscript 2. The closure of M in the operator norm we shall denote by C. Since M is a *-algebra so C is a σ-weakly dense C^*-subalgebra in $\mathcal{M}$.

The classes of right (left) bounded elements in $\mathcal{H}$ we shall denote by $\mathcal{U}'$ and $\mathcal{U}''$ respectively, i.e.

$$\mathcal{U}' = \{\eta \in D_F : \exists c > 0 \ \|\pi(\xi)\eta\|_2 \le c\|\xi\|_2 \ \forall \xi \in \mathcal{U}\},$$

$$\mathcal{U}'' = \{\xi \in D_S : \exists c > 0 \ \|\pi'(\eta)\xi\|_2 \le c\|\eta\|_2 \ \forall \eta \in \mathcal{U}'\},$$

where $\pi'(\eta)\xi = \pi(\xi)\eta, \xi \in \mathcal{U}$. Because algebra $\mathcal{U}$ is achieved so $\mathcal{U} = \mathcal{U}''$.

Finally, the injection of M into $\mathcal{M}_*$, $x \to \phi_x$, given by

$$\phi_x(u) = \sum_{i=1}^{n} < J\pi(u)^* J\Lambda(y_i), \Lambda(z_i) >, \tag{3}$$

where $x = \sum_i z_i^* y_i$, z_i, $y_i \in N$, and $u \in \mathcal{M}$, we shall denote by Φ. As was shown in [63] ϕ_x is well defined, i.e. independent of the representation of x, and Φ is an injective linear positive map onto a norm dense subspace in $\mathcal{M}_*$. In the following we use also another expressions for ϕ_x given in [60]:

$$\phi_x(z^* y) = < J\pi(x)^* J\Lambda(y), \Lambda(z) >, \quad y, z \in N, \tag{4}$$

$$\phi_x(y) = < \Lambda(y), J\Lambda(x) >, \quad y \in N, \tag{5}$$

$$\phi_x(y) = \phi_y(x), \quad y \in M. \tag{6}$$

The closure of M in the norm $\|x\|_L = \max\{\|x\|_\infty, \|\phi_x\|_1\}$ we shall denote by L. As was shown in [60] Φ extends to an injection from L into $\mathcal{M}_*$.

2.2 Dynamics in the Markovian Regime

Since the generalization to the time continuous case is straightforward we restrict our considerations to the discrete time case, i.e. we consider a semigroup $(T^n)_{n\ge0}$. Suppose that a bounded linear operator $T : \mathcal{M} \to \mathcal{M}$ satisfies the following assumptions:

A1) T is two-positive,
A2) T is normal with the predual $T_{*1} : \mathcal{M}_* \to \mathcal{M}_*$,
A3) $T(\mathbf{1}) \le \mathbf{1}$, where $\mathbf{1}$ is the identity operator in $\mathcal{M}$,
A4) $\phi \circ T \le \phi$, i.e. $\phi(Tx) \le \phi(x)$ for all $x \in \mathcal{M}_+$,
A5) $T \circ \sigma_t = \sigma_t \circ T \ \forall t \in \mathbf{R}$,
A6) T_{*1} preserves the space $\Phi(M)$, i.e. $T_{*1} : \Phi(M) \to \Phi(M)$.

Let us comment on the above conditions. By Theorem 1, the first five of the above assumptions only generalize properties of the reduced dynamics. The last assumption is a technical one. It is a consequence of arbitrariness of the weight ϕ, and is superfluous in the case when ϕ is a tracial weight as we will see in Section 2.5.

Suppose that T satisfies A1-A6. Then, by A1 and A3, T satisfies the Schwarz inequality $T(x)^*T(x) \leq T(x^*x)$, and so is contractive in the operator norm. By the Schwarz inequality and A4, $T : N \to N$. The map $T_2\Lambda(x) = \Lambda(Tx)$, $x \in N$, is contractive in $\| \cdot \|_2$, and so extends to the whole space $\mathcal{H}$. This extension we denote also by T_2. By A4, $T : M \to M$, and so the map $T_1 : \Phi(M) \to \Phi(M)$ given by $T_1(\phi_x) = \phi_{Tx}$, $x \in M$, is well defined.

Theorem 2. T_1 extends to a linear two-positive and contractive operator on $\mathcal{M}_*$ (denoted also by T_1).

Proof. Step 1. First we show that for any $x \in M$, $\|T_1\phi_x\|_1 \leq 2\|\phi_x\|_1$. If $x \in M_h$, then $Tx \in M_h$, and so

$$\|\phi_{Tx}\|_1 = \inf\{\phi(h) + \phi(k) : h - k = Tx, \, h, k \in M_+\}$$

$$\leq \inf\{\phi(Ty) + \phi(Tz) : y - z = x, \, y, z \in M_+\}$$

$$\leq \inf\{\phi(y) + \phi(z) : y - z = x, \, y, z \in M_+\} = \|\phi_x\|_1.$$

The formula for the norm of ϕ_{Tx} follows from the Remark on page 165 in [63]. Hence, using the Hermitian decomposition and the property $\|\phi_{x^*}\|_1 = \|\phi_x\|_1$, we arrive at $\|\phi_{Tx}\|_1 \leq 2\|\phi_x\|_1$, for all $x \in M$. The bounded extension of T_1 onto $\mathcal{M}_*$ we denote also by T_1.

Step 2. We show that Φ is completely positive. Suppose $\tilde{x} \in M \otimes M_{n \times n}$ and $\tilde{x} \geq 0$, where $M_{n \times n}$ is the algebra of $n \times n$ matrices. A functional $(\phi_{x_{ij}})_{i,j} = \Phi \otimes \mathrm{id}(\tilde{x})$ is positive on $M \otimes M_{n \times n}$ if and only if for any $y_1, ..., y_n \in N$ there is

$$\sum_{i,j=1}^{n} \phi_{x_{ij}}(y_j^*y_i) \geq 0.$$

Let $\tilde{\xi} = (J\Lambda(y_1), ..., J\Lambda(y_n)) \in \mathcal{H} \otimes \mathbf{C}^n$. Because $\pi \otimes \mathrm{id}(\tilde{x})$ is positive so

$$< \pi \otimes \mathrm{id}(\tilde{x})\tilde{\xi}, \, \tilde{\xi} > = \sum_{i,j=1}^{n} < J\pi(x_{ij})^* J\Lambda(y_i), \, \Lambda(y_j) > = \sum_{i,j=1}^{n} \phi_{x_{ij}}(y_j^*y_i) \geq 0.$$

Step 3. Because Φ is completely positive so T_1 is two-positive. Hence the dual operator $T_1^* : \mathcal{M} \to \mathcal{M}$ is also bounded and two-positive. By step 1, $\|T_1\psi\|_1 \leq \|\psi\|_1$ for all $\psi \in \mathcal{M}_{*+}$. Hence $\psi(1) - \psi(T_1^*1) \geq 0$, and so $T_1^*1 \leq 1$. Thus T_1^* satisfies the Schwarz inequality and so is contractive in the operator norm. Hence $\|T_1\psi\|_1 \leq \|\psi\|_1$ for all $\psi \in \mathcal{M}_*$. $\square$

Let us next consider the dual operator $T_1^* : \mathcal{M} \to \mathcal{M}$. Suppose that $x \in M$. Then, for any $y \in M$,

$$\phi_y(T_1^*x) = \phi_{Ty}(x) = \phi_x(Ty) = (T_{*1}\phi_x)(y).$$

By Proposition 7 in [60], we infer that $T_1^*x \in L$ and $\phi_{T_1^*x} = T_{*1}\phi_x$. By assumption A6, $T_1^*x \in M$. If $x \in M_+$, then

$$\phi(T_1^* x) \, = \, < J\pi(1)^* J\Lambda(\sqrt{T_1^* x}), \, \Lambda(\sqrt{T_1^* x}) > = \, \phi_{T_1^* x}(1)$$

$$= \, \|\phi_{T_1^* x}\|_1 \, = \, \|T_{*1}\phi_x\|_1 \, \leq \, \|\phi_x\|_1 \, = \, \phi(x),$$

and so $\phi \circ T_1^* \leq \phi$. In particular, $T_1^* : M \to M$ and the map $(T_1^*)_1(\phi_x) = \phi_{T_1^* x}$, $x \in M$, is well defined and extends to a bounded operator on $\mathcal{M}_*$. Let $x \in M$ and $y \in \mathcal{M}$. Then

$$(T_1^*)_1(\phi_x)(y) \, = \, \phi_{T_1^* x}(y) \, = \, \phi_x(Ty) \, = \, (T_{*1}\phi_x)(y).$$

Because $\Phi(M)$ is norm dense in $\mathcal{M}_*$ so $(T_1^*)_1 = T_{*1}$. This identification allows us to denote the operator T_1^* on $\mathcal{M}$ by T_*. Using the same argument as for the operator T we obtain that operator $T_{*2} : \Lambda(N) \to \Lambda(N)$, $T_{*2}\Lambda(x) = \Lambda(T_* x)$, extends to a contraction on $\mathcal{H}$ which we denote also by T_{*2}.

Theorem 3. $T_2^* = T_{*2}$, where T_2^* is the adjoint operator of T_2.

Proof. Step 1. Suppose that $x, y \in M$. Then

$$(T_{*1}\phi_x)(y) \, = \, \phi_x(Ty) \, = \, < T_2\Lambda(y), \, J\Lambda(x) > .$$

On the other hand

$$(T_{*1}\phi_x)(y) \, = \, \phi_{T_* x}(y) \, = \, < \Lambda(y), \, JT_{*2}\Lambda(x) > .$$

Because $\Lambda(M)$ is dense in $\mathcal{H}$ so $JT_{*2}J = T_2^*$, and hence $(T_{*2})^* = JT_2 J$.

Step 2. $T_2 : D(\Delta^{1/2}) \to D(\Delta^{1/2})$ and $T_2 \circ \Delta^{1/2} = \Delta^{1/2} \circ T_2|_{D(\Delta^{1/2})}$.

Suppose $x \in N \cap N^*$. Then $\pi(\Delta^{it}\Lambda(x)) = \pi(\sigma_t x)$. Hence $\Delta^{it}\Lambda(x) \in \mathcal{U}$ and $\Delta^{it}\Lambda(x) = \Lambda(\sigma_t x)$. Thus, by assumption A5,

$$T_2\Delta^{it}\Lambda(x) \, = \, \Lambda(T\sigma_t x) \, = \, \Lambda(\sigma_t Tx) \, = \, \Delta^{it}T_2\Lambda(x).$$

Because Δ^{it} and T_2 are contractions and $\mathcal{U}$ is dense in $\mathcal{H}$ so $T_2\Delta^{it} = \Delta^{it}T_2$. Let $A = \ln \Delta$ and let $A = \int \lambda dE(\lambda)$ be its spectral decomposition. Then $T_2 E(B) = E(B)T_2$ for any Borel set $B \subset \mathbf{R}$. Because $\Delta^{1/2} = \int e^{\lambda/2}dE(\lambda)$, the assertion follows.

Step 3. Suppose again that $x \in N \cap N^*$. Then $(T_2 S)\Lambda(x) = (ST_2)\Lambda(x)$. Because $S = J\Delta^{1/2}$ and $D(\Delta^{1/2}) = D_S$ so

$$(T_2 J\Delta^{1/2})\Lambda(x) \, = \, (J\Delta^{1/2}T_2)\Lambda(x) \, = \, (JT_2\Delta^{1/2})\Lambda(x).$$

Because $\Delta^{1/2}\Lambda(x) = J\Lambda(x^*)$, the set $\Delta^{1/2}\mathcal{U}$ is dense in $\mathcal{H}$. Thus $T_2 J = JT_2$ and $T_{*2}J = JT_{*2}$. By step 1, $T_{*2} = T_2^*$ which completes the proof. $\square$

Summing up this subsection: In each space $\mathcal{M}$, $\mathcal{M}_*$ and $\mathcal{H}$ there is a pair of contractive (in the corresponding norms) operators

$$T, T_* : \mathcal{M} \to \mathcal{M} \qquad \text{two-positive and normal,} \tag{7}$$

$$T_1, T_{*1} : \mathcal{M}_* \to \mathcal{M}_* \qquad \text{two-positive,} \tag{8}$$

$$T_2, T_{*2} : \mathcal{H} \to \mathcal{H}, \tag{9}$$

such that T is dual to T_{*1}, T_* is dual to T_1, and T_2 and T_{*2} are adjoint in $\mathcal{H}$. Moreover, both T and T_* leave the space M invariant, and

$$\phi \circ T \leq \phi, \qquad \phi \circ T_* \leq \phi. \tag{10}$$

2.3 The Unitary Decomposition of T_2

Suppose $S_{2,m} = T_{*2}^m T_2^m$, $m \in \mathbf{N}$, and let $K^m = \{\xi \in \mathcal{H} : \; S_{2,m}\xi = \xi\}$. It is clear that K^m is a closed linear subspace in $\mathcal{H}$. If $\xi \in K^m$ and $m \geq 2$, then

$$\|\xi\|_2 = \|S_{2,m}\xi\|_2 \leq \|T_2^{m-1}\xi\|_2 \leq \|\xi\|_2.$$

Hence $\|T_2^{m-1}\xi\|_2 = \|\xi\|_2$. Because

$$\|S_{2,(m-1)}\xi - \xi\|_2^2 = \|S_{2,(m-1)}\xi\|_2^2 - 2\|T_2^{m-1}\xi\|_2^2 + \|\xi\|_2^2 \leq 0$$

so $S_{2,(m-1)}\xi = \xi$. Thus (K^m) is a decreasing sequence of closed subspaces in $\mathcal{H}$ and we define $K^\infty = \bigcap_{m=1}^\infty K^m$. Let $P_{2,m}$ and P_2 be the orthogonal projections in $\mathcal{H}$ onto K^m and K^∞ respectively. It is well known, see for example [36], that for any $\xi \in \mathcal{H}$,

$$P_{2,m}\xi = \lim_{n \to \infty} \frac{1}{n} \sum_{k=0}^{n-1} S_{2,m}^k \xi,$$

and the limit exists in the norm in $\mathcal{H}$. From step 3 in the proof of Theorem 3 we obtain that $P_{2,m} \circ J = J \circ P_{2,m}$, and $P_2 \circ J = J \circ P_2$.

Theorem 4. Suppose $S_m = T_*^m T^m$. Then, for any $x \in M$, the mean average limit converges σ-strongly* to a two-positive and contractive in the operator norm projection $P_m : M \to M$,

$$P_m x = \lim_{n \to \infty} \frac{1}{n} \sum_{k=0}^{n-1} S_m^k x.$$

Moreover, $\phi(P_m x) \leq \phi(x)$ for all $x \in M_+$.

Proof. Let us define

$$x_n = \frac{1}{n} \sum_{k=0}^{n-1} S_m^k x, \qquad \xi_n = \Lambda(x_n) = \frac{1}{n} \sum_{k=0}^{n-1} S_{2,m}^k \Lambda(x).$$

Then $x_n \in M$ and $\xi_n \in \mathcal{U}$. Let $\xi = \lim_{n \to \infty} \xi_n = P_{2,m}\Lambda(x)$.

Step 1. ξ, $\xi^\sharp \in \mathcal{U}$. Because $S(S_{2,m}^k \Lambda(x)) = S_{2,m}^k(\Lambda(x^*))$ so $\xi_n^\sharp$ converges in $\mathcal{H}$. Since S is closed, $\xi \in D_S$ and $\xi_n^\sharp \to \xi^\sharp$. Hence $\pi(\xi)$ is affiliated to $\pi(\mathcal{M})$. Suppose that $\eta \in \mathcal{U}'$. Then

$$\|\pi'(\eta)\xi_n\|_2 = \|\pi(\xi_n)\eta\|_2 \leq \|\pi(\xi_n)\|_{op}\|\eta\|_2.$$

However, $\|\pi(\xi_n)\|_{op} = \|x_n\|_\infty \leq \|x\|_\infty$. Hence $\|\pi'(\eta)\xi_n\|_2 \leq \|x\|_\infty\|\eta\|_2$. Because $\xi_n \to \xi$ and $\pi'(\eta)$ is bounded so also $\|\pi'(\eta)\xi\|_2 \leq \|x\|_\infty\|\eta\|_2$. Thus $\pi(\xi)$ is bounded which implies that $\xi \in \mathcal{U}$. Replacing x by x^* one may check in the same way that $\xi^\sharp \in \mathcal{U}$.

Step 2. Let $y = \Lambda^{-1}(\xi)$. Then $x_n \to y$ in the σ-strong* topology. Since $\pi : \mathcal{M} \to$

$\pi(\mathcal{M})$ is a homeomorphism with respect to the σ-strong* topology so it is sufficient to consider operators $\pi(x_n)$ and $\pi(y)$. Suppose that $\eta \in \mathcal{U}'$. Then

$$\|\pi(x_n)\eta - \pi(y)\eta\|_2 = \|\pi'(\eta)(\xi_n - \xi)\|_2 \to 0.$$

Because $\|\pi(x_n)\|_{op} \leq \|x\|_\infty$ and $\mathcal{U}'$ is dense in $\mathcal{H}$ so $\pi(x_n) \to \pi(y)$ strongly. Since $\xi_n^\sharp \to \xi^\sharp \in \mathcal{U}$ and $\pi(x_n)^*\eta = \xi_n^\sharp \eta$, $\pi(y)^*\eta = \xi^\sharp \eta$ for all $\eta \in \mathcal{H}$, so $\pi(x_n) \to \pi(y)$ strongly*. In particular, $\|\pi(y)\|_{op} \leq \|x\|_\infty$. Since σ-strong* and strong* topologies coincides on bounded sets, the assertion follows.

Step 3. $y \in M$. Suppose that $z \in M$ and let $z_n = \frac{1}{n}\sum_{k=0}^{n-1} S_m^k z$. Then

$$|\phi_z(y)| = | < \Lambda(y), J\Lambda(z) > | = | < P_{2,m}\Lambda(x), J\Lambda(z) > |$$

$$= | < \Lambda(x), JP_{2,m}\Lambda(z) > | = \lim_{n\to\infty} | < \Lambda(x), J\Lambda(z_n) > | = \lim_{n\to\infty} |\phi_{z_n}(x)|$$

$$= \lim_{n\to\infty} |\phi_x(z_n)| \leq \|\phi_x\|_1 \|z\|_\infty.$$

Hence, by Corollary 17 in [60], $y \in L$. Because $x = x_1 - x_2 + i(x_3 - x_4)$, where $x_j \in M_+$, so $y_j := \Lambda^{-1}(P_{2,m}x_j) \in L_+$, and $(x_j)_n \to y_j$ σ-strongly. Since the weight ϕ is σ-weakly lower continuous, $\phi(y_j) \leq \liminf \phi((x_j)_n)$. However, by formula (10),

$$\phi((x_j)_n) = \frac{1}{n}\sum_{k-0}^{n-1} \phi(S_m^k x_j) \leq \phi(x_j).$$

Hence $y_j \in M_+$ and so, by linearity, $y \in M$.
Step 4. Finally, let us define a map $P_m : M \to M$, by $P_m x = \Lambda^{-1}(P_{2,m}\Lambda(x))$. Then P_m is a positive and contractive in the operator norm projection such that $P_m x = \lim_{n\to\infty} x_n$ in the σ-strong* topology. Moreover, $\phi(P_m x) \leq \phi(x)$ for all $x \in M_+$. Thus only two-positivity remained to be shown. Let $\tilde{M} = M \otimes M_{2\times 2}$ and let $\tilde{\phi} = \phi \otimes \text{Tr}$, where Tr is the standard trace on $M_{2\times 2}$, the algebra of 2×2 matrices. Replacing T and T_* by $\tilde{T} = T \otimes \text{id}$ and $\tilde{T}_* = T_* \otimes \text{id}$, and noting that $\tilde{S}_{2,m} = \tilde{T}_{*2}^m \tilde{T}_2^m$ is contractive in the corresponding Hilbert space and $\tilde{S}_m = \tilde{T}_*^m \tilde{T}^m$ is positive on $\tilde{M} = M \otimes M_{2\times 2}$, we conclude that $P_m \otimes \text{id}$ is positive, and so P_m is two-positive. $\square$

Suppose now that $x, y \in M$. Then

$$\phi_{P_m x}(y) = < \Lambda(y), JP_{2,m}\Lambda(x) > = < \Lambda(P_m y), J\Lambda(x) > = \phi_x(P_m y) \quad (11)$$

because $P_{2,m}$ commutes with J. Using the above property we define a contractive projection $P_{1,m}$ on $\mathcal{M}_*$ as follows. Let $P_{1,m}\phi_x = \phi_{P_m x}$, $x \in M$. Then

$$\|P_{1,m}\phi_x\|_1 = \sup_{y\in M, \|y\|_\infty \leq 1} |\phi_{P_m x}(y)| = \sup_{y\in M, \|y\|_\infty \leq 1} |\phi_x(P_m y)| \leq \|\phi_x\|_1.$$

Because $\Phi(M)$ is norm dense in $\mathcal{M}_*$ so $P_{1,m}$ extends to a two-positive contractive projection on $\mathcal{M}_*$ which we denote also by $P_{1,m}$. Let $P_{1,m}^* : \mathcal{M} \to \mathcal{M}$ be the dual projection. By formula (11), $P^*_{1,m}|_M = P_m$. Hence there exists a unique extension of P_m to a two-positive normal and contractive in the operator norm projection on $\mathcal{M}$ which we denote also by P_m. Using the properties of projections $P_{2,m}$ and the fact that $P_2\xi = \lim_{m\to\infty} P_{2,m}\xi$ one may show in the same way as above the existence of a two-positive and contractive (in the corresponding norms) projections $P_1 : \mathcal{M}_* \to \mathcal{M}_*$, and its dual $P : \mathcal{M} \to \mathcal{M}$ associated with the orthogonal projection P_2 in $\mathcal{H}$. Two-positivity follows actually from the fact that for all $x \in M$, $P_m(x) \to P(x)$ in the σ-strong* topology, and so

$$\sum_{i,j=1}^{2} y^*_j P(x^*_j x_i) y_i = \lim_{m\to\infty} \left(\sum_{i,j=1}^{2} y^*_j P_m(x^*_j x_i) y_i \right) \leq 0$$

for any x_i, $y_i \in M$, $i = 1, 2$. It is also clear that $P : M \to M$ and $\phi \circ P \leq \phi$. For $x \in M$, Px is given by $Px = \Lambda^{-1}(P_2\Lambda(x))$.

Suppose now that $R_{2,m} = P_2 T^m_2 T^m_{*2} P_2$, $m \in \mathbf{N}$. Clearly, $R_{2,m}$ is a contraction in $\mathcal{H}$, and let K_m be its fixed point space, i.e.

$$K_m = \{\xi \in \mathcal{H} : R_{2,m}\xi = \xi\}.$$

Then $K_{m+1} \subset K_m \subset K^\infty$. Let $K = \bigcap_{m=1}^{\infty} K_m$.

Proposition 5. K is a unitary space for T_2.

Proof. Suppose $\xi \in K$. Because $\xi \in K^\infty$ so $\|T^m_2\xi\|_2 = \|\xi\|_2$ for all $m \in \mathbf{N}$. Because $\xi \in K_m$ so $\|T^m_{*2}P_2\xi\|_2 = \|\xi\|_2$. However, $P_2\xi = \xi$ and $T_{*2} = T^*_2$ which implies that $\|T^{*m}_2\xi\|_2 = \|\xi\|_2$ for all $m \in \mathbf{N}$. Conversely, suppose that for all $m \in \mathbf{N}$ there is $\|T^m_2\xi\|_2 = \|T^{*m}_2\xi\|_2 = \|\xi\|_2$. Then $T^m_{*2}T^m_2\xi = T^m_2 T^m_{*2}\xi = \xi$, and so $P_2\xi = \xi$ which implies that $R_{2,m}\xi = \xi$. Hence $\xi \in K$. □

Let $Q_{2,m}$ and Q_2 be the orthogonal projections of $\mathcal{H}$ onto K_m and K respectively. Let $R_m = PT^m T^m_* P$. Since operators $R_m : M \to M$ possess the same properties as S_m, so by repeating the arguments of Theorem 4 and the discussion after it we arrive at the following result which we state without a proof.

Theorem 6. There are two-positive and contractive (in the corresponding norms) projections $Q_1 : \mathcal{M}_* \to \mathcal{M}_*$ and its dual $Q : \mathcal{M} \to \mathcal{M}$, and an orthogonal projection Q_2 from $\mathcal{H}$ onto K such that

$$Q_1\Phi(x) = \Phi(Qx) \quad \forall x \in M, \tag{12}$$

$$Q_2\Lambda(x) = \Lambda(Qx) \quad \forall x \in N \cap N^*, \tag{13}$$

$$Q : M \to M \quad \text{and} \quad \phi \circ Q \leq \phi. \tag{14}$$

Proposition 7. $Q(1) \leq 1$. $Q(1) = 1$ if and only if the projection Q is ϕ-compatible.

Proof. Assume that $Q(1) = 1$. Then, for any $x \in M_+$,

$$\phi(Qx) = \phi_{Qx}(1) = (Q_1\phi_x)(1) = \phi_x(Q(1)) = \phi_x(1) = \phi(x).$$

Moreover, $\phi|_{Q(\mathcal{M})}$ is semifinite. Conversely, suppose that $\phi(Qx) = \phi(x)$ for all $x \in M_+$. Then $\phi_x(Q(1)) = \phi_x(1)$ for all $x \in M$. Since $\Phi(M)$ is norm dense in $\mathcal{M}_*$, the assertion follows. The inequality $Q(1) \le 1$ follows from formula (14). $\square$

2.4 The Isometric-Sweeping Decomposition

We start with the following observation.
Proposition 8.
$$QT = TQ, \quad QT_* = T_*Q,$$
$$Q_1T_1 = T_1Q_1, \quad Q_1T_{*1} = T_{*1}Q_1.$$

Proof. By duality, it is sufficient to check only the first line. But this follows from the property $Q_2T_2 = T_2Q_2$ (since K is the unitary space for T_2), formula (13), the fact that all three operators Q, T and T_* are normal maps, and M is σ-weakly dense in $\mathcal{M}$. $\square$

We are now in a position to formulate our main results.

Theorem 9. $\mathcal{M}_* = \mathcal{M}_{*1} \oplus \mathcal{M}_{*2}$, where $\mathcal{M}_{*1}$ and $\mathcal{M}_{*2}$ are norm closed T_{*1} and T_1-invariant *-subspaces. The restriction of T_1 to $\mathcal{M}_{*1}$ is an invertible isometry and $T_{*1}|_{\mathcal{M}_{*1}}$ is its inverse. For any $\psi \in \mathcal{M}_{*2}$ and all $x \in C$ there is

$$\lim_{n \to \infty} (T_1^n \psi)(x) = \lim_{n \to \infty} (T_{*1}^n \psi)(x) = 0. \tag{15}$$

If $\{T_{*1}^n\}$ is relatively compact in the strong operator topology, then

$$\lim_{n \to \infty} \|T_{*1}^n \psi\|_1 = 0 \tag{16}$$

for all $\psi \in \mathcal{M}_{*2}$.
Remark. If $\psi \in \mathcal{M}_{*2}$, then for any ϕ-finite projection $e \in \mathcal{M}$ we have $(T_{*1}^n \psi)(e) \to 0$, which justifies the name of sweeping. $\mathcal{M}_{*1}$ is called the isometric part.
Proof. Let $\mathcal{M}_{*1}$ be the image of projection Q_1 and let $\mathcal{M}_{*2} = \{\psi - Q_1\psi : \psi \in \mathcal{M}_*\}$. The first part of the theorem is clear. If $x \in M$ and $Qx = x$, then

$$\Lambda(T_*Tx) = T_{*2}T_2\Lambda(x) = \Lambda(x) = \Lambda(TT_*x).$$

Hence $T_*Tx = TT_*x = x$, and so $T_{*1}T_1\phi_x = T_1T_{*1}\phi_x = \phi_x$. Because the space $\{\phi_x : Qx = x\}$ is norm dense in $\mathcal{M}_{*1}$ so $T_{*1}|_{\mathcal{M}_{*1}}T_1|_{\mathcal{M}_{*1}} = T_1|_{\mathcal{M}_{*1}}T_{*1}|_{\mathcal{M}_{*1}} = \mathrm{id}|_{\mathcal{M}_{*1}}$. Since both T_1 and T_{*1} are contractive, they are isometric operators on $\mathcal{M}_{*1}$. Suppose now that $x, z \in M$. Then, $Q_2^{\perp}\Lambda(x) \in K^{\perp}$, the completely nonunitary subspace for T_2, and so

$$(T_1^n(\phi_x - Q_1\phi_x))(z) = (T_1^n\phi_{x-Qx})(z) = < \Lambda(z), J\Lambda(T^n(x - Qx)) >$$

$$= < \Lambda(z), JT_2^n Q_2^{\perp}\Lambda(x) > \to 0$$

when $n \to \infty$. The case of T_{*1} may be handled in the same way. Since the space $\{\phi_x - Q_1\phi_x\}$ is norm dense in $\mathcal{M}_{*2}$, and M is norm dense in C, the formula (15) follows. Finally, suppose that $\{T_{*1}^n\}$ is relatively compact in the strong operator topology. Then, for any $\psi \in \mathcal{M}_{*2}$, the set $\{T_{*1}^n\psi\}_0^\infty$ is relatively compact in the norm topology in $\mathcal{M}_*$. Let ψ_0 be a strong accumulation point of this set. Then there exists a subsequence (m_n) of natural numbers such that

$$\lim_{n\to\infty} \|T_{*1}^{m_n}\psi - \psi_0\|_1 = 0.$$

Hence, for any $x \in M$, $(T_{*1}^{m_n}\psi)(x) \to \psi_0(x)$. By formula (15), $\psi_0(x) = 0$. However, ψ_0 is normal and M is σ-weakly dense in $\mathcal{M}$ so $\psi_0 = 0$. Thus $\|T_{*1}^{m_n}\psi\|_1 \to 0$. Since T_{*1} is norm contractive, the formula (16) follows, and the proof is complete. $\square$

By duality, one can obtain a similar decomposition of the algebra $\mathcal{M}$.

Theorem 10. $\mathcal{M} = \mathcal{M}_1 \oplus \mathcal{M}_2$, where $\mathcal{M}_1$ is a σ-weakly closed *-subalgebra and $\mathcal{M}_2$ is a σ-weakly closed linear *-subspace in $\mathcal{M}$. Both $\mathcal{M}_1$ and $\mathcal{M}_2$ are T and T_*-invariant. The restriction of T to $\mathcal{M}_1$ is a *-automorphism, whereas

$$\lim_{n\to\infty} \psi(T^n x) = 0 \tag{17}$$

for all $\psi \in \mathcal{M}_*$ and all $x \in \mathcal{M}_2 \cap C$. If the predual semigroup $\{T_{*1}^n\}$ is relatively compact in the strong operator topology, then

$$\lim_{n\to\infty} \psi(T^n x) = 0 \tag{18}$$

for all $x \in \mathcal{M}_2$, uniformly on bounded sets in $\mathcal{M}_2$. If $\mathcal{M}_1 \neq 0$, then the projection from $\mathcal{M}$ onto $\mathcal{M}_1$ is a conditional expectation, and it is a ϕ-compatible conditional expectation whenever $1 \in \mathcal{M}_1$.

Proof. Let $\mathcal{M}_1 = \{Qx : x \in \mathcal{M}\}$ and let $\mathcal{M}_2 = \{x - Qx : x \in \mathcal{M}\}$. Because projection Q is two-positive normal and commutes with both T and T_* so the decomposition of $\mathcal{M}$ follows. To proceed further we first show that $M_1 = \{Qx : x \in M\}$ is a *-algebra. By linearity, it is sufficient to check that $x^*x \in M_1$ whenever $x \in M_1$. Suppose that $x \in M_1$. Then $Q(x) = x$ and $Q(x^*) = x^*$. By Proposition 7 and two-positivity, projection Q satisfies the Schwarz inequality. Hence $Q(x^*x) \geq Q(x^*)Q(x) = x^*x$, and so $\phi(Q(x^*x)) \geq \phi(x^*x)$. On the other hand $\phi(Q(x^*x)) \leq \phi(x^*x)$. Thus $\phi(Q(x^*x) - x^*x) = 0$ and so, by faithfulness of ϕ, $Q(x^*x) = x^*x$. Hence $x^*x \in M_1$ which implies that M_1 is a *-algebra. Since $\mathcal{M}_1$ is a σ-weak closure of of M_1, it is also a *-algebra. Because $T(x^*x) \geq (Tx)^*Tx$ so, for any $x \in M_1$,

$$0 \leq \phi(T(x^*x)) - \phi((Tx)^*Tx) \leq \|\Lambda(x)\|_2^2 - \|T_2\Lambda(x)\|_2^2 = 0.$$

Hence, by faithfulness of ϕ, $T(x^*x) = (Tx)^*Tx$. This implies that for any $\psi \in \mathcal{M}_{*+}$, the positive form b on M given by

$$b(x, y) = \psi(T(x^*y) - T(x^*)T(y)),$$

vanishes on M_1. Because ψ was arbitrary we conclude that $T(xy) = T(x)T(y)$ for all $x, y \in M_1$. However, σ-weak continuity of T implies that $T(xy) = T(x)T(y)$ for all $x, y \in \mathcal{M}_1$. If $x \in M_1$, then $Q(x^*x) = x^*x$, and so $T_*T(x^*x) = TT_*(x^*x) = x^*x$. Again, by σ-weak continuity of T and T_*, this equality holds for all $x \in \mathcal{M}_1$. Thus $T|_{\mathcal{M}_1}$ is a *-automorphism. If $y \in \mathcal{M}_2 \cap C$, then formula (15) yields (17). Finally, suppose that (T_{*1}^n) is relatively compact in the strong operator topology. Then, by formula (16), for any $y \in \mathcal{M}_2$ with $\|y\|_\infty \nleq 1$, and all $\psi \in \mathcal{M}_*$ there is

$$\lim_{n\to\infty} |\psi(T^m y)| = \lim_{n\to\infty} |\psi(T^n(\mathrm{id} - Q)y)| = \lim_{n\to\infty} |(T_{*1}^n(\mathrm{id} - Q_1)\psi)(y)|$$

$$\leq \lim_{n\to\infty} \|T_{*1}^n(\mathrm{id} - Q_1)\psi\|_1 = 0$$

since $(\mathrm{id} - Q_1)\psi \in \mathcal{M}_{*2}$. If $\mathcal{M}_{*1} \neq 0$, the Q is a norm one projection, and so it is a conditional expectation onto the algebra $\mathcal{M}_1$. If $1 \in \mathcal{M}_1$, then $\mathcal{M}_1$ is a von Neumann algebra, i.e. $\mathcal{M}_1'' = \mathcal{M}_1$, and, by Proposition 7, Q is a ϕ-compatible conditional expectation onto $\mathcal{M}_1$. $\square$

It is worth noting that this result is optimal in the following sense. There exists a σ-weakly continuous semigroup of operators on $B(\mathcal{H}_S)$ satisfying A1-A6 and such that $\mathcal{M}_1 = 0, \mathcal{M}_2 = B(\mathcal{H}_S), \psi(T_t x) \to 0$ for all $\psi \in \mathrm{Tr}(\mathcal{H}_S)$ and all $x \in K(\mathcal{H}_S)$, and $\psi(T_t 1) = 1$ for all $t \geq 0$ [7]. Here $\mathrm{Tr}(\mathcal{H}_S)$ is the Banach space of trace class operators, the predual space to $B(\mathcal{H}_S)$, and $K(\mathcal{H}_S)$ stands for the C^*-algebra of compact operators.

2.5 Remarks

Assumptions A1-A6 are necessary if one wants to consider arbitrary von Neumann algebras, especially those of type III. In special cases they may be simplified.

Proposition 11. Suppose $\mathcal{M}$ is a semifinite von Neumann algebra, i.e. of type I or II, with a faithful normal and semifinite trace τ. Suppose further that operator $T : \mathcal{M} \to \mathcal{M}$ satisfies assumptions A1-A4 with $\phi = \tau$. Then A5 and A6 follows.

Proof. Since $\phi = \tau$, $\sigma_t = \mathrm{id}$. Hence only assumption A6 needs to be shown. By A4, $\|Tx\|_1 = \tau(Tx) \leq \tau(x) = \|x\|_1$, if $x \in M_+$. Hence T is bounded in the norm $\| \cdot \|_1$ and so extends to a bounded operator T_1 on $\mathcal{M}_* = L^1(\mathcal{M})$. Let $T_1^* : \mathcal{M} \to \mathcal{M}$ be the dual operator. Clearly, T_1^* is also two-positive. Moreover, for any $\psi \in L_+^1(\mathcal{M})$,

$$\psi(T_1^* 1) = (T_1 \psi)(1) = \|T_1 \psi\|_1 \leq \|\psi\|_1 = \psi(1),$$

which implies that $T_1^*(1) \leq 1$. Hence T_1^* satisfies the Schwarz inequality and so is contractive in the operator norm. Thus T_1 is contractive in the $\| \cdot \|_1$ norm. Suppose now that $x \in M_+$. Then, for any $y \in \mathcal{M}$,

$$\Phi(x)(y) = \langle J\pi(y)^* J\Lambda(x^{1/2}), \Lambda(x^{1/2}) \rangle = \tau(xy) \equiv \tau_x(y).$$

So, by linearity, $\Phi(x) = \tau_x$ for all $x \in M$. By Proposition 1 in [65], $\tau(T_1^* x) \leq \tau(x)$, for all $x \in M_+$, and so $T_1^* : M \to M$. Because $\Phi(T_1^* x) = T_{*1} \Phi(x)$, for all $x \in M$, so $T_{*1} : \Phi(M) \to \Phi(M)$. $\square$

The above decomposition $\mathcal{M} = \mathcal{M}_1 \oplus \mathcal{M}_2$ (called the isometric-sweeping decomposition) is obviously related to the asymptotic properties of the semigroup (T^n). Such properties for positive or completely positive semigroups having a faithful normal stationary state (or a faithful family of subinvariant normal states) have been studied by many authors. For example, in [24] and [40] the problem of the approach to equilibrium was addressed. See also the contribution of Fagnola and Rebolledo in this volume. In [26, 30, 37, 64, 65] the existence of the mean ergodic projection on a von Neumann algebra $\mathcal{M}$ was considered. Such a projection being a conditional expectation onto the fixed point subalgebra $\mathcal{M}^T$ provides another decomposition, namely $\mathcal{M} = \mathcal{M}^T \oplus \mathcal{N}$, with the obvious inclusion $\mathcal{M}^T \subset \mathcal{M}_1$. However, the evolution restricted to $\mathcal{M}^T$ is trivial, while on $\mathcal{M}_1$ it is given by a group of automorphisms. Moreover, the restriction of the dynamics to $\mathcal{N}$ cannot be controlled in general. For a partial result in this direction see [25]. From this point of view the isometric-sweeping decomposition for states (see Theorem 9) is closer to the so-called Jacobs-deLeeuw-Glicksberg splitting onto the so-called reversible and flight parts which holds whenever the semigroup is relatively compact in the weak operator topology, see for example [36]. However, in such a case there is no clear physical interpretation of the flight vectors which are characterized by the property that 0 is a weak accumulation point of their trajectory. It should be also pointed out that the isometric-sweeping decomposition may exist, even when the Jacobs-deLeeuw-Glicksberg splitting fails to hold [43]. This is due to the fact that our assumption A4 about the existence of a subinvariant faithful normal weight, contrary to the existence of a subinvariant faithful normal state, has no direct topological consequences for the semigroup (T^n) or its predual. However, in a special case of $\mathcal{M} = B(\mathcal{H}_S)$, $\mathcal{M}_* = \text{Tr}(\mathcal{H}_S)$ and $\phi = \text{Tr}$, if the Jacobs-deLeeuw-Glicksberg splitting exists, then it coincides with the isometric-sweeping decomposition. To this end, let us define

$$\text{Tr}(\mathcal{H}_S)_r = \overline{\text{Lin}}\{\psi \in \text{Tr}(\mathcal{H}_S) : T_{*1}\psi = e^{i\alpha}\psi \text{ for some } \alpha \in \mathbf{R}\}$$

and

$$\text{Tr}(\mathcal{H}_S)_0 = \{\psi \in \text{Tr}(\mathcal{H}_S) : 0 \text{ is a weak limit point of } \{T_{*1}^n \psi\}\},$$

where T_{*1} is the predual operator of T, see assumption A2. The Jacobs-deLeeuw-Glicksberg theorem states that if T_{*1} is relatively compact in the weak operator topology, then

$$\text{Tr}(\mathcal{H}_S) = \text{Tr}(\mathcal{H}_S)_r \oplus \text{Tr}(\mathcal{H}_S)_0.$$

Theorem 12. Suppose $\mathcal{M} = B(\mathcal{H}_S)$ and assumptions A1-A4 hold with $\phi = \text{Tr}$. If T_{*1} is relatively compact in the weak operator topology, then $\text{Tr}(\mathcal{H}_S)_1 = \text{Tr}(\mathcal{H}_S)_r$ and $\text{Tr}(\mathcal{H}_S)_2 = \text{Tr}(\mathcal{H}_S)_0$.

Proof. By Proposition 11, all assumptions A1-A6 are satisfied and so the isometric-sweeping decomposition takes place. To prove the theorem we proceed by steps.

Step1. $\mathrm{Tr}(\mathcal{H}_S)_r \subset \mathrm{Tr}(\mathcal{H}_S)_1$. Suppose $T_{*1}\psi = e^{i\alpha}\psi$, $\psi \in \mathrm{Tr}(\mathcal{H}_S)$. Then also $T_{*2}\psi = e^{i\alpha}\psi$, where $T_{*2} : \mathrm{HS}(\mathcal{H}_S) \to \mathrm{HS}(\mathcal{H}_S)$, and $\mathrm{HS}(\mathcal{H}_S)$ stands for the Hilbert space of Hilbert-Schmidt operators acting on $\mathcal{H}_S$. Hence

$$\|e^{-i\alpha}\psi - T_{*2}^*\psi\|_2^2 = \|\psi\|_2^2 - <e^{-i\alpha}\psi, T_{*2}^*\psi>_{HS} - <T_{*2}^*\psi, e^{-i\alpha}\psi>_{HS}$$
$$+ \|T_{*2}^*\psi\|_2^2 \leq 0$$

and so $T_{*2}^*\psi = e^{-i\alpha}\psi$. Therefore, by Proposition 5, $\psi \in K$ and hence $\psi \in \mathrm{Tr}(\mathcal{H}_S)_1$.

Step 2. By the very construction, $\mathrm{Tr}(\mathcal{H}_S)_1 \perp \mathrm{Tr}(\mathcal{H}_S)_2$ in the following sense: $\mathrm{Tr}(\psi_1\psi_2) = 0$ for all $\psi_1 \in \mathrm{Tr}(\mathcal{H}_S)_1$ and all $\psi_2 \in \mathrm{Tr}(\mathcal{H}_S)_2$. We show that $\mathrm{Tr}(\mathcal{H}_S)_r \perp \mathrm{Tr}(\mathcal{H}_S)_0$ also. Since T_2 and T_{*2} are contractions in $\mathrm{HS}(\mathcal{H}_S)$, so both $\{T_2^n\}$ and $\{T_{*2}^n\}$ are wo-relatively compact in $B(\mathrm{HS}(\mathcal{H}_S))$. Let $\mathcal{T}_2(\mathcal{T}_2^*)$ denote the closure in the weak operator topology of $\{T_2^n\}(\{T_{*2}^n\})$ in $B(\mathrm{HS}(\mathcal{H}_S))$ and $Q_2(\tilde{Q}_2)$ be the unit in the kernel of $\mathcal{T}_2(\mathcal{T}_2^*)$, respectively. Since $(\mathcal{T}_2)^* = \mathcal{T}_2^*$ so $\tilde{Q}_2 = Q_2^*$. However, the reversible parts of T_2 and T_{*2} in $\mathrm{HS}(\mathcal{H}_S)$ coincide, hence $\mathrm{im}Q_2 = \mathrm{im}\tilde{Q}_2 = \mathrm{im}Q_2^*$. Because $Q_2^2 = Q_2$, $Q_2^{*2} = Q_2^*$, so $Q_2^*Q_2 = Q_2$ and $Q_2Q_2^* = Q_2^*$. Therefore $(Q_2 - Q_2^*)^2 = 0$ which implies that $Q_2 = Q_2^*$. Thus, for any $x \in \mathrm{HS}(\mathcal{H}_S)_r$ and $y \in \mathrm{HS}(\mathcal{H}_S)_0$ we have

$$< x, y >_{HS} = < Q_2x, (\mathrm{id} - Q_2)y >_{HS} = 0$$

However, $\mathrm{Tr}(\mathcal{H}_S)_r \subset \mathrm{HS}(\mathcal{H}_S)_r$ and $\mathrm{Tr}(\mathcal{H}_S)_0 \subset \mathrm{HS}(\mathcal{H}_S)_0$, hence the assertion follows. Moreover, if $\psi \in \mathrm{Tr}(\mathcal{H}_S)$ and $\psi \perp \mathrm{Tr}(\mathcal{H}_S)_r$, then $\psi \in \mathrm{Tr}(\mathcal{H}_S)_0$. To prove this suppose that $\psi = \psi_1 + \psi_2$, where $\psi_1 \in \mathrm{Tr}(\mathcal{H}_S)_r$ and $\psi_2 \in \mathrm{Tr}(\mathcal{H}_S)_0$. By the assumption, $\mathrm{Tr}\psi\mathrm{Tr}(\mathcal{H}_S)_r = 0$. Because $\psi_1^* \in \mathrm{Tr}(\mathcal{H}_S)_r$, where ψ^* stands for Hermitian conjugate, and $\mathrm{Tr}\psi_2\psi_1^* = 0$, so $\mathrm{Tr}\psi_1\psi_1^* = 0$. Hence $\psi_1 = 0$.

Step 3. A set $K_\psi = \{T_{*1}^n\psi\}_{n \geq 0}$ is relatively compact for any $\psi \in \mathrm{Tr}(\mathcal{H}_S)$ in the weak topology on $\mathrm{Tr}(\mathcal{H}_S)$. Suppose that $\psi \geq 0$. By the Eberlein-Šmulian theorem K_ψ is weakly sequentially compact. Let $\{\psi_m\}$ be an arbitrary sequence in K_ψ. Then there exists a subsequence $\{\psi_{m_n}\}$ such that w-$\lim \psi_{m_n} = \psi_0$, where $\psi_0 \in \mathrm{Tr}(\mathcal{H}_S)_+$. However, $\psi_{m_n} \in \mathrm{Tr}(\mathcal{H}_S)_+$ so, by Corollary 5.11 in [59], $\lim_{n\to\infty} \|\psi_{m_n} - \psi_0\|_1 = 0$. This implies that K_ψ is sequentially compact, and so it is relatively compact in the trace norm topology. Suppose now that $\psi \in \mathrm{Tr}(\mathcal{H}_S)$. Then $\psi = \psi_1 - \psi_2 + i\psi_3 - i\psi_4$, where all $\psi_j \in \mathrm{Tr}(\mathcal{H}_S)_+$. Each set $K_j = \{T_{*1}^n\psi_j\}_{n\geq 0}$ is relatively compact. Because the function $f(\psi_1, \psi_2, \psi_3, \psi_4) = \psi_1 - \psi_2 + i\psi_3 - i\psi_4$, $\psi_j \in \mathrm{Tr}(\mathcal{H}_S)_+$, is norm continuous so the set $f(\times K_j)$ is compact in $\mathrm{Tr}(\mathcal{H}_S)$. However, for all $n \geq 0$, $T_{*1}^n\psi \in f(\times \overline{K_j})$, which implies that $\overline{K_\psi}$ is compact. Hence, the semigroup (T_{*1}^n) is relatively compact in the strong operator topology.

Step 4. By Lemma 4.2 in [42], for any $\psi\mathrm{Tr}(\mathcal{H}_S)_0$, $\lim_{n\to\infty} \|T_{*1}^n\psi\|_1 = 0$. Because $\psi \in \mathrm{Tr}(\mathcal{H}_S)_2 \Rightarrow \mathrm{Tr}\psi\mathrm{Tr}(\mathcal{H}_S)_1 = 0$ and, by step 1, $\mathrm{Tr}(\mathcal{H}_S)_r \subset \mathrm{Tr}(\mathcal{H}_S)_1$ so

$$\psi \in \mathrm{Tr}(\mathcal{H}_S)_2 \Rightarrow \mathrm{Tr}\psi\mathrm{Tr}(\mathcal{H}_S)_r = 0 \Rightarrow \psi \in \mathrm{Tr}(\mathcal{H}_S)_0.$$

The last implication follows from step 2. Hence $\mathrm{Tr}(\mathcal{H}_S)_2 \subset \mathrm{Tr}(\mathcal{H}_S)_0$. Suppose now that $\mathrm{Tr}(\mathcal{H}_S)_2 \neq \mathrm{Tr}(\mathcal{H}_S)_0$. We take $\psi \in \mathrm{Tr}(\mathcal{H}_S)_0$ such that $\psi \notin \mathrm{Tr}(\mathcal{H}_S)_2$. Let $\psi = \psi_1 + \psi_2$ be its isometric-sweeping decomposition, i.e. $\psi_1 \in \mathrm{Tr}(\mathcal{H}_S)_1$ and $\psi_2 \in \mathrm{Tr}(\mathcal{H}_S)_2$ with $\psi_1 \neq 0$. Then $\psi_1 = \psi - \psi_2 \in \mathrm{Tr}(\mathcal{H}_S)_0$ and so $\lim_{n\to\infty} \|T_{*1}^n \psi_1\|_1 = 0$. On the other hand, by Theorem 9, $\|T_{*1}^n \psi_1\|_1 = \|\psi_1\|_1 > 0$, the contradiction. Therefore $\mathrm{Tr}(\mathcal{H}_S)_2 = \mathrm{Tr}(\mathcal{H}_S)_0$. Thus equality $\mathrm{Tr}(\mathcal{H}_S)_r = \mathrm{Tr}(\mathcal{H}_S)_1$ also holds. $\square$

In the next section we illustrate these general results by physically motivated examples.

3 Review of Decoherence Effects in Infinite Spin Systems

3.1 Infinite Spin Systems

The ideal infinite quantum spin system S consists of an array of noninteracting spin-$\frac{1}{2}$ particles fixed at positions $n = 1, 2, \ldots$ and exposed to a magnetic field. The algebra $\mathcal{M}$ of its bounded observables is given by the σ-weak closure of $\pi_0(\otimes_1^\infty M_{2\times 2})$, where π_0 is a (faithful) GNS representation with respect to a tracial state tr on the Glimm algebra $\otimes_1^\infty M_{2\times 2}$, and $M_{2\times 2}$ is the algebra generated by Pauli matrices. It is worth pointing out that such a tracial state corresponds to the infinite temperature case. Let us also point out that $\mathcal{M}$ is not a "big" matrix algebra. It is a continuous algebra (factor of type II_1) in which there are no pure normal states. In fact, any projection $e \in \mathcal{M}$ contains a nontrivial subprojection $f \in \mathcal{M}$. It is worth noting that the absence of minimal projections is a new feature which may be present only in systems in the thermodynamic limit. Since the algebra $\mathcal{M}$ is a finite factor so the theorem about the isometric-sweeping decomposition can be strengthen in the following way.

Theorem 13. Suppose that a σ-weakly continuous semigroup $(T_t)_{t\geq 0}$, $T_t : \mathcal{M} \to \mathcal{M}$, satisfies:

A1) T_t are two-positive,

A3) $T_t(1) \leq 1$,

A4) $\mathrm{tr} \circ T_t \leq \mathrm{tr}$,

see Section 2.2. Then $\mathcal{M} = \mathcal{M}_1 \oplus \mathcal{M}_2$, where

(i) $\mathcal{M}_1$ is a von Neumann subalgebra of $\mathcal{M}$ and the evolution T_t when restricted to $\mathcal{M}_1$ is reversible, given by a one parameter group of *-automorphisms of $\mathcal{M}_1$.

(ii) $\mathcal{M}_2$ is a linear space (closed in the norm topology) such that for any observable $B = B^* \in \mathcal{M}_2$ and any statistical state $\rho \in \mathcal{M}_*$ of the system there is

$$\lim_{t\to\infty} \langle T_t B \rangle_\rho = 0, \tag{19}$$

where $\langle A \rangle_\rho = \mathrm{tr}(\rho A)$ stands for the expectation value of an observable A in the state ρ.

Proof. Using the same argument as in the proof of Proposition 11 one can show that all T_t can be extended to contractive operators $(T_t)_1$ on $\mathcal{M}_* = L^1(\mathcal{M})$. Their dual operators $(T_t)_1^* : \mathcal{M} \to \mathcal{M}$ also satisfy A1, A3 and A4. Because $\mathcal{M} \subset L^1(\mathcal{M})$ so for any $A, B \in \mathcal{M}$,

$$\text{tr}[((T_t)_1^*)_1^* A] B = \text{tr} A[(T_t)_1^* B] = \text{tr}(T_t A)B.$$

Since $\mathcal{M}$ is dense in $L^1(\mathcal{M})$ in the $\| \cdot \|_1$-norm so $((T_t)_1^*)_1^* = T_t$. Hence T_t are normal operators. By Proposition 11, operators T_t satisfy all assumptions A1-A6 from Section 2.2, and so the isometric-sweeping decomposition follows. Finally, we prove the equation (19). Suppose that $\psi \in L^2(\mathcal{M}) \subset L^1(\mathcal{M})$, where $L^2(\mathcal{M})$ denotes the Hilbert space of square summable with respect to the trace tr operators [54]. From the very definition of the sweeping part it follows that for any $B \in \mathcal{M}_2$,

$$\lim_{t \to \infty} \text{tr}(\psi T_t B) = 0.$$

Because for any $\phi \in L^1(\mathcal{M})$ and any $\epsilon > 0$ there exists $\psi \in L^2(\mathcal{M})$ such that $\|\phi - \psi\|_1 < \epsilon$ so

$$|\text{tr}(\phi T_t B)| \leq \epsilon\|B\| + |\text{tr}(\psi T_t B)|.$$

Hence $\lim_{t \to \infty} |\text{tr}(\phi T_t B)| \leq \epsilon\|B\|$. Since ϵ was arbitrary, the proof is complete. $\square$

The above result means that any observable A of the system may be written as a sum $A = A_1 + A_2$, $A_i \in \mathcal{M}_i$, $i = 1, 2$, and all expectation values of the second term A_2 are beyond experimental resolution after the decoherence time. Therefore, if decoherence is efficient then almost instantaneously what we can observe are observables contained in the subalgebra $\mathcal{M}_1$. In other words we apply Borel's 0th axiom: Events with very small probability never occur. Hence all possible outcomes of the process of decoherence can be directly expressed by the description of this subalgebra and its reversible time evolution.

3.2 Continuous Pointer States [10]

Suppose that the infinite spin system S described by the algebra $\mathcal{M}$ (see the previous subsection) interacts with its environment E. The reservoir is chosen to consists of noninteracting phonons of an infinitely extended one dimensional harmonic crystal at the inverse temperature $\beta = \frac{1}{kT}$. The Hilbert space $\mathcal{H}$ representing pure states of a single phonon is (in the momentum representation) $\mathcal{H} = L^2(\mathbf{R}, dk)$. A phonon energy operator is given by the dispersion relation $\omega(k) = |k|$ ($\hbar = 1$, $c = 1$). It follows that the Hilbert space of the reservoir is $\mathcal{F} \otimes \mathcal{F}$, where $\mathcal{F}$ is the symmetric Fock space over $\mathcal{H}$. A phonon field $\phi(f) = \frac{1}{\sqrt{2}}(a^*(f) + a(f))$, where $a^*(f)$ and $a(f)$ are given by the Araki-Woods representation [4]:

$$a^*(f) = a_F^*((1+\rho)^{1/2}f) \otimes I + I \otimes a_F(\rho^{1/2}\overline{f}), \tag{20}$$

$$a(f) = a_F((1+\rho)^{1/2}f) \otimes I + I \otimes a_F^*(\rho^{1/2}\overline{f}). \tag{21}$$

Here $a_F^*(a_F)$ denotes respectively creation (annihilation) operators in the Fock space, and ρ is the thermal equilibrium distribution related to the phonons energy according to the Planck law

$$\rho(k) = \frac{1}{e^{\beta\omega(k)} - 1}.$$

Since the phonons are noninteracting, their dynamics is completely determined by the energy operator $H_E = H_0 \otimes I - I \otimes H_0$, where $H_0 = d\Gamma(\omega) = \int \omega(k)a_F^*(k)a_F(k)dk$ describes the dynamics of the reservoir at zero temperature. The reference state of the reservoir is taken to be a gauge-invariant quasi-free thermal state given by

$$\omega_E(a^*(f)a(g)) = \int \rho(k)\overline{g}(k)f(k)dk.$$

Clearly, ω_E is invariant with respect to the free dynamics of the environment.

The joint system $S + E$ evolves unitarily with the Hamiltonian H consisting of three parts

$$H = H_S \otimes 1 + 1 \otimes H_E + H_I. \tag{22}$$

We assume that $H_S = 0$ and that the coupling between the system and the environment is linear, i.e. $H_I = \lambda Q \otimes \phi(g)$, where

$$Q = \pi_0 \left(\sum_{n=1}^{\infty} \frac{1}{2^n} \sigma_n^3 \right), \tag{23}$$

σ_n^3 is the third Pauli matrix in the n-th site, and $\lambda > 0$ is a coupling constant. The factor $\frac{1}{2^n}$ in equation (23) reflects the property that interaction between spin particles and the reservoir decreases as $n \to \infty$. Its form was chosen to simplify further calculations. The test function $g(k) = |k|^{1/2}\chi(k)$, where $\chi(k)$ is an even and real valued function such that:
(i) χ is differentiable with bounded derivative,
(ii) for large $|k|$, $|\chi(k)| \leq \frac{C}{k^{2+\epsilon}}$, $C > 0$, $\epsilon > 0$,
and $\chi(0) = 1$. The behavior of the test function g at the origin and the asymptotic bound (ii) are taken to ensure that H is essentially self-adjoint. Properties (i) and (ii) will also secure that the two-point thermal correlation function of the field operator is integrable.

The reduced dynamics of an observable $X \in \mathcal{M}$ is given by

$$T_t(X) = \Pi^{\omega_E}(e^{itH}(X \otimes I)e^{-itH}), \tag{24}$$

where Π^{ω_E} is a conditional expectation (the dual operation to the partial trace) with respect to the reference state ω_E of the reservoir. Because the thermal correlation function

$$\langle \phi_t(g)\phi(g) \rangle = \omega_E(e^{itH_E}\phi(g)e^{-itH_E}\phi(g)) = \omega_E(\phi(e^{it\omega}g)\phi(g))$$

is integrable, we use the so-called singular coupling limit [2, 48] which states that $T_t = e^{tL}$ is a quantum Markov semigroup with the generator L given by a master equation in the standard (Gorini-Kossakowski-Sudershan-Lindblad) form

$$L(X) = ib[X, Q^2] + \lambda a(QXQ - \frac{1}{2}\{Q^2, X\}).$$

(25)

Parameters $a > 0$ and $b \in \mathbf{R}$ are determined by

$$\int_0^\infty < \phi_t(g)\phi(g) > dt = \frac{a}{2} + ib.$$

(26)

Direct calculations show that

$$< \phi_t(g)\phi(g) > = \sqrt{2\pi}\mathcal{F}(f_1)(t) + \frac{\sqrt{2\pi}}{2}\mathcal{F}(f_2)(t) + \frac{\sqrt{2\pi}}{2}\mathcal{F}(f_3)(t),$$

where

$$f_1(k) = \frac{|k|\chi^2(k)}{e^{\beta|k|} - 1}, \quad f_2(k) = |k|\chi^2(k), \quad f_3(k) = k\chi^2(k),$$

and $\mathcal{F}$ stands for the Fourier transform. Hence, by the inverse Fourier formula, one has

$$a = 2\pi f_1(0) + \pi f_2(0) - \frac{2\pi}{\beta},$$

(27)

and

$$b = \frac{\sqrt{2\pi}}{2} \int_0^\infty \left(\frac{d}{dt}F(\chi^2)(t)\right) dt = -\int_0^\infty \chi^2(k)dk.$$

(28)

The master equation (25) consists of two terms. The first one is a Hamiltonian term $H'_S = bQ^2$, and the second is a dissipative operator

$$L_D(X) = \frac{2\pi\lambda}{\beta}(QXQ - \frac{1}{2}\{Q^2, X\}).$$

(29)

Because these two parts commute, it follows from the Trotter-Kato product formula that

$$T_t(X) = e^{itH'_S}(e^{tL_D}X)e^{-itH'_S}.$$

(30)

We now describe the effect of dissipation. Because $\mathcal{M}$ is a limit (in the σ-weak topology) of local algebras $M_{2^n \times 2^n} = \otimes_1^\infty M_{2\times2}$, and e^{tL_D} preserves each local algebra so we may assume that $X = (x_{ij}) \in M_{2^n \times 2^n}$. Then

$$L_D(X)_{ij} = -\frac{\pi\lambda}{\beta} \cdot \frac{(j - i)^2}{4^{n-1}} x_{ij},$$

(31)

$i, j \in \{1, ..., 2^n\}$, and so

$$e^{tL_D}(X)_{ij} = x_{ij}\exp\left(-\gamma t\frac{(j-i)^2}{4^{n-1}}\right),\tag{32}$$

where $\gamma = \pi k\lambda T$. It follows from equation (32) that the loss of coherence is faster for coefficients which are more distant to the diagonal, and it increases with reservoir temperature similarly as in the model of a harmonic oscillator linearly coupled to an infinite bath of harmonic oscillators [61]. Suppose $\mathcal{A}$ is a von Neumann algebra generated by $\pi_0(\sigma_n^3)$, $n \in \mathbf{N}$. Then $\mathcal{A}$ is a maximal commutative subalgebra in $\mathcal{M}$, and let $P : \mathcal{M} \to \mathcal{A}$ be the normal norm one projection onto the von Neumann subalgebra $\mathcal{A}$ [55]. Since $[H'_S, \pi_0(\sigma_n^3)] = 0$, it follows from equations (30) and (32) that all observables from $\mathcal{A}$ are T_t-invariant.

Theorem 14. For any statistical state ρ of the spin system and any spin observable X

$$\lim_{t\to\infty}\langle T_t(X)\rangle_\rho = \langle P(X)\rangle_\rho,\tag{33}$$

Proof. Because the semigroup $(T_t)_{t\geq 0}$ satisfies all assumptions of Theorem 13 so the isometric-sweeping decomposition follows. Hence it is enough to show that $\mathcal{A} = \mathcal{M}_1$. The inclusion $\mathcal{A} \subset \mathcal{M}_1$ is obvious. Suppose that $X \in \mathcal{M}_1$ and $X \notin \mathcal{A}$, i.e. $P(X) \neq X$. Let $Y = X - P(X)$. Then $Y \in \mathcal{M}_1$ and $Y \neq 0$. We may assume that $\|Y\|_2 =, 1$, where $\|\cdot\|_2$ is the norm in the Hilbert space $L^2(\mathcal{M})$. Let us take a sequence (A_n), where $A_n \in M_{2^n \times 2^n} \subset \mathcal{M}$, such that $A_n \to X$ in $L^2(\mathcal{M})$. Then $B_n = A_n - P(A_n) \in M_{2^n \times 2^n}$ and $B_n \to Y$. Hence, there exists $n_0 \in \mathcal{N}$ such that $\|Y - B_{n_0}\|_2 < \frac{1}{4}$. Because $\|T_t B_{n_0}\|_2 \to 0$, when $t \to \infty$, so there exists $t_0 > 0$ such that $\|T_{t_0} B_{n_0}\|_2 < \frac{1}{4}$. Thus

$$1 = \|T_{t_0}Y\|_2 \leq \|T_{t_0}(Y - B_{n_0})\|_2 + \|T_{t_0}B_{n_0}\|_2 < \frac{1}{2},$$

which is a contradiction. $\square$

Finally, we describe the algebra $\mathcal{A}$. Since $\mathcal{A}$ is commutative, it is isomorphic to an algebra of functions on some configuration space Ω. In the sequel we identify an operator $X \in \mathcal{A}$ with the corresponding function $X(\eta)$, $\eta \in \Omega$. Let P_n^+ and P_n^- be spectral projections of σ_n^3, i.e. $\sigma_n^3 = P_n^+ - P_n^-$. An infinite product $P_1^\sharp P_2^\sharp \cdots$, where $\sharp$ stands for + or -, defines a state on the subalgebra of continuous functions in $\mathcal{A}$, and so corresponds to a point in the configuration space. Thus $\Omega = \{(i_1, i_2, ...) : i_n = \pm\}$ or, in other words, each point of Ω describes a configuration of up and down spins located at $n = 1, 2, ...$. If μ_0 is a probability measure on $\{-1, 1\}$ which assigns value one half to both $\uparrow$ and $\downarrow$ spin positions, and if $\mu = \otimes_1^\infty \mu_0$ is the corresponding product probability measure on Ω, then for any $X \in \mathcal{A}$

$$\text{tr}(X) = \int X(\eta)d\mu(\eta),\tag{34}$$

and so the induced pointer states form an uncountable family. More precisely, for any $s \in [0, 1]$ there exists a projection $e \in \mathcal{A}$ such that $\text{tr}(e) = s$. Thus, since normalization to the unit interval is not essential, the decoherence induced pointer states of the presented model indeed correspond to a pointer with continuous readings.

3.3 Decoherence-Induced Spin Algebra [6]

Suppose that the system S and the environment E are the same as in the previous subsection. The evolution of the system is given by a free Hamiltonian which corresponds to the interaction of the spins with an external magnetic field parallel to the z-axis and of strength $H(n)$ at the site n

$$H_S = \pi_0 \left(-g\mu_B \sum_{n=1}^{\infty} H(n)\sigma_n^3 \right), \tag{35}$$

where g is the Landé factor, μ_B is the Bohr magneton and σ_n^3 is the third Pauli matrix in the nth site. We assume that the magnetic field decreases as $H(n) \sim (\frac{1}{q})^n$ for some $q \geq 2$. Since the coefficients $H(n)$ are summable, the Hamiltonian H_S is bounded.

The Hamiltonian H of the joint system is given by formula (22) with $H_I = \lambda Q \otimes \phi(g)$, where

$$Q = \pi_0 \left(\sum_{n=1}^{\infty} a_n \sigma_n^1 \right), \tag{36}$$

σ_n^1 stands for the first Pauli matrix in the nth site, $\lambda > 0$ is a coupling constant, and $a_n \sim (\frac{1}{p})^n$ for some $p \geq 2$. Again, since the coefficients a_n are summable, the spin part of the coupling operator Q is bounded. We assume that $g(k) = |k|^{1/2}\chi(k)$, where $\chi(k)$ satisfies the same assumptions as in the previous subsection. The reduced dynamics of the system is given by equation (24) which in the Markovian approximation leads to the following formula for the corresponding generator

$$L(A) = i[H_S - bQ^2, A] + L_D(A), \tag{37}$$

where

$$L_D(A) = \frac{2\pi\lambda}{\beta}(QAQ - \frac{1}{2}\{Q^2, A\}), \tag{38}$$

and $b = \int_0^{\infty} \chi^2(k)dk > 0$. The first part in equation (37) is the commutator with a new collective Hamiltonian $H_C = H_S - bQ^2$, while the second term is a dissipative operator. The collective Hamiltonian

$$H_C = -\pi_0 \left(g\mu_B \sum_{n=1}^{\infty} H(n)\sigma_n^3 \right) - \pi_0 \left(b \sum_{n,m=1}^{\infty} J(n+m)\sigma_n^1 \sigma_m^1 \right), \tag{39}$$

where $J(n) = a_n$, is nothing else but the Hamiltonian of the Ising model with an infinite range interaction. However, the potentials $H(n)$ and $J(n)$ are not translationally invariant.

Theorem 15. For the semigroup $T_t = e^{tL}$ the isometric-sweeping decomposition holds with $\mathcal{M}_1 = \mathbf{C} \cdot \mathbf{1}_S$.

Proof. We proceeds by steps.

Step 1. It is clear from the form of the generator L, see equation (37), that it generates a semigroup T_t which satisfies all assumptions of Theorem 13. Hence the isometric-sweeping decomposition follows.

Step 2. The subalgebra $\mathcal{M}_1$ is defined by the property $T_t^* T_t x = T_t T_t^* x = x$ for all $t \geq 0$. Hence

$$\bigcap_{l=0}^{\infty} \ker(L_D \circ \delta_{H_C}^l) \subset \mathcal{M}_1,$$

where $\delta_{H_C}(\cdot) = i[H_C, \cdot]$. We prove now the reverse inclusion. Suppose that $x \in \mathcal{M}_1$. Then, by differentiating the equation $T_t^* T_t x = x$ at time $t = 0$, we get $\mathcal{M}_1 \subset \ker L_D$. Assume that

$$\mathcal{M}_1 \subset \bigcap_{l=0}^{n-1} \ker L_D \circ \delta_{H_C}^l$$

for some $n \geq 1$. Because

$$\frac{d^{n+1}}{dt^{n+1}} T_t^* T_t x|_{t=0} = 0$$

so

$$\frac{d^{n+1}}{dt^{n+1}} T_t^* T_t x|_{t=0} = (-\delta_{H_C} + L_D)^{n+1}(x)$$

$$+ \sum_{m=1}^{n} \binom{n+1}{m} (-\delta_{H_C} + L_D)^{n+1-m} \circ (\delta_{H_C} + L_D)^m (x) + (\delta_{H_C} + L_D)^{n+1}(x)$$

$$= (-1)^{n+1} \delta_{H_C}^{n+1}(x) + (-1)^n L_D \circ \delta_{H_C}^n(x) + \delta_{H_C}^{n+1}(x) + L_D \circ \delta_{H_C}^n(x)$$

$$+ \sum_{m=1}^{n} \binom{n+1}{m} (-\delta_{H_C} + L_D)^{n+1-m} \circ \delta_{H_C}^m(x)$$

$$= (-1)^{n+1} \delta_{H_C}^{n+1}(x) + (-1)^n L_D \circ \delta_{H_C}^n(x) + \delta_{H_C}^{n+1}(x) + L_D \circ \delta_{H_C}^n(x)$$

$$+ \sum_{m=1}^{n} \binom{n+1}{m} (-1)^{n+1-m} \delta_{H_C}^{n+1}(x) + \sum_{m=1}^{n} \binom{n+1}{m} (-1)^{n-m} L_D \circ \delta_{H_C}^n(x)$$

$$= \delta_{H_C}^{n+1}(x) \sum_{m=0}^{n+1} \binom{n+1}{m} (-1)^{n+1-m}$$

$$+ \{1 + (-1)^n [\sum_{m=0}^{n+1} \binom{n+1}{m} (-1)^m - (-1)^{n+1}]\} L_D \circ \delta_{H_C}^n(x)$$

$$= 2 L_D \circ \delta_{H_S}^n(x) = 0.$$

Hence, by induction,

$$\mathcal{M}_1 \subset \bigcap_{l=0}^{\infty} \ker L_D \circ \delta_{H_C}^l.$$

Step 3. Let C_1 (respectively C_3) be a C^*-subalgebra in the Glimm algebra generated by $\{\sigma_1^{i_1}...\sigma_n^{i_n}\}$, where $i_k = 0, 1$ ($i_k = 0, 3$ respectively), and $n \in \mathbf{N}$. Then both $\pi_0(C_1)''$ and $\pi_0(C_3)''$, where $\pi_0(C_1)''$ denotes the bicommutant of the algebra $\pi_0(C_1)$, are maximal Abelian self-adjoint algebras (m.a.s.a in short) in $\mathcal{M}$ such that $\pi_0(C_1)'' \cap \pi_0(C_3)'' = \mathbf{C} \cdot \mathbf{1}_S$. The choice of coefficients $(H(n))$ and (a_n) guarantees that $L^\infty(Q) = \pi_0(C_1)''$ and $L^\infty(H_S) = \pi_0(C_3)''$, where $L^\infty(Q)$ is the von Neumann algebra generated by operator Q. Hence $L^\infty(Q) \cap L^\infty(H_S) = \mathbf{C} \cdot \mathbf{1}_S$.
Step 4. We show now that if $[Q, [Q, x]] = 0$ for some $x \in \mathcal{M}$, then $x \in L^\infty(Q)$. Let us define the derivation $\delta_x(\cdot) = i[\cdot, x]$. If $[Q, [Q, x]] = 0$, then $[Q, x] \in L^\infty(Q)$ since, by step 3, $L^\infty(Q)$ is a m.a.s.a. Suppose that W is a polynomial. Then

$$\delta_x(W(Q)) = i[Q, x]W'(Q) \in L^\infty(Q).$$

This implies that $\delta_x(L^\infty(Q)) \subset L^\infty(Q)$ since δ_x is continuous in the weak operator topology. Because $L^\infty(Q)$ is commutative so $\delta_x|_{L^\infty(Q)} = 0$, and hence $[Q, x] = 0$. Because $L^\infty(Q)$ is a m.a.s.a so $x \in L^\infty(Q)$.
Step 5. Next we show that $\ker L_D \cap L^\infty(H_C)' = \mathbf{C} \cdot \mathbf{1}_S$. Here $L^\infty(H_C)'$ stands for the commutant in $\mathcal{M}$ of the algebra $L^\infty(H_C)$. Suppose that $x \in \ker L_D \cap L^\infty(H_C)'$. Then $[Q, [Q, x]] = 0$ and $[H_C, x] = 0$. By step 4, $x \in L^\infty(Q)$ which implies that $[H_S, x] = [H_C + bQ^2, x] = 0$. Hence $x \in L^\infty(H_S)$. Because, by step 3, $L^\infty(Q) \cap L^\infty(H_S) = \mathbf{C} \cdot \mathbf{1}_S$ so $x = z\mathbf{1}_S$, where $z \in \mathbf{C}$.
Step 6. By step 2, $\delta_{H_C}(\mathcal{M}_1) \subset \mathcal{M}_1$. Hence, the derivation $\delta_1 := \delta_{H_C}|_{\mathcal{M}_1}$ is well defined and bounded. Thus $\delta_1(\cdot) = i[H_1, \cdot]$, where $H_1 = H_1^* \in \mathcal{M}_1$ [52]. By step 2 again, $H_1 \in \ker L_D$. On the other hand

$$[H_C, H_1] = -i\delta_1(H_1) = [H_1, H_1] = 0.$$

Hence $H_1 \in L^\infty(H_C)'$ and so, by step 5, H_1 is proportional to the identity operator. Suppose now that $x \in \mathcal{M}_1$. Then $[H_C, x] = -i\delta_1(x) = 0$, and so $x \in L^\infty(H_C)'$. Because $x \in \ker L_D$ so, by step 5, x is proportional to the identity operator. Hence $\mathcal{M}_1 = \mathbf{C} \cdot \mathbf{1}_S$. $\square$

It follows that the system is ergodic, i.e. all expectation values of an observable $T_t(A)$, where $A = \pi_0(\sigma_1^{i_1}...\sigma_n^{i_n})$, tend to zero when $t \to \infty$, if at least one $i_k \neq 0$. We use now this dynamical destruction of all coherent terms for reducing the number of degrees of freedom.
Corollary 16. Suppose that the first site does not interact with the reservoir, i.e. we put in equation (36) $a_1 = 0$. Then $\mathcal{M}_1 = M_{2\times2}$ and for any $A \in \mathcal{M}_1$

$$T_t(A) = e^{ith_1\sigma^3}Ae^{-ith_1\sigma^3}, \tag{40}$$

where $h_1 = H(1)$.
Proof. Suppose that in equation (36) the coefficient $a_1 = 0$. The corresponding semigroup we shall denote by T_t^1. Let $\mathcal{A}$ be a subalgebra in $\mathcal{M}$ generated by $\{\pi_0(\sigma_1^k) : k = 0, 1, 2, 3\}$. Suppose that $x \in \mathcal{M}$. Then

$$x = \sum_{k=0}^{3} \pi_0(\sigma_1^k)x_k,$$

where operators x_k belong to $\mathcal{A}'$, the commutant in $\mathcal{M}$ of algebra $\mathcal{A}$. Let S_t be a semigroup on $\mathcal{M}$ with a generator L_0 given by the following Markov master equation

$$L_0(A) = i[H_S^0 - bQ^2, A] + L_D(A),$$

where L_D is defined in equation (38), and

$$H_S^0 = \pi_0\left(-g\mu_B \sum_{n=2}^{\infty} H(n)\sigma_n^3\right).$$

Note that the summation index ranges from 2 to infinity. Then

$$T_t^1(x) = \sum_{k=0}^{3} \pi_0(U_t^* \sigma_1^k U_t)S_t(x_k),$$

where $U_t = e^{-ith_1\sigma_1^3}$. Since operators $\pi_0(U_t^*\sigma_1^k U_t)$, $k = 0, 1, 2, 3$, are orthogonal in $L^2(\mathcal{M})$ so

$$\|T_t^1(x)\|_{L^2}^2 = \sum_{k=0}^{3} \|S_t(x_k)\|_{L^2}^2.$$

Let us notice that the semigroup S_t restricted to the commutant $\mathcal{A}'$ has the same properties as the semigroup T_t. Hence, if any of x_k is not proportional to the identity operator, then, by the above theorem, $\|S_t(x_k)\|_{L^2} < \|x_k\|_{L^2}$ for all $t > 0$. Thus $\|T_t^1(x)\|_{L^2} < \|x\|_{L^2}$, too, which implies that such an operator cannot belong to $\mathcal{M}_1$. Hence, if $x \in \mathcal{M}_1$, then $x_k = z_k 1_S$, $z_k \in \mathbf{C}$, for all $k = 0, 1, 2, 3$, and so $x \in \mathcal{A}$. It follows that $\mathcal{M}_1 = \mathcal{A} = M_{2\times 2}$, and the dynamics on it is given by unitary operators U_t. $\square$

This result shows that the infinite quantum spin system, subjected to a specific interaction with the phonon field, after the decoherence time may be effectively described as a quantum system with only one degree of freedom. In other words, the environment forces the spin particles to behave in a collective way what allows introduction of three collective observables which satisfy the standard commutation relations of spin momenta. Generalization to a finite number of degrees of freedom is straightforward: If it happens that $a_1 = a_2 = ... = a_n = 0$, then $\mathcal{M}_1 = M_{2^n \times 2^n}$.

3.4 From Quantum to Classical Dynamical Systems [38]

The origin of deterministic laws that govern the classical domain of our everyday experience has also attracted much attention in recent years. In particular, the emergence of classical mechanics described by differential, and hence local, equations of motion from the evolution of delocalized quantum states was at the center of this issue. For example, the question in which asymptotic regime non-relativistic quantum mechanics reduces to its ancestor, i.e. Hamiltonian mechanics, was addressed in [27]. It was shown there that for very many bosons with weak two-body interactions there is a class of states for which time evolution of expectation values of certain operators in these states is approximately described by a non-linear Hartree equation. The problem under what circumstances such an equation reduces to the Newtonian mechanics of point particles was also discussed in that paper. A different point of view was taken in a seminal paper by Gell-Mann and Hartle [28]. They gave a thorough analysis of the role of decoherence in the derivation of phenomenological classical equations of motion. Various forms of decoherence (weak, strong) and realistic mechanisms for the emergence of various degrees of classicality were also presented. In the same spirit it was shown in [38] that an infinite quantum system subjected to a specific interaction with another quantum system may be effectively described as a simple classical dynamical system. More precisely, the effective observables of the system were parameterized by a single collective variable which underwent a continuous periodic evolution. We shall say that decoherence induces classical behavior of the quantum system if $\mathcal{M}_1$ is commutative and its evolution is given by a continuous flow on the configuration space of the algebra $\mathcal{M}_1$. In that example $\mathcal{M}_1$ was the same as in Section 3.2, i.e. $\mathcal{M}_1 = L^\infty(\mathcal{C}, d\mu)$, where $\mathcal{C}$ is the Cantor set. However, contrary to the continuous pointer case, the evolution was given by a one parameter periodic group of automorphisms α_t. Let us describe briefly this evolution. Suppose that $S^1 = \{e^{ia}, a \in \mathbf{R}\}$ and let $\lambda : \mathcal{C} \to S^1$ be given by

$$\lambda(i_1, i_2, \ldots) = \exp\left(2\pi i \sum_{n=1}^{\infty} \frac{i_n}{2^{n+1}}\right),$$

where $i_j \in \{0, 2\}$. It induces an isomorphism $\hat{\lambda}$ of the corresponding algebras

$$\hat{\lambda} : L^\infty(S^1, da) \to L^\infty(\mathcal{C}, d\mu),$$

where da denotes the normalized Lebesgue measure on the Borel σ-algebra of the circle S^1. Then the group of automorphism $\hat{\alpha}_t$ of the algebra $L^\infty(S^1, da)$, $\hat{\alpha}_t = \hat{\lambda}^{-1} \circ \alpha_t \circ \hat{\lambda}$, is induced by a continuous flow on the underlying configuration space $e^{ia} \to e^{i(a+2\pi t)}$, i.e. a uniform rotation of the circle.

An even more transparent example of an environment induced classical dynamical system was presented in [39]. A quantum dynamical semigroup $(T_t)_{t \geq 0}$ of a quantum system represented by a factor of type II_∞ such that $(\mathcal{M}_1, T_t|_{\mathcal{M}_1})$ was isomorphic with a classical system $(L^\infty(\mathbf{R}^3, d\mathbf{x}), g_t)$, where g_t was a uniform motion, i.e. with a constant velocity, in $\mathbf{R}^3$ was constructed there. It should be noted, however, that such a factor has no direct physical interpretation. In order to deal with realistic infinite quantum systems like Bose gases one has to consider abstract C^*-algebras of canonical commutation relations (CCR) and their temperature representations which lead to type III factors. The discussion of irreversible dynamics of such algebras is less advanced and concentrated mainly on the so-called quasi-free dynamical semigroups [1, 21, 23], see also [18, 19], where a general form of quasi-free dynamical semigroups was given. In the next section we address a question how one can construct a class of quasi-free dynamical semigroups on the CCR algebras which would combine deterministic and stochastic evolutions on the one particle space.

4 Dynamical Semigroups on CCR Algebras

4.1 Algebras of Canonical Commutation Relations (CCR)

The aim of this subsection is to set up a formalism in which we can discuss systematically the canonical commutation relations for a system with finite or infinite number of degrees of freedom. A comprehensive discussion of this subject can be found in [14]. Suppose S is a real linear space equipped with a nondegenerate symplectic bilinear form $\sigma : S \times S \to \mathbf{R}$. Moreover, we assume the existence of a linear operator J on S with the following properties

$$\sigma(Jf, g) = -\sigma(f, Jg), \quad J^2 = -\mathrm{id}.$$

With the help of J and σ one can introduce a complex pre-Hilbert space with the scalar multiplication and the scalar product defined by

$$(\lambda_1 + i\lambda_2)f = \lambda_1 f + \lambda_2 Jf, \quad \lambda_i \in \mathbf{R},$$

$$< f, g > = \sigma(f, Jg) + i\sigma(f, g),$$

where $f, g \in S$. Its norm completion will be denoted by $\mathcal{H}$. On S there is usually defined its own topology τ, which is stronger than the norm topology and which makes S a real locally convex topological vector space, for the definition see the next subsection. The case in which S is infinite dimensional is typical for field theories and many-body problems, whereas finite dimensional S corresponds to quantum mechanics of finite number of particles. Let $\Delta(S)$ be the space of formal linear combinations

$$\Delta(S) = \{\overset{\text{finite}}{\sum} z_k W(f_k)\},$$

where $z_k \in \mathbf{C}$, $f_k \in S$, and $W(f_k)$ being abstract symbols called the Weyl operators. Clearly $\Delta(S)$ is a complex linear space. The product of two Weyl operators is defined as

$$W(f)W(g) = e^{-i\sigma(f, g)/2}W(f + g),$$

while the *-operation as $W(f)^* = W(-f)$, and they are next extended to $\Delta(S)$ by linearity (anti-linearity) respectively. The completion $\Delta_1(S)$ of $\Delta(S)$ with respect to the norm $\| \sum z_k W(f_k) \|_1 = \sum |z_k|$, is a Banach *-algebra. We now define a new norm on $\Delta_1(S)$ by

$$\|R\| = \sup_{\pi} \|\pi(R)\|, \quad R \in \Delta_1(S),$$

where the supremum is taken over all nondegenerate representations π of $\Delta_1(S)$ for which $\pi(W(\lambda f))$, $f \in S$, is continuous in $\lambda \in \mathbf{R}$ with respect to the σ-weak topology on $B(\mathcal{H}_\pi)$. The completion of $\Delta(S)$ with respect to this norm is a C^*-algebra, say $\mathcal{W}(S)$, which we refer to as the C^*-algebra of the canonical commutation relations [20]. It is worth noting that such an algebra is simple.

4.2 Promeasures on Locally Convex Topological Vector Spaces

Suppose E is a locally convex topological vector space over $\mathbf{R}$, i.e. such that its topology is defined by a family of seminorms separating points. It is clear that the topology of a locally convex space is always Hausdorff. By E' we denote the topological dual, and by E^* the algebraic dual of the space E. Let I be the set of all closed linear subspaces V in E such that $\dim(E/V) < \infty$, and let $p_V : E \to E/V$ be the canonical projection. We say that $V \leq W$, $V, W \in I$, if $W \subset V$. For any $V \leq W$ we define a surjective linear map $p_{VW} : E/W \to E/V$ by $p_{VW}(p_W f) = p_V f$, $f \in E$. Then $(E/V, p_{VW}, I)$ is a projective net of finite dimensional locally convex (and hence locally compact) topological vector spaces. The projective limit of this net is canonically isomorphic to the topological space E'^* equipped with the $\sigma(E'^*, E')$-topology. Let $M(E/V)$ denote the set of all complex measures on E/V with finite variations. $M(E/V)$, when equipped with the natural sum and multiplication by scalars, the multiplication given by convolution $*$, and the norm $\|\mu_V\| = |\mu_V|(E/V)$, is a Banach algebra. As a Banach space $M(E/V)$ is the dual space to $C_0(M/V)$, the Banach space of continuous functions on E/V vanishing at infinity and equipped with the sup-norm. By definition, see [12], a promeasure on E is an arbitrary projective net $(\mu_V, (p_{VW})_*, I)$, where μ_V is a positive finite measure on E/V, $(p_{VW})_* : M(E/W) \to M(E/V)$ is the induced algebraic homomorphism, and $(p_{VW})_*(\mu_W) = \mu_V$ for all $V \leq W$. It is worth noting that in general the projective limit $\lim_\leftarrow \mu_V$ may not exist on E'^*. However, if $\dim E < \infty$, then any promeasure on E can be identified with a measure in an obvious way. To simplify notation we shall denote a promeasure by $(\mu_V)_{V \in I}$ or just by μ, if there is no risk of confusion. Since for all $V \leq W$,

$$\mu_V(E/V) = (p_{VW})_*\mu_W(E/V) = \mu_W(p_{VW}^{-1}(E/V)) = \mu_W(E/W),$$

we conclude that $\|\mu_{V_1}\| = \|\mu_{V_2}\|$ for all $V_1, V_2 \in I$. This common value uniquely associated with the promeasure $\mu = (\mu_V)_{V \in I}$, is called its total mass and will be denoted by $\|\mu\|$. If the total mass is equal to one, we shall say that $(\mu_V)_{V \in I}$ is a probability promeasure on E.

Suppose now that $(\mu_V)_{V \in I}$ and $(\nu_V)_{V \in I}$ are promeasures on E. Because

$$(p_{VW})_*(\mu_W * \nu_W) = (p_{VW})_*(\mu_W) * (p_{VW})_*(\nu_W) = \mu_V * \nu_V$$

so $\mu * \nu = (\mu_V * \nu_V)_{V \in I}$ is again a promeasure on E, which we shall call the convolution of promeasures $(\mu_V)_{V \in I}$ and $(\nu_V)_{V \in I}$. It is clear that convolution of probability promeasures is also a probability promeasure. If $T : E \to E$ is $\mathbf{R}$-linear and continuous, then for any $V \in I$ also $T^{-1}(V) \in I$ and so the linear operator

$$T_V : E/T^{-1}(V) \to E/V, \quad T_V(p_{T^{-1}(V)}f) = p_V(Tf),$$

$f \in E$, is well defined. Moreover, it is injective. Let $(T_V)_* : M(E/T^{-1}(V)) \to M(E/V)$ be the induced homomorphism of measure algebras.

Proposition 17. Suppose $(\mu_V)_{V \in I}$ is a promeasure on E. Then $\nu = (\nu_V)_{V \in I}$, where $\nu_V = (T_V)_*(\mu_{T^{-1}(V)})$, is also a promeasure on E, which we shall denote by $T_*(\mu)$. Moreover, $\|T_*(\mu)\| = \|\mu\|$.

Proof. If $V \le W$, then

$$(p_{VW})_* \nu_W = (p_{VW})_*(T_W)_*(\mu_{T^{-1}(W)}) = (p_{VW} \circ T_W)_*(\mu_{T^{-1}(W)}).$$

Because $p_{VW} \circ T_W = T_V \circ p_{T^{-1}(V)T^{-1}(W)}$ so

$$(p_{VW} \circ T_W)_*(\mu_{T^{-1}(W)}) = (T_V)_*(p_{T^{-1}(V)T^{-1}(W)})_*(\mu_{T^{-1}(W)})$$
$$= (T_V)_*(\mu_{T^{-1}(V)}) = \nu_V.$$

By definition, $\|T_*(\mu)\| = \|\nu_V\|$ for any $V \in I$. Hence

$$\|T_*(\mu)\| = \mu_{T^{-1}(V)}(T_V^{-1}(E/V)) = \|\mu_{T^{-1}(V)}\| = \|\mu\|,$$

which completes the proof. $\Box$

Suppose now that $x' \in E'$. If $\mu = (\mu_V)_{V \in I}$ is a promeasure on E, then $\mu_{x'} = (x')_*(\mu)$ is a finite measure on $\mathbf{R}$. Hence

$$\mathcal{F}(\mu)(x') = \int\limits_{-\infty}^{\infty} e^{it} \mu_{x'}(dt),$$

is a function on E' which we shall call the Fourier transform of the promeasure μ. It is a positive definite function which is continuous on every finite dimensional subspace of E' [12]. Let us recall that E' as the topological dual space is equipped with the $\sigma(E', E)$-topology.

Proposition 18. If $\mu = (\mu_V)_{V \in I}$ and $\nu = (\nu_V)_{V \in I}$ are promeasures on E, then for all $x' \in E'$,

$$\mathcal{F}(\mu * \nu)(x') = \mathcal{F}(\mu)(x') \cdot \mathcal{F}(\nu)(x').$$

Proof.

$$\mathcal{F}(\mu * \nu)(x') = \int\limits_{-\infty}^{\infty} e^{it}(x')_*(\mu * \nu)(dt) = \int\limits_{-\infty}^{\infty} e^{it}(\mu_{x'} * \nu_{x'})(dt)$$

$$= \int\limits_{-\infty}^{\infty}\int\limits_{-\infty}^{\infty} e^{i(t+s)} \mu_{x'}(dt)\nu_{x'}(ds) = \mathcal{F}(\mu)(x') \cdot \mathcal{F}(\nu)(x'). \quad \Box$$

Proposition 19. If $T : E \to E$ is $\mathbf{R}$-linear and continuous, then for any promeasure μ on E, $\mathcal{F}(T_*(\mu)) = \mathcal{F}(\mu) \circ T'$, where $T' : E' \to E'$ is the dual operator.

Proof. Let $x' \in E'$. Then

$$\mathcal{F}(T_*(\mu))(x') = \int_{-\infty}^{\infty} e^{it}(x')_*(T_*\mu)(dt) = \int_{-\infty}^{\infty} e^{it}(x' \circ T)_*(\mu)(dt)$$

$$= \int_{-\infty}^{\infty} e^{it}(T'(x'))_*(\mu)(dt) = \mathcal{F}(\mu)(T'(x')). \quad \square$$

Combining Propositions 18 and 19 we obtain the following:

$$\mathcal{F}(\mu * (T_*\nu)) = \mathcal{F}(\mu) \cdot (\mathcal{F}(\nu) \circ T'). \tag{41}$$

4.3 Perturbed Convolution Semigroups of Promeasures

Suppose that $(S_t)_{t\geq0}$ is a semigroup of **R**-linear and continuous contractions in E, i.e. $S_0 = \text{id}$ and $S_t \circ S_s = S_{t+s}$ for all $s,\, t \geq 0$. Let $\mu_t = (\mu_V(t))_{V\in I}, t \geq 0$, be a family of probability promeasures on E. We shall say that μ_t is an S_t-perturbed convolution semigroup if $\mu_0 = \delta_{\bar{0}}$, where $\delta_{\bar{0}}$ denotes the Dirac point measure with support in the zero vector of E, and for all $s,\, t \geq 0$,

$$\mu_t * [(S_t)_*\mu_s] = \mu_{s+t}, \tag{42}$$

i.e. for all $V \in I$ the following equality holds

$$\mu_V(t) * [(S_{t,V})_*\mu_{S_t^{-1}(V)}(s)] = \mu_V(s + t). \tag{43}$$

Let $r,\, s,\, t \geq 0$. The following calculations

$$\mu_{t+s} * [(S_{t+s})_*\mu_r] = [\mu_t * (S_t)_*\mu_s] * [(S_t)_*(S_s)_*\mu_r]$$

$$= \mu_t * [(S_t)_*(\mu_s * (S_s)_*\mu_r)] = \mu_t * [(S_t)_*\mu_{s+r}],$$

show that this notion is well defined. Construction of perturbed semigroups is similar to that of typical convolution semigroups. For example, we prove the following.

Theorem 20. Let ν be a probability promeasure on E. Suppose that the function $t \to (S_{t,V})_*\nu_{S_t^{-1}(V)} \in M(E/V)$ is weakly*-measurable for all $V \in I$. Then there exists a Poisson S_t-perturbed convolution semigroup μ_t of probability promeasures on E.

Proof. We proceed by steps.

Step 1. Because the function $t \to (S_{t,V})_*\nu_{S_t^{-1}(V)}$ is weakly*-measurable and $\|(S_{t,V})_*\nu_{S_t^{-1}(V)}\| = 1$ for all $t \geq 0$, so there exists a measure $\nu_V(t) \in M(E/V)$ such that for any $f \in C_0(E/V)$,

$$\nu_V(t)(f) = \int_0^t (S_{s,V})_*\nu_{S_s^{-1}(V)}(f)ds.$$

We show that $\nu_t = (\nu_V(t))_{V \in I}$ is a promeasure on E. Let $V \leq W$. Then, for any $t \geq 0$,

$$(p_{VW})_* \nu_W(t) = (p_{VW})_* \left[\int_0^t (S_{s,W})_* \nu_{S_s^{-1}(W)} ds \right] = \int_0^t (p_{VW} \circ S_{s,W})_* \nu_{S_s^{-1}(W)} ds$$

$$= \int_0^t (S_{s,V} \circ p_{S_s^{-1}(V) S_s^{-1}(W)})_* \nu_{S_s^{-1}(W)} ds = \int_0^t (S_{s,V})_* \nu_{S_s^{-1}(V)} ds = \nu_V(t).$$

Step 2. If ν is a promeasure on E, then $\mu = (\mu_V)_{V \in I}$ given by $\mu_V = e^{\nu_V}$ is also a promeasure on E. By definition,

$$\mu_V = \delta_{\bar{0}} + \nu_V + \frac{\nu_V * \nu_V}{2!} + \ldots$$

Because $\|\nu_V * \nu_V\| = \|\nu_V\|^2$ so the series is norm convergent in $M(E/V)$. Suppose that $V \leq W$. Because $(p_{VW})_*$ is norm continuous so

$$(p_{VW})_* \mu_W = \lim_{n \to \infty} (p_{VW})_* \sum_{k=0}^n \frac{(\nu_W)^k}{k!} = \mu_V.$$

Step 3. For a probability promeasure ν on E we define

$$\mu_t = e^{-at} \exp[a \int_0^t (S_s)_* \nu ds],$$

where $a > 0$. By steps 1 and 2, μ_t is a probability promeasure on E such that $\mu_0 = \delta_{\bar{0}}$. We show that $(\mu_t)_{t \geq 0}$ is an S_t-perturbed convolution semigroup. Let $V \in I$. Then

$$\mu_V(t) * [(S_{t,V})_* \mu_{S_t^{-1}(V)}(s)] = e^{-a(t+s)} \exp[a(\nu_V(t) + (S_{t,V})_* \nu_{S_t^{-1}(V)}(s))]$$

$$= e^{-a(t+s)} \exp[a(\int_0^t (S_{r,V})_* \nu_{S_r^{-1}(V)} dr + \int_0^s (S_{t,V})_* (S_{r,S_t^{-1}(V)})_* \nu_{S_r^{-1}(S_t^{-1}V)} dr)].$$

Because $S_{t,V} \circ S_{r,S_t^{-1}(V)} = S_{r+t,V}$ so

$$\mu_V(t) * [(S_{t,V})_* \mu_{S_t^{-1}(V)}(s)]$$

$$= e^{-a(t+s)} \exp[a(\int_0^t (S_{r,V})_* \nu_{S_r^{-1}(V)} dr + \int_0^s (S_{t+r,V})_* \nu_{S_{t+r}^{-1}(V)} dr)]$$

$$e^{-a(t+s)}\exp[a(\int_0^{t+s}(S_{r,V})_*\nu_{S_r^{-1}(V)}dr)] = \mu_V(t+s).$$

Guided by the theory of stochastic processes we shall call such a semigroup the Poisson S_t-perturbed convolution semigroup. $\square$

Let $(\mu_t)_{t\geq 0}$ be an S_t-perturbed convolution semigroup. If $\Gamma_t = \mathcal{F}(\mu_t)$, then by formula (41),

$$\Gamma_{t+s}(x') = \Gamma_t(x')\cdot\Gamma_s(S_t'x'), \tag{44}$$

for all $x' \in E'$. We use this property in the next subsection to construct a quantum dynamical semigroup on the CCR algebra $\mathcal{W}$.

4.4 Quantum Dynamical Semigroups on CCR Algebras

In order to construct a quantum dynamical semigroup on $\mathcal{W} = \mathcal{W}(S)$ we combine a deterministic evolution given by a semigroup of injective algebraic homomorphisms of $\mathcal{W}$ and a stochastic evolution represented by a perturbed convolution semigroup of probability promeasures. Let $(S_t)_{t\geq 0}$ be a semigroup of $\mathbf{R}$-linear and continuous operators on (S, τ) such that $\sigma(S_t f, S_t g) = \sigma(f, g)$ for all $f, g \in S$ and all $t \geq 0$. It was shown in [41] that with such a semigroup one can associate a semigroup of algebraic and unital *-homomorphisms $\alpha_t : \mathcal{W} \to \mathcal{W}$ being the extension of the map $\alpha_t(W(f)) - W(S_t f)$, $f \in S$. It is worth pointing out that since $\mathcal{W}$ is simple so all α_t are injective. It is obvious that such a semigroup generalizes the notion of automorphic evolution. Since all α_t are injective and map unitary operators from $\mathcal{W}$ into unitary operators we shall say that $(\alpha)_{t\geq 0}$ represents a deterministic evolution of the system.

Now let $E = S'$, where S' is the topological dual space to (S, τ). S' with the $\sigma(S', S)$-topology is a locally convex topological vector space over $\mathbf{R}$ such that $S'' = S$. Since the topology τ is stronger than the norm topology so the following inclusion $S \subset \mathcal{H} \subset S'$ holds. Let $(S_t')_{t\geq 0}$, $S_t' : S' \to S'$, be the dual semigroup. By definition, S_t' are $\mathbf{R}$-linear and continuous operators on S' such that $(S_t')' = S_t$.

Theorem 21. Suppose that $(\mu_t)_{t\geq 0}$ is an S_t'-perturbed convolution semigroup of probability promeasures on S'. Then there exists a unique quantum dynamical semigroup $(T_t)_{t\geq 0}$, $T_t : \mathcal{W} \to \mathcal{W}$, such that

$$T_t W(f) = \mathcal{F}(\mu_t)(f)W(S_t f), \tag{45}$$

for all $f \in S$. If the following conditions
a) $\lim_{t\to 0+}\sigma(S_t f, g) = \sigma(f, g)$ for all $f, g \in S$,
b) $\lim_{t\to 0+}\omega(W(S_t f - f)) = 1$, $f \in S$, for some state $\omega \in \mathcal{S}$,
c) $\lim_{t\to 0+}\mu_V(t) = \delta_{\bar{0}} \in M(E/V)$ in the vague topology for all $V \in I$,
hold, then the semigroup $(T_t)_{t\geq 0}$ is ω-continuous.

Proof. Let $T_t^0 W(f) = \Gamma_t(f)W(f)$, where $\Gamma(f) = \mathcal{F}(\mu_t)(f)$, and $f \in S$. We show that operator T_t^0 can be extended to a completely positive norm contractive and unital operator on $\mathcal{W}$. By linearity, $T_t^0 : \Delta(S) \to \Delta(S)$. The space S when equipped with the discrete topology is an Abelian group whose dual group (the group of characters) $\hat{S}$ is a commutative compact group. The pairing between S and $\hat{S}$ we denote by $(f, \hat{f})$, $f \in S$ and $\hat{f} \in \hat{S}$. With a character $\hat{f}$ one can associate a $*$-automorphism $\beta_{\hat{f}}$ of $\Delta(S)$ defined by $\beta_{\hat{f}} W(f) = (f, \hat{f}) W(f)$, and then extended by linearity to $\Delta(S)$. Since Γ_t is a positive definite function on the group $\dot{S}$ and $\Gamma_t(\vec{0}) = 1$ so for any $t \geq 0$ there exists a probability Borel measure $\hat{\mu}_t$ on $\hat{S}$ such that

$$\Gamma_t(f) = \int_{\hat{S}} (f, \hat{f}) \hat{\mu}_t(d\hat{f}),$$

see for example [51]. Hence, for any $x \in \Delta(S)$,

$$T_t^0(x) = \int_{\hat{S}} \beta_{\hat{f}}(x) \hat{\mu}_t(d\hat{f}). \tag{46}$$

Because $\|\beta_{\hat{f}}(x)\| = \|x\|$ so $\|T_t^0(x)\| \leq \|x\|$, and T_t^0 can be extended to a contractive operator on the algebra $\mathcal{W}$. It is also clear that $T_t^0(1) = 1$. Since the formula (46) holds for any $A \in \mathcal{W}$, so for all $A_1, ..., A_n$ and $B_1, ..., B_n$ from $\mathcal{W}$ we get

$$\sum_{i,j=1}^n B_j^*(T_t^0(A_j^* A_i))B_i = \sum_{i,j=1}^n B_j^*[\int_{\hat{S}} \beta_{\hat{f}}(A_j)^* \beta_{\hat{f}}(A_i) \hat{\mu}_t(d\hat{f})]B_i$$

$$= \int_{\hat{S}} (\sum_{j=1}^n \beta_{\hat{f}}(A_j)B_j)^* (\sum_{i=1}^n \beta_{\hat{f}}(A_i)B_i) \hat{\mu}_t(d\hat{f}) \geq 0.$$

Thus T_t^0 is completely positive for all $t \geq 0$.

Let us now define $T_t : \mathcal{W} \to \mathcal{W}$, $T_t = \alpha_t \circ T_t^0$, $t \geq 0$. By definition, T_t is a norm contractive completely positive and unital operator on $\mathcal{W}$. Let us check that $(T_t)_{t \geq 0}$ is a semigroup. Clearly, it is enough to show the semigroup property on Weyl operators. Suppose that $s, t \geq 0$ and $f \in S$. Then

$$(T_s \circ T_t)W(f) = (\alpha_s \circ T_s^0)(\alpha_t \circ T_t^0)W(f) = (\alpha_s \circ T_s^0)(\Gamma_t(f)W(S_t f))$$

$$= \Gamma_t(f)\Gamma_s(S_t f)W(S_{s+t} f).$$

Because $(\mu_t)_{t \geq 0}$ is an S_t'-perturbed convolution semigroup on S' so, by formula (44), $\Gamma_t(f)\Gamma_s(S_t f) = \Gamma_{s+t}(f)$. Hence $(T_s \circ T_t)W(f) = T_{s+t}W(f)$.

Finally, we show that the mapping $t \to \pi_\omega(T_t A)$, $A \in \mathcal{W}$, where π_ω is the GNS representation associated with the state ω, is σ-strongly continuous. Because operators T_t are contractive it is enough to check that for any $f \in S$,

$$\lim_{t \to 0^+} \pi_\omega(T_t W(f)) = \pi_\omega(W(f))$$

in the strong topology. Suppose that $g \in S$. Then, by assumptions, a) and b)

$$\|(\pi_\omega(W(S_t f)) - \pi_\omega(W(f)))\pi_\omega(W(g))\Omega\|^2$$

$$= 2 - 2\text{Re} < \pi_\omega(W(S_t f))\pi_\omega(W(g))\Omega, \ \pi_\omega(W(f))\pi_\omega(W(g))\Omega >$$

$$= 2 - 2\text{Re}[e^{i\sigma(S_t f, g)} e^{-i\sigma(f,g)} e^{i\sigma(S_t f, f)/2} \omega(W(S_t f - f))],$$

tends to zero when $t \to 0^+$. Let $V \subset S'$, $V \in I$, and let V^0 be the subspace of S orthogonal to V. Because $S = \bigcup_{V \in I} V^0$, so for any $f \in S$ there exists $V \in I$ such that $f \in V^0$. However, for any $g \in V^0$,

$$\mathcal{F}(\mu_t)(g) = \int_{S'/V} e^{i f'(g)} \mu_V(t)(df').$$

Let us notice that since $g \in V^0$, the action $f'(g)$, $f' \in S'/V$, is well defined. By assumption c) $\lim_{t \to 0^+} \Gamma_t(f) = 1$. Hence for any $g \in S$,

$$\lim_{t \to 0^+} \|(\pi_\omega(T_t W(f)) - \pi_\omega(W(f)))\pi_\omega(W(g))\Omega\|^2 = 0,$$

and so the semigroup $(T_t)_{t \geq 0}$ is ω-continuous. $\square$

By means of S'_t-perturbed convolution semigroups of probability promeasures on S' one can construct a general class of quantum dynamical semigroups on the CCR algebras. In the next subsection we present a simple example, known to physicists as the quantum Brownian motion in the large mass limit, which illustrates the above framework.

4.5 Example: Quantum Brownian Motion

Let us consider a quantum particle in $\mathbf{R}^3$. For such a quantum system $S = \mathbf{R}^6 \equiv \mathbf{R}^3 \times \mathbf{R}^3$. The CCR algebra $\mathcal{W} = \mathcal{W}(S)$ is generated by Weyl operators $W(\vec{a}, \vec{b})$, $\vec{a}, \vec{b} \in \mathbf{R}^3$. A remarkable theorem due to Stone and von Neumann shows that there is essentially only one faithful representation π of $\mathcal{W}$, the so-called Schrödinger representation in the Hilbert space $L^2(\mathbf{R}^3, d\vec{x})$ of square integrable functions:

$$\pi(W(\vec{a}, \vec{b})) = \prod_{j=1}^{3} e^{i a_j b_j / 2} U_j(a_j) V_j(b_j), \tag{47}$$

where $\vec{a} = (a_1, a_2, a_3)$, $\vec{b} = (b_1, b_2, b_3)$, and the unitary operators U_j and V_j are given by

$$U_j(a_j)\psi(x_1, x_2, x_3) = e^{i a_j x_j}\psi(x_1, x_2, x_3), \tag{48}$$

$$V_j(b_j)\psi(x_1, x_2, x_3) = \psi(x_1 - \delta_{1j}b_j, x_2 - \delta_{2j}b_j, x_3 - \delta_{3j}b_j). \tag{49}$$

The generators of the one parameter groups of unitary operators U_j (V_j) we denote by $\hat{x}_j$ ($\hat{p}_j$) respectively. Dynamical de-quantization of such a system means that due to the interaction with an environment the effective (or decoherence free) observables of the system form a commutative algebra of functions on the configuration space $\mathbf{R}^3$. Guided by the discussion of the previous subsection we construct a quantum dynamical semigroup on $\mathcal{W}$ using a convolution semigroup of probability measures on $S = S'$. To this end suppose that the deterministic evolution of the system is trivial, i.e.that $S_t = \text{id}$ for all $t \geq 0$. Suppose further that $(\mu_t)_{t\geq 0}$ are probability distributions of a Gaussian process X_t on S with the mean vector $\vec{m}(t) = t\vec{m}$, $\vec{m} = (m_1, ..., m_6)$, and the covariance matrix $K(s, t) = \min(s, t) \cdot K$, where $K = [K_{jk}]$ is a positive definite matrix in $\mathbf{R}^6$. Then the Fourier transform of μ_t, the so-called characteristic function of the process X_t is given by

$$\Gamma_t(\vec{x}) = \exp[i < \vec{x}, \vec{m}(t) > - \frac{1}{2} < K(t, t)\vec{x}, \vec{x} >]$$

$$= \exp\left[it \sum_{j=1}^{6} x_j m_j - \frac{t}{2} \sum_{j,k=1}^{6} K_{jk} x_j x_k\right], \tag{50}$$

where for simplicity we replaced a vector $(\vec{a}, \vec{b})$ by $\vec{x} \in \mathbf{R}^6$. If the process X_t takes values in $\{\vec{0}\} \times \mathbf{R}^3$, i.e. $K_{jk} = 0$ whenever $j \leq 3$ or $k \leq 3$, and $\vec{m} = (0, 0, 0, m_1, m_2, m_3)$, then

$$\Gamma_t(\vec{a}, \vec{b}) = \exp\left[it \sum_{j=1}^{3} b_j m_j - \frac{t}{2} \sum_{j,k=1}^{3} K_{jk}^0 b_j b_k\right], \tag{51}$$

where the matrix $K^0 : \mathbf{R}^3 \to \mathbf{R}^3$ is defined by $K_{jk}^0 = K_{(j+3)(k+3)}$. It follows from equation (51) that the unitary operators $W(\vec{a}, \vec{0})$ are T_t-invariant while the other operators tend to zero when $t \to \infty$. Hence, for any $A \in \mathcal{W}$

$$\lim_{t \to \infty} T_t(A) = P(A), \tag{52}$$

in the norm topology, where $P : \mathcal{W} \to \mathcal{C}$ is a projection onto a commutative subalgebra generated by the unitary operators $W(\vec{a}, \vec{0})$, $\vec{a} \in \mathbf{R}^3$. More precisely

$$P(W(\vec{a}, \vec{b})) = \int_{\overline{\mathbf{R}^3}} \mu_H(d\chi)\tau_\chi(W(\vec{a}, \vec{b})), \tag{53}$$

where μ_H is a normalized Haar measure on the Bohr compactification $\overline{\mathbf{R}^3}$ of $\mathbf{R}^3$, and τ_χ is an automorphism of $\mathcal{W}$ given by

$$\tau_\chi(W(\vec{a}, \vec{b})) = \chi(\vec{b})W(\vec{a}, \vec{b}). \tag{54}$$

Because

$$\int_{\overline{\mathbf{R}^3}} \mu_H(d\chi)\chi(\vec{b}) = 0$$

if $\vec{b} \neq \vec{0}$ so $P(W(\vec{a}, \vec{b})) = 0$ whenever $\vec{b} \neq \vec{0}$. Hence the isometric-sweeping decomposition is given by $A = P(A) + (\mathrm{id} - P)(A)$, $A \in \mathcal{W}$. Because the algebra $\mathcal{C}$ is isomorphic to the C^*-algebra of quasi-periodic functions on $\mathbf{R}^3$ so the semigroup $(T_t)_{t \geq 0}$ indeed leads to the dynamical de-quantization of the system.

Finally, we derive the formula for the generator L of the semigroup $(T_t)_{t \geq 0}$ in the Schrödinger representation π. Since

$$T_t \pi(W(\vec{a}, \vec{b})) = \Gamma_t(\vec{a}, \vec{b}) \pi(W(\vec{a}, \vec{b})), \tag{55}$$

where $\Gamma_t(\vec{a}, \vec{b})$ is given by equation (51), and

$$[\hat{x}_j, \pi(W(\vec{a}, \vec{b}))] = b_j \pi(W(\vec{a}, \vec{b}))$$

so

$$L\pi(W(\vec{a}, \vec{b})) = i \sum_{j=1}^{3} m_j [\hat{x}_j, \pi(W(\vec{a}, \vec{b}))] - \frac{1}{2} \sum_{j,k=1}^{3} K_{jk}^0 [\hat{x}_j, [\hat{x}_k, \pi(W(\vec{a}, \vec{b}))]].$$

If $m_j = 0$ for all $j \in \{1, 2, 3\}$ and if the matrix K_{jk}^0 is diagonal, then L is the generator of the quantum Brownian motion in the large mass limit [11, 69].

5 Outlook

The example presented in the previous subsection is rather simple. To construct the semigroup of quantum Brownian motion one assumes that the system has a finite number of degrees of freedom and that the corresponding convolution semigroup is that of a Gaussian process. It is obvious, however, that it is a very specific case. Possible generalizations include for example application of arbitrary infinitely divisible processes like the α-stable and Poisson processes.

The main advantage, however, of the proposed framework is based on the possibility of construction quantum dynamical semigroups of systems in the thermodynamic limit. CCR algebras with infinite dimensional spaces S and their temperature representations describe Bose gases. It is believed that such systems strongly interact with their environment and so they show up efficient decoherence phenomena. For example decoherence due to environmental coupling should be transparent when a Bose-Einstein condensate is present in the trap. Many features of Bose-Einstein condensates are well described by the mean-field theory. In the mean-field picture all atoms of the condensate have the same macroscopic function satisfying the Gross-Pitajevski equation. This equation is so successful because short-range correlations between the bosons induced by the interatomic potential may be neglected. However, such a system cannot be isolated in practice. Even at low temperatures the interaction between atoms of the condensate and those outside the trap leads to ejection of atoms from the condensate suggesting rich microscopic dynamic not captured by the mean-field theory. Therefore, a new rigorous model of interacting continuous quantum systems is necessary for the description of such phenomena. We hope that the proposed framework of irreversible dynamics of infinite quantum systems will play an essential role in future investigations of this subject.

Acknowledgements

One of the authors (R. O.) has been supported by the Polish Ministry of Scientific Research and Information Technology under the grant No PBZ-MIN-008/P03/2003. We both thank the Referee for his valuable comments and remarks.

References

1. R. Alicki, Rep. Math. Phys. **14**, 27 (1978)
2. R. Alicki and K. Lendi, *Quantum Dynamical Semigroups and Applications*, LNP 289, Springer, Berlin, 1987
3. A. Amann, Fortschr. Phys. **34**, 167 (1986)
4. H. Araki and E. J. Woods, J. Math. Phys. **4**, 637 (1963)
5. Ph. Blanchard et al. (eds.), *Decoherence: Theoretical, Experimental and Conceptual Problems*, Springer, Berlin, 2000
6. Ph. Blanchard, P. Ługiewicz and R. Olkiewicz, Phys. Lett. A **314**, 29 (2003)
7. Ph. Blanchard and R. Olkiewicz, J. Stat. Phys. **94**, 933 (1999)
8. Ph. Blanchard and R. Olkiewicz, Phys. Lett. A. **273**, 223 (2000)
9. Ph. Blanchard and R. Olkiewicz, Rev. Math. Phys. **15**, 217 (2003)
10. Ph. Blanchard and R. Olkiewicz, Phys. Rev. Lett. **90**, 010403 (2003)
11. Ph. Blanchard and R. Olkiewicz, International J. Modern Phys. B **18**, 501 (2004)
12. N. Bourbaki, *Elements de mathematique. Integration*, Chapitre IX, Hermann, Paris, 1969
13. O. Bratteli and D.W. Robinson, *Operator Algebras and Quantum Statistical Mechanics I*, Springer, New York, 1979
14. O. Bratteli and D.W. Robinson, *Operator Algebras and Quantum Statistical Mechanics II*, Springer, New York, 1981
15. D. Braun, F. Haake and W. T. Strunz, Phys. Rev. Lett. **86**, 2913 (2001)
16. M. Brune et al., Phys. Rev. Lett. **77**, 4887 (1996)
17. E. B. Davies, Commun. Math. Phys. **39**, 91 (1974)
18. B. Demoen, P. Vanheuverzwijn and A. Verbeure, Lett. Math. Phys. **2**, 161 (1977)
19. B. Demoen, P. Vanheuverzwijn and A. Verbeure, Rep. Math. Phys. **15**, 27 (1979)
20. G. G. Emch, *Algebraic Methods in Statistical Mechanics and Quantum Field Theory*, Wiley Interscience, New York, 1972
21. G. G. Emch, S. Albeverio and J.-P. Eckmann, Rep. Math. Phys. **13**, 73 (1978)
22. G. G. Emch and J. C. Varilly, Lett. Math. Phys. **3**, 113 (1979)
23. D. E. Evans, J. T. Lewis, J. Funct. Anal. **26**, 369 (1977)
24. A. Frigerio, Lett. Math. Phys. **2**, 79 (1977)
25. A. Frigerio, Commun. Math. Phys. **63**, 269 (1978)
26. A. Frigerio, M. Veri, Math. Z. **180**, 275 (1982)
27. J. Fröhlich, T. Tsai and H. Yau, Commun. Math. Phys. **225**, 223 (2002)
28. M. Gell-Mann, J.B. Hartle, Phys. Rev. D **47**, 3345 (1993)
29. D. Giulini et al., *Decoherence and the Appearance of a Classical World in Quantum Theory*, 2nd Edn., Springer, Berlin, 2003
30. U. Groh, In: *One-Parameter Semigroups of Positive Operators*, LNM Vol. 1184, R. Nagel (ed.), Springer, Berlin, 1986
31. R. Haag, *Local Quantum Physics*, 2nd Edn., Springer, Berlin, 1996
32. R. Haag and D. Kastler, J. Math. Phys. **7**, 848 (1964)

33. D. Home, *Conceptual Foundations of Quantum Physics*, Plenum Press, New York, 1997
34. J. K. Hornberger et al., Phys. Rev. Lett. **90**, 160401 (2003)
35. E. Joos and H. D. Zeh, Z. Phys. B **59**, 223 (1985)
36. U. Krengel, *Ergodic Theorems*, Walter de Gruyter, Berlin, 1985
37. B. Kümmerer, R. Nagel, Acta Sci. Math. **41**, 151 (1979)
38. P. Ługiewicz and R. Olkiewicz, J. Phys. A **35**, 6695 (2002)
39. P. Ługiewicz and R. Olkiewicz, Commun. Math. Phys. **239**, 241 (2003)
40. W. A. Majewski, J. Stat. Phys. **55**, 417 (1989)
41. J. Manuceau, Ann. Inst. Henri Poincaré **8**, 139 (1968)
42. R. Nagel, In: *Semigroups of linear and nonlinear operations and applications*, G. R. Goldstein and J. A. Goldstein (eds.), Kluwer Academic Publishers, 1993
43. R. Olkiewicz, Commun. Math. Phys. **208**, 245 (1999)
44. R. Omnès, Rev. Mod. Phys. **64**, 339 (1992)
45. R. Omnès, *The Interpretation of Quantum Mechanics*, Princeton University Press, New Jersey, 1994
46. R. Omnès, *Understanding Quantum Mechanics*, Princeton University Press, New Jersey, 1999
47. R. Omnès, Phys. Rev. A **65**, 052119 (2002)
48. P. F. Palmer, J. Math. Phys. **18**, 527 (1977)
49. H. Primas, *Asymptotically Disjoint Quantum States*, in [5], pp. 161-178
50. J. M. Raimond, M. Brune and S. Haroche, Rev. Mod. Phys. **73**, 565 (2001)
51. W. Rudin, *Fourier Analysis on Groups*, Interscience Publisher, New York, 1962
52. S. Sakai, *Operator Algebras in Dynamical Systems*, Cambridge University Press, Cambridge, 1991
53. E. Schrödinger, Naturwissenschaft **23**, 807 (1935)
54. I. E. Segal, Ann. of Math. **57**, 401 (1952)
55. S. Strătilă and D. Voiculescu, *Representations of AF-Algebras and of the Group $U(\infty)$*, Berlin, Springer, 1975
56. S. Strătilă, *Modular Theory in Operator Algebras*, Edituria Academiei, Abacus Press, 1981
57. S. Strătilă and L Zsidó, *Lectures on von Neumann Algebras*, Abacus Press, Tunbridge Wells, 1979
58. V. S. Sunder, *An Invitation to von Neumann Algebras*, Springer, New York, 1987
59. M. Takesaki, *Theory of Operator Algebras I*, Springer, New York, 1979
60. M. Terp, J. Operator. Th. **8**, 327 (1982)
61. J. Twamley, Phys. Rev. D **48**, 5730 (1993)
62. A. Uhlmann, Rep. Math. Phys. **9**, 273 (1976)
63. M.E. Walter, Math. Scand. **37**, 145 (1975)
64. S. Watanabe, Hokkaido Math. J. **8**, 176 (1979)
65. F. J. Yeadon, J. London Math. Soc. (2) **16**, 326 (1977)
66. H. D. Zeh, *The Physical Basis of The Direction of Time*, 4th Edn., Springer, Berlin, 2001
67. W. H. Zurek, Phys. Rev. D **26**, 1862 (1982)
68. W. H. Zurek, Progr. Theor. Phys. **89**, 281 (1993)
69. W. H. Zurek, S. Habib and J. P. Paz, Phys. Rev. Lett. **70**, 1187 (1993)

Notes on the Qualitative Behaviour of Quantum Markov Semigroups*

Franco Fagnola[1] and Rolando Rebolledo[2]

[1] Politecnico di Milano, Dipartimento di Matematica "F. Brioschi"
Piazza Leonardo da Vinci 32, 20133 Milano, Italy
e-mail: franco.fagnola@polimi.it
[2] Facultad de Matemáticas, Universidad Católica de Chile
Casilla 306 Santiago 22, Chile
e-mail: rrebolle@uc.cl

*Partially supported by FONDECYT grant 1030552 and CONICYT/ECOS exchange program

1 Introduction

Within these notes we provide a survey of results connected with the large time behavior of Quantum Markov Semigroups. More precisely, given a QMS $\mathcal{T} = (\mathcal{T}_t)_{t \geq 0}$ we explore below a number of conditions on its generator $\mathcal{L}$ to have the following list of properties of the semigroup:

- Existence of stationary states;
- Existence of a *faithful* stationary state;
- Convergence towards the equilibrium;
- Recurrence and Transience.

It is worth noticing several peculiarities of this mathematical study. To have a notion of semigroup broad enough to include both, quantum dynamics ("Master Equations") as well as the classical Markov structure, one sacrifices strong topological properties. This leads to difficult problems like that of characterizing the semigroup from its generator, a problem which has been solved for an important class of Quantum Markov semigroups (see [43], [20], [21], [39]) but which is far from being closed. As we will see below, in most cases the generator of such a semigroup is given as a densely defined sesquilinear form on a Hilbert space. This forces to construct the Quantum Markov semigroup following the procedure used in Classical Probability to built up the minimal semigroup (see, for instance, [16]). In this case, the preservation of the identity, which is equivalent to the characterization of the domain of the generator, is often a non trivial problem. Tools for solving this problem have been developed by Chebotarev and Fagnola in [14] (see also [26], [15]).

As a compensation of weak topological properties, a strong algebraic condition is assumed, namely the property of *complete positivity*. This notion is the key feature which allowed the development of the current theory of Quantum Markov Semigroups.

Let be given a von Neumann algebra $\mathfrak{M}$ of operators over a complex separable Hilbert space $\mathfrak{h}$, endowed with a trace $\mathrm{tr}(\cdot)$. We follow the previous text [52] to introduce a *Quantum Markov Semigroup* as a w^*-continuous semigroup $\mathcal{T} = (\mathcal{T}_t)_{t \in \mathbb{R}^+}$ of normal completely positive linear maps on $\mathfrak{M}$ satisfying that $\mathcal{T}_0(\cdot)$ is the identity map of $\mathfrak{M}$ and such that $\mathcal{T}_t(\mathbf{1}) = \mathbf{1}$, for all $t \geq 0$, where $\mathbf{1}$ denotes the unit operator in $\mathfrak{M}$.

In general these semigroups are termed as *Quantum Dynamical Semigroups* too, however, we will reserve that name for the most general category of semigroups including both Markov and sub-Markov semigroups for which the weaker condition $\mathcal{T}_t(\mathbf{1}) \leq \mathbf{1}$ holds.

The infinitesimal generator of $\mathcal{T}$ is the operator $\mathcal{L}(\cdot)$ with domain $D(\mathcal{L})$ which is the vector space of all elements $a \in \mathfrak{M}$ such that the w^*–limit of $t^{-1}(\mathcal{T}_t(a) - a)$ exists. For $a \in D(\mathcal{L}(\cdot))$, $\mathcal{L}(a)$ is defined as the limit above.

We recall here for easy reference the following well-known examples.

Example 1.1. The dynamics of *closed systems* is defined via a group of linear unitary transformations $U_t : \mathfrak{h} \to \mathfrak{h}$, $(t \in \mathbb{R})$, with a generator H which is self-adjoint. Indeed, it suffices to consider the von Neumann algebra $\mathfrak{M} = \mathcal{L}(\mathfrak{h})$ of all bounded linear operators acting on $\mathfrak{h}$, the quantum Markov semigroup being defined as

$$\mathcal{T}_t(a) = U_t^* a U_t = e^{itH} a e^{-itH},$$

for all $t \geq 0$.

If H is bounded, the quantum Markov semigroup is uniformly strongly continuous, that is

$$\lim_{t \to 0} \sup_{\|a\| \leq 1} \|\mathcal{T}_t(a) - a\| = 0.$$

In this case the infinitesimal generator is

$$\mathcal{L}(a) = i[H, a], \ (a \in \mathfrak{M}).$$

Example 1.2. A slight modification of the above example consists in taking a general strongly continuous semigroup $(P_t)_{t \in \mathbb{R}^+} \mathbb{R}^+$ acting on a complex separable Hilbert space $\mathfrak{h}$, with generator G. Define

$$\mathcal{T}_t(a) = P_t^* a P_t,$$

for all $t \geq 0$, $a \in \mathcal{L}(\mathfrak{h})$. If G is bounded, the generator is

$$\mathcal{L}(a) = G^* a + a G, \ (a \in \mathcal{L}(\mathfrak{h})).$$

If G is unbounded, the above expression needs to be interpreted as a sesquilinear form:

$$\mathcal{L}(a)(v, u) = \langle Gv, au \rangle + \langle v, aGu \rangle,$$

for all u, v in the domain $D(G)$ of G, $a \in \mathfrak{M}$.

Definition 1.1. *A state ω on the given von Neumann algebra $\mathfrak{M}$ is normal it is $\sigma-$ weakly continuous, or equivalently if $\omega(\bigvee_\alpha a_\alpha) = \bigvee_\alpha \omega(a_\alpha)$ for any increasing net $(a_\alpha)_\alpha$ of positive elements in $\mathfrak{M}$, where $\bigvee_\alpha$ is the symbol for the least upper bound of a net.*

ω is faithful if $\omega(a) > 0$ for all non-zero positive element $a \in \mathfrak{M}$.

Given a quantum Markov semigroup $\mathcal{T}$ in $\mathfrak{M}$, a state ω is invariant with respect to the semigroup if $\omega(\mathcal{T}_t(a)) = \omega(a)$, for any $a \in \mathfrak{M}$, $t \geq 0$.

A von Neumann algebra $\mathfrak{M}$ is σ-finite if and only if there exists a normal faithful state on $\mathfrak{M}$ (see [11], Proposition 2.5.6). In particular, any von Neumann algebra on a separable Hilbert space is σ-finite. Within this framework we will be placed throughout this paper.

The notes are organized as follows: first we introduce some preliminary notations and concepts. We then analyze ergodic type theorems. Next, we establish a
criterion on the existence of stationary states, depending on conditions on the generator of the semigroup. We continue analyzing faithfulness of stationary states and
giving a result on the convergence towards the equilibrium of the QMS, the analysis of recurrence and transience being the final lecture. We have made an effort to
keep the text as simple as possible, sacrificing in many cases the proof of statements
requiring more sophisticated concepts which the interested reader can follow in the
references. Namely, this is the case for a number of results whose proof is based
on dilations of QMS by means of quantum cocycles, so called *quantum dilations of
Markov semigroups*. This subject uses in force the results of the chapter on quantum stochastic differential equations written by Franco Fagnola. Along the last five
years we have been invited to lecture on the subject of these notes several times, so
that besides the papers containing the original results of our joint work, the reader is
addressed to the previous surveys [33], [34] to complement the current text with additional examples and the use of quantum dilations to study the convergence towards
the equilibrium of QMS with unbounded generators.

1.1 Preliminaries

We start by fixing notations which will be used throughout the remains of the current
article. We write $\mathfrak{h}$ a complex separable Hilbert space, endowed with a scalar prod-
uct $\langle \cdot, \cdot \rangle$ antilinear in the first variable, linear in the second. $\mathfrak{L}(\mathfrak{h})$ denotes the von
Neumann algebra of all the bounded linear operators in $\mathfrak{h}$. The w^* or σ-weak topol-
ogy of $\mathfrak{L}(\mathfrak{h})$ is the weaker topology for which all maps $x \mapsto \mathrm{tr}(\rho x)$ are continuous,
where $\rho \in \mathfrak{I}_1(\mathfrak{h})$ and $\mathrm{tr}(\cdot)$ denotes the trace.

The *predual space* of a von Neumann algebra $\mathfrak{M}$, which is the Banach space
of all σ-weakly continuous linear functionals on $\mathfrak{M}$, is denoted $\mathcal{A}_*$, in particular,
$\mathfrak{L}(\mathfrak{h})_* = \mathfrak{I}_1(\mathfrak{h})$, the space of *trace-class* operators.

Any quantum Markov semigroup $\mathcal{T}$ on $\mathfrak{M}$ induces a predual semigroup $\mathcal{T}_*$ on
$\mathcal{A}_*$ defined by

$$\mathcal{T}_{*t}(\omega)(a) = \omega\left(\mathcal{T}_t(a)\right),\tag{1}$$

for all $\omega \in \mathcal{A}_*$, $a \in \mathfrak{M}$, $t \geq 0$.

The cone of positive elements in the algebra $\mathfrak{M}$ is denoted $\mathfrak{M}^+$. The space of
normal states is $\mathfrak{S} = \{\omega \in \mathcal{A}_*^+ : \omega(\mathbf{1}) = 1\}$, where $\mathbf{1}$ denotes the identity in $\mathfrak{M}$.

A function $t : D(t) \times D(t) \to \mathbb{C}$, where $D(t)$ is a subspace of $\mathfrak{h}$, is a *sesquilinear
form* over a Hilbert space $\mathfrak{h}$ if $t(v, u)$ is antilinear in v and linear in u. The set of all
sesquilinear forms over $\mathfrak{h}$ is denoted $\mathfrak{F}(\mathfrak{h})$. The form is said to be *densely defined* if
its domain $D(t)$ is dense, *symmetric* if $t(v, u) = \overline{t(u, v)}$, for all $u, v \in D(t)$, and
positive if $t(u, u) \geq 0$, for all $u \in D(t)$. We follow Bratteli and Robinson [12] to
recall some useful properties of forms. A *quadratic form* $u \mapsto t(u, u)$ is associated
to each sesquilinear form t. This quadratic form determines t by polarization. A
positive quadratic form is said to be *closed* whenever the conditions

1. $u_n \in D(t)$,
2. $\|u_n - u\| \to 0$,

imply that $u \in D(t)$ and $t(u_n - u, u_n - u) \to 0$.

Moreover, the quadratic form $u \mapsto t(u, u)$ is densely defined, positive and closed if and only if there exists a unique positive selfadjoint operator T such that $D(t) = D(T^{1/2})$ and

$$t(v, u) = \langle T^{1/2} v, T^{1/2} u \rangle,$$

for all $u, v \in D(t)$. In that case it holds in particular that $t(v, u) = \langle v, Tu \rangle$, for any $u \in D(T)$, $v \in D(t)$.

2 Ergodic Theorems

We start analyzing the simpler of our problems which is the convergence of Cesàro means of a given quantum Markov semigroup. Indeed this problem is simpler due to the following well known fact.

Proposition 2.1. *The unit ball of $\mathfrak{L}(\mathfrak{h})$ is w^*-compact.*

Proof. Indeed this follows from Alaoglu-Bourbaki theorem and the fact that $\mathfrak{L}(\mathfrak{h})$ is the topological dual of the Banach space $\mathfrak{J}_1(\mathfrak{h})$ endowed with the norm $T \mapsto \mathrm{tr}(|T|)$ (see [11], Proposition 2.4.3, p.68).

Thus, given $a \in \mathfrak{M}$, the ball $B_a = \{x \in \mathfrak{L}(\mathfrak{h}) : \|x\| \leq \|a\|\}$ is w^*-compact. Moreover, given any quantum Markov semigroup $\mathcal{T}$, the *orbit* of $a \in \mathfrak{M}$ is

$$\mathcal{T}(a) = \{\mathcal{T}_t(a) \in \mathfrak{M} : t \geq 0\}.$$

We call $\mathbf{co}(\mathcal{T}(a))$ the convex hull of the orbit and we denote $\overline{\mathbf{co}(\mathcal{T}(a))}$ its w^*-closure. As a result we have

Corollary 2.1. *Given any quantum Markov semigroup on $\mathfrak{M}$, and any $a \in \mathfrak{M}$, the set $\overline{\mathbf{co}(\mathcal{T}(a))}$ is compact in the w^*-topology.*

Proof. Indeed, since $\|\mathcal{T}_t(a)\| \leq \|a\|$ we have that $\overline{\mathbf{co}(\mathcal{T}(a))} \subseteq B_a$, is a closed subset of a compact set.

Here however, we are looking for more applicable results, thus, the use of sequences (or, say, criteria for sequential w^*-compactness) will be sufficient for our purposes.

Proposition 2.2. *For any $a \in \mathfrak{M}$ the w^*-limit of any sequence*

$$\left(\frac{1}{t_n} \int_0^{t_n} \mathcal{T}_s(a) ds; \ n \geq 0 \right),$$

with $t_n \to \infty$, is $\mathcal{T}_t$-invariant, for all $t \geq 0$.

Proof. Call b one of these limit points, i.e.

$$b = w^* - \lim_n \frac{1}{t_n} \int_0^{t_n} \mathcal{T}_s(a)ds.$$

Then

$$\mathcal{T}_r(b) = w^* - \lim_n \frac{1}{t_n} \int_0^{t_n} \mathcal{T}_{s+r}(a)ds$$

$$= w^* - \lim_n \frac{1}{t_n} \int_r^{t_n+r} \mathcal{T}_s(a)ds$$

$$= w^* - \lim_n \frac{1}{t_n} \left(\int_0^{t_n} \mathcal{T}_s(a)ds + \int_{t_n}^{t_n+r} \mathcal{T}_s(a)ds - \int_0^r \mathcal{T}_s(a)ds \right)$$

$$= w^* - \lim_n \frac{1}{t_n} \int_0^{t_n} \mathcal{T}_s(a)ds.$$

Now, we would like to say a word about the convergence of states, which requires some additional notions.

Definition 2.1. *A sequence of states $(\omega_n)_n$ is said to converge* narrowly *to $\omega \in \mathfrak{S}$ if it converges in the weak topology of the Banach space $\mathcal{A}_*$ i.e.*

$$\lim_{n \to \infty} \omega_n(x) = \omega(x)$$

for all $x \in \mathfrak{M}$.

A sequence of states $(\omega_n)_n$ is tight *if for any $\epsilon > 0$ there exists a finite rank projection $p \in \mathfrak{M}$ and $n_0 \in \mathbb{N}$ such that*

$$\omega_n(p) \geq 1 - \epsilon,$$

for all $n \geq n_0$.

Theorem 2.1. *Any tight sequence of states on $\mathfrak{L}(\mathfrak{h})$ admits a narrowly convergent subsequence.*

The reader is referred to [19] Lemma 4.3 p.291 or [49] Theorem 2 p.27 (see also [44] Appendix 1.4) for the proof. A detailed exposition of this kind of results is contained in [17].

Corollary 2.2. *Suppose that $\mathfrak{M} = \mathfrak{L}(\mathfrak{h})$. If for each state ω the family*

$$\frac{1}{t} \int_0^t \mathcal{T}_{*s}(\omega)ds, \quad t > 0$$

is tight, then each sequential limit point of the family is invariant under $\mathcal{T}_*$.

The proof of this corollary is completely similar to Corollary 2.2 and it is omitted.

We denote $\mathfrak{F}(T)$ the set of fixed points of T in $\mathfrak{M}$. A straightforward generalization of Proposition 2.2, shows that any limit point of the family of Cesàro means of the orbit $T(a)$ (indeed any limit point of $\overline{\mathrm{co}(T(a))}$) belongs to $\mathfrak{F}(T)$, for any $a \in \mathfrak{M}$. To prove the convergence of the whole family it suffices to prove that $\overline{\mathrm{co}(T(a))} \cap \mathfrak{F}(T)$ is reduced to a single point.

If the existence of a faithful, normal, stationary state ω is assumed, then $\mathfrak{F}(T)$ becomes a von Neumann subalgebra of $\mathfrak{M}$. Moreover, this subalgebra is globally invariant under the modular automorphism σ_t^ω introduced in the theory of Tomita and Takesaki (see [11, 12], [38]), since $\sigma_t^\omega(\cdot)$ and $T_t(\cdot)$ commute. Therefore, there exists a faithful normal conditional expectation $E^{\mathfrak{F}}(T)$ which satisfies

CE1 $E^{\mathfrak{F}}(T) : \mathfrak{M} \to \mathfrak{F}(T)$ is linear, w^*–continuous, completely positive,
CE2 $E^{\mathfrak{F}}(T)(1) = 1$,
CE3 $\omega \circ E^{\mathfrak{F}}(T) = \omega$, and $E^{\mathfrak{F}}(T)(aE^{\mathfrak{F}}(T)(b)) = E^{\mathfrak{F}}(T)(a)E^{\mathfrak{F}}(T)(b)$, for all $a, b \in \mathfrak{M}$.

The above characterization contains the Ergodic Theorem for QMS. Indeed $E^{\mathfrak{F}}(T)$ is unique since, given any other map E which satisfies **CE1**, **CE2**, **CE3**, it follows that $E^{\mathfrak{F}}(T) = E \circ E^{\mathfrak{F}}(T) = E^{\mathfrak{F}}(T) \circ E = E$. More precisely,

Theorem 2.2. *If ω is a faithful, normal state which is invariant under T, then there exists a unique normal conditional expectation $E^{\mathfrak{F}}(T)$ onto $\mathfrak{F}(T)$. In addition, $E^{\mathfrak{F}}(T) \circ T_t(\cdot) = E^{\mathfrak{F}}(T)(\cdot)$ for all $t \geq 0$; for any element $a \in \mathfrak{M}$, $E^{\mathfrak{F}}(T)(a)$ belongs to the w^*–closure $\overline{\mathrm{co}(T(a))}$ of the convex hull of the orbit $T(a) = (T_t(a))_{t \geq 0}$. Moreover, invariant states under the action of the predual semigroup $(T_{*t})_{t \geq 0}$, are elements of the form $\varphi \circ E^{\mathfrak{F}}(T)$ of the predual $\mathfrak{M}_*$, where φ runs over all states defined on the algebra $\mathfrak{F}(T)$.*

Since the paper of S.Ch.Moy [45], several authors have studied the construction of conditional expectations in a non-commutative framework, in particular Umegaki [58], Takesaki [57], Accardi and Cechini [1] (see eg. the survey included in the work of Petz [48]). As we have pointed out along this section, the concept of a conditional expectation is crucially related to the existence of invariant states and Cesàro (or Abel) limits.

3 The Minimal Quantum Dynamical Semigroup

Throughout this chapter, which is merely expository, we consider the von Neumann algebra $\mathfrak{M} = \mathcal{L}(\mathfrak{h})$ and we rephrase, for easier reference the crucial result which allows to construct a quantum dynamical semigroup starting from a generator given as a sesquilinear form. For further details on this matter we refer to [26], section 3.3, see also [15].

Let G and L_ℓ, $(\ell \geq 1)$ be operators in $\mathfrak{h}$ which satisfy the following hypothesis:

- **(H-min)** G is the infinitesimal generator of a strongly continuous contraction semigroup in $\mathfrak{h}$, $D(G)$ is contained in $D(L_\ell)$, for all $\ell \geq 1$, and, for all $u, v \in D(G)$, we have

$$\langle Gv, u \rangle + \sum_{\ell=1}^{\infty} \langle L_\ell v, L_\ell u \rangle + \langle v, Gu \rangle = 0.$$

Under the above assumption (**H-min**), for each $x \in \mathcal{L}(\mathfrak{h})$ let $\mathfrak{L}(x) \in \mathfrak{F}(\mathfrak{h})$ be the sesquilinear form with domain $D(G) \times D(G)$ defined by

$$\mathfrak{L}(x)(v, u) = \langle Gv, xu \rangle + \sum_{\ell=1}^{\infty} \langle L_\ell v, x L_\ell u \rangle + \langle v, xGu \rangle. \tag{2}$$

It is well-known (see e.g. [20] Sect.3, [26] Sect. 3.3) that, given a domain $D \subseteq D(G)$, which is a core for G, it is possible to built up a quantum dynamical semigroup, called the *minimal* QDS, satisfying the equation:

$$\langle v, T_t(x)u \rangle = \langle v, xu \rangle + \int_0^t \mathfrak{L}(T_s(x))(v, u)ds, \tag{3}$$

for $u, v \in D$.

This equation, however, in spite of the hypothesis (**H-min**) and the fact that D is a core for G, does not necessarily determine a unique semigroup. The minimal QDS is characterized by the following property: for any w^*-continuous family $(T_t)_{t\geq 0}$ of positive maps on $\mathcal{L}(\mathfrak{h})$ satisfying (3) we have $T_t^{(\min)}(x) \leq T_t(x)$ for all positive $x \in \mathcal{L}(\mathfrak{h})$ and all $t \geq 0$ (see e.g. [26] Th. 3.21).

Let $\mathcal{T}_*^{(\min)}$ denote the predual semigroup on $\mathfrak{I}_1(\mathfrak{h})$ with infinitesimal generator $\mathcal{L}_*^{(\min)}$. It is worth noticing here that $\mathcal{T}_*^{(\min)}$ is a *weakly* continuous semigroup on the Banach space $\mathfrak{I}_1(\mathfrak{h})$, hence it is *strongly* continuous. The linear span $\mathcal{V}$ of elements of $\mathfrak{I}_1(\mathfrak{h})$ of the form $|u\rangle\langle v|$ is contained in the domain of $\mathcal{L}_*^{(\min)}$. Thus we can write the equation (3) as follows

$$\mathrm{tr}(|u\rangle\langle v|T_t(x)) = \mathrm{tr}(|u\rangle\langle v|x) + \int_0^t \mathrm{tr}(\mathcal{L}_*^{(\min)}(|u\rangle\langle v|)T_s(x))ds.$$

This equation reveals that the solution to (3) is unique whenever the linear manifold $\mathcal{L}_*^{(\min)}(\mathcal{V})$ is big enough. Indeed, the following characterization holds.

Proposition 3.1. *Under the assumption (**H-min**) the following conditions are equivalent:*

(i) the minimal QDS is Markov (i.e. $T_t^{(\min)}(1) = 1$),

(ii) $(T_t^{(\min)})_{t\geq 0}$ is the unique w^*-continuous family of positive contractive maps on $\mathcal{L}(\mathfrak{h})$ satisfying (3) for all positive $x \in \mathcal{L}(\mathfrak{h})$ and all $t \geq 0$,

(iii) the domain $\mathcal{V}$ is a core for $\mathcal{L}_*^{(\min)}$,

(iv) for all $\lambda > 0$, the unique solution x to $\mathfrak{L}(x) = \lambda x$, in the form sense, is $x = 0$.

The equivalence of (i) and (iii) (resp. (i) and (ii)) have been proved in [20] Th. 3.2 and also in [26] Prop. 3.31 (resp. [26] Th. 3.21). For the sake of completeness we sketch this proof below.

Proof.   For any $u, v \in \mathcal{V}$, $x \in \mathfrak{L}(\mathfrak{h})$, the equation (3) can be written as

$$\mathrm{tr}(x T_{*t}^{(\mathrm{min})}(|u\rangle\langle v|)) = \mathrm{tr}(x|u\rangle\langle v|) + \int_0^t \left\langle v, \mathfrak{L}(T_s^{(\mathrm{min})}(x))u \right\rangle ds.$$

So that,

$$\frac{1}{t}\mathrm{tr}(x\left(T_{*t}^{(\mathrm{min})}(|u\rangle\langle v|) - |u\rangle\langle v|\right)) = \frac{1}{t}\int_0^t [\left\langle Gv, T_s^{(\mathrm{min})}(x)u \right\rangle$$
$$+ \sum_{\ell=1}^\infty \left\langle L_\ell v, T_s^{(\mathrm{min})}(x) L_\ell u \right\rangle$$
$$+ \left\langle v, T_s^{(\mathrm{min})}(x) Gu \right\rangle] ds.$$

The functions of s appearing in the right-hand side of the previous equation are all continuous and (**H-min**) implies that

$$\left|\left\langle L_\ell v, T_s^{(\mathrm{min})}(x) L_\ell u \right\rangle\right| \le \|L_\ell v\| \|L_\ell u\|,$$

for all $\ell \ge 1$. Moreover,

$$\sum_{\ell=1}^\infty \|L_\ell v\| \|L_\ell u\| \le \left(\sum_{\ell=1}^\infty \|L_\ell v\|^2\right)^{1/2} \left(\sum_{\ell=1}^\infty \|L_\ell u\|^2\right)^{1/2}$$
$$= (-2\mathrm{Re}\langle v, Gv\rangle)^{1/2}(-2\mathrm{Re}\langle u, Gu\rangle)^{1/2},$$

so that the series of the left-hand side converges and the function

$$s \mapsto \sum_{\ell=1}^\infty \left\langle L_\ell v, T_s^{(\mathrm{min})}(x) L_\ell u \right\rangle,$$

is continuous by Lebesgue's theorem on dominated convergence. So that,

$$\lim_{t \to 0} \frac{1}{t}\mathrm{tr}(x\left(T_{*t}^{(\mathrm{min})}(|u\rangle\langle v|) - |u\rangle\langle v|\right)) = \langle v, \mathfrak{L}(x)u\rangle.$$

Therefore, since the weak and the strong generators of $T_*^{(\mathrm{min})}$ coincide, the rang-one operator $|u\rangle\langle v|$ belongs to the domain of $\mathcal{L}_*$ for all $u, v \in D(G)$ and

$$\mathcal{L}_*(|u\rangle\langle v|) = |Gu\rangle\langle v| + \sum_{\ell=1}^\infty |L_\ell u\rangle\langle L_\ell v| + |u\rangle\langle Gv|. \tag{4}$$

We now go into the proof of the equivalence of all conditions (i) to (iv).

Notice that, under the hypothesis (**H-min**), one has $\langle v, \mathfrak{L}(1)u \rangle = 0$ for all $u, v \in D$, so that the constant family of operators $t \mapsto 1$ solves the equation (3). Therefore, (i) follows immediately from (ii). On the other hand, if (i) holds, let $\mathcal{T}$ be another contractive quantum dynamical semigroup solving (3). Then, for any $x \in \mathfrak{L}(\mathfrak{h})$ such that $0 \leq x \leq 1$ and all $t \geq 0$ we have,

$$
\begin{aligned}
\mathcal{T}_t^{(\min)}(x) &\leq \mathcal{T}_t(x) \\
&= \mathcal{T}_t(1) - \mathcal{T}_t(1 - x) \\
&\leq 1 - \mathcal{T}_t(1 - x) \\
&\leq 1 - \mathcal{T}_t^{(\min)}(1 - x) \\
&= \mathcal{T}_t^{(\min)}(x).
\end{aligned}
$$

Since each operator y in $\mathfrak{L}(\mathfrak{h})$ can be written as a linear combination of four such operators x before, it follows that $\mathcal{T}_t^{(\min)}(y) = \mathcal{T}_t(y)$ for any $y \in \mathfrak{L}(\mathfrak{h})$ and (i) is equivalent to (ii).

Since $\mathcal{V}$ is trace-norm dense in the Banach space of trace-class operators, $\mathcal{V}$ is a core for $\mathcal{L}_*$ if and only if the ortogonal complement of the linear manifold $(\lambda I - \mathcal{L}_*)(\mathcal{V})$ in $\mathfrak{L}(\mathfrak{h})$ is trivial for some $\lambda > 0$ (see e.g. [24], Prop. 3.1). Suppose (iv), then if x is an element of the above orthogonal complement, it holds

$$
\mathrm{tr}((\lambda I - \mathcal{L}_*)(|u\rangle\langle v|)x) = 0,
$$

that is, $\mathfrak{L}(x) = \lambda x$ in the form sense. So that $x = 0$ by (iv) and $\mathcal{V}$ is a core for $\mathcal{L}_*$, that is (iv) implies (iii). Reciprocally assume (iii) to be true. If there exists a solution $x \neq 0$ of $\mathfrak{L}(x) = \lambda x$ for a $\lambda > 0$, then $(\lambda I - \mathcal{L}_*)(\mathcal{V})$ is not trivial, so that $\mathcal{V}$ is not a core for the predual generator, contradicting (iii). Thus, (iii) and (iv) are equivalent.

For every $\rho \in \mathcal{V}$, we have $\mathrm{tr}(\mathcal{L}_*(\rho)) = 0$. Under condition (iii) the above identity holds for any ρ in the domain of $\mathcal{L}_*$ and

$$
\frac{d}{dt}\mathrm{tr}(\mathcal{T}_{*t}^{(\min)}(\rho)) = \mathrm{tr}(\mathcal{L}_*(\mathcal{T}_{*t}^{(\min)}(\rho))) = 0,
$$

thus, $\mathrm{tr}(\mathcal{T}_{*t}^{(\min)}(\rho)) = \mathrm{tr}(\rho)$, for all $t \geq 0$ and every trace class operator ρ because $\mathcal{V}$ is dense with respect to the trace norm. Thus, $\mathcal{T}^{(\min)}$ is Markov. That is, (iii) implies (i).

The proof that (i) implies (iv) is more technically involved and the interested reader is referred to [26], Prop.3.30 to conclude. However, just to give a flavor of the main idea used in this part of the proof, suppose that $\mathcal{L}(x)$ is a bounded operator. If (iv) does not hold, there exists $\lambda > 0$ and $x \in B(\mathcal{H}_S)$ such that $\mathcal{L}(x) = \lambda x$. In this case $\mathcal{T}_t(x) = \exp(\lambda t)x$. Notice that

$$
-2\,\|x\|\,1 \leq (x + x^*) \leq 2\,\|x\|\,1, \tag{5}
$$

$$
-2\,\|x\|\,1 \leq i(x - x^*) \leq 2\,\|x\|\,1. \tag{6}
$$

If we apply $T_t(\cdot)$ to (5) and use the fact that the semigroup is Markov, we obtain

$$-2\,\|x\|\,1 \le e^{\lambda t}(x + x^*) \le 2\,\|x\|\,1,$$

for all $t \ge 0$. This is a contradiction, so that necessarily $x + x^* = 0$, working similarly with (6) yields the conclusion that $x = 0$ and (iv) holds.

Assume that the minimal QDS is Markov and call $(P_t)_{t \in \mathbb{R}^+}\,\mathbb{R}^+$ the semigroup of contractions generated by G, then the equation (3) may be written in an equivalent form as

$$\langle v, T_t(x)u \rangle = \langle P_t v, x P_t u \rangle + \sum_{\ell \ge 1} \int_0^t \langle L_\ell P_{t-s} v, T_t(x) L_\ell P_{t-s} u \rangle ds, \qquad (7)$$

Moreover, the solution to (7) is obtained as the supremum of an approximating sequence $(T^{(n)})_{n \in \mathbb{N}}$ defined recursively as follows on positive elements $x \in \mathfrak{L}(\mathfrak{h})$:

$$T_t^{(0)}(x) = P_t^* x P_t \qquad (8)$$

$$\langle u, T_t^{(n+1)}(x)u \rangle = \langle P_t u, x P_t u \rangle \qquad (9)$$

$$+ \sum_{\ell \ge 1} \int_0^t \langle L_\ell P_{t-s} u, T_t^{(n)}(x) L_\ell P_{t-s} u \rangle ds, \ (u \in D(G)).$$

For any positive element $x \in \mathfrak{L}(\mathfrak{h})$ the above sequence is increasing with n. This allows to define $T_t(x) = \sup_n T_t^{(n)}(x)$. $\mathcal{T}$ is the *Minimal Quantum Markov semi-group*, (see for instance [26]).

The previous discussion has shown the importance of getting a minimal quantum dynamical semigroup which preserves the identity, that is, a quantum Markov semigroup. Recently A.M. Chebotarev and the first author have obtained easier criteria to verify the Markov property. For instance, the following result ([14] Th. 4.4 p.394) which will be good enough to be applied in our framework.

Proposition 3.2. *Under the hypothesis* **(H-min)** *suppose that there exists a self-adjoint operator C in $\mathfrak{h}$ with the following properties:*

(a) the domain of G is contained in the domain of $C^{1/2}$ and is a core for $C^{1/2}$,
(b) the linear manifold $L_\ell(D(G^2))$ is contained in the domain of $C^{1/2}$,
(c) there exists a self-adjoint operator Φ, with $D(G) \subseteq D(\Phi^{1/2})$ and $D(C) \subseteq D(\Phi)$, such that, for all $u \in D(G)$, we have

$$-2\mathrm{Re}\langle u, Gu \rangle = \sum_\ell \|L_\ell u\|^2 = \|\Phi^{1/2}u\|^2,$$

(d) for all $u \in D(C^{1/2})$ we have $\|\Phi^{1/2}u\| \le \|C^{1/2}u\|$,
(e) for all $u \in D(G^2)$ the following inequality holds

$$2\mathrm{Re}\langle C^{1/2}u, C^{1/2}Gu \rangle + \sum_{\ell=1}^\infty \|C^{1/2}L_\ell u\|^2 \le b\|C^{1/2}u\|^2 \qquad (10)$$

where b is a positive constant depending only on G, L_ℓ, C.

Then the minimal QDS is Markov.

As shown in [26] the domain of G^2 can be replaced by a linear manifold D which is dense in $\mathfrak{h}$, is a core for $C^{1/2}$, is invariant under the operators P_t of the contraction semigroup generated by G, and enjoys the properties:

$$R(\lambda; G)(D) \subseteq D(C^{1/2}), \qquad L_\ell\left(R(\lambda; G)(D)\right) \subseteq D(C^{1/2})$$

where $R(\lambda; G)$ ($\lambda > 0$) are the resolvent operators. Moreover the inequality (10) must be satisfied for all $u \in R(\lambda; G)(D)$.

We finish this section giving a useful criterion to characterize the domain of the generator, assuming two hypothesis: (**H-min**) of section 3 and (**H-Markov**) that the minimal QDS is Markov. We recall that, as in section 3, D is a core for G. Moreover, $\mathcal{V}$ the linear manifold generated by the rank-one operators $|u\rangle\langle v|$, where $u, v \in D$, is a core for $\mathcal{L}_*$.

Lemma 3.1. *Under the above hypotheses the domain of $\mathcal{L}$ is given by all the elements $X \in \mathcal{B}(\mathfrak{h})$ for which the application $(v, u) \mapsto \pounds(X)(v, u)$ is norm–continuous in the product Hilbert space.*

Proof. We remark that $X \in D(\mathcal{L})$ if and only if the linear form $\rho \mapsto \mathrm{tr}(\mathcal{L}_*(\rho)X)$, defined on $D(\mathcal{L}_*)$, is continuous for the norm $\|\cdot\|_1$ of $\mathfrak{I}_1(\mathfrak{h})$, since $\mathcal{L} = (\mathcal{L}_*)^*$. So that the essential of the proof consists in establishing the equivalence of the above property with the continuity of $(v, u) \mapsto \pounds(X)(v, u)$ as stated. Moreover, the reader will agree that the latter is a necessary condition for X being an element of $D(\mathcal{L})$, so that it remains to prove the sufficiency.

Indeed, if $\mathcal{L}(X)$ is bounded then it is represented by $Y = \mathcal{L}(X) \in \pounds(\mathfrak{h})$. Then for any $\rho \in \mathcal{V}$, the computation of $\mathrm{tr}(()\mathcal{L}_*(\rho)X)$ yields

$$|\mathrm{tr}(\mathcal{L}_*(\rho)X)| = |\mathrm{tr}(\rho Y)| \leq \|\rho\|_1 \|Y\|.$$

The proof is then completed by a standard argument based on the core property of $\mathcal{V}$.

4 The Existence of Stationary States

This section is aimed at finding a criterion for the existence of stationary states for a quantum Markov semigroup whose generator is unbounded and known as a sesquilinear form. We will be concerned with the case $\mathfrak{M} = \pounds(\mathfrak{h})$ again.

4.1 A General Result

We begin by introducing a notation for truncated operators in $\pounds(\mathfrak{h})$.

Definition 4.1. *For each self-adjoint operator Y, bounded from below, we denote by $Y \wedge r$ the truncated operator*

$$Y \wedge r = Y E_r + r E_r^\perp \tag{11}$$

where E_r denotes the spectral projection of Y associated with the interval $]-\infty, r]$.

We are now in a position to prove our first result on the existence of normal stationary states.

Theorem 4.1. *Let $\mathcal{T}$ be a quantum Markov semigroup. Suppose that there exist two self-adjoint operators X and Y with X positive and Y bounded from below and with finite dimensional spectral projections associated with bounded intervals such that*

$$\int_0^t \langle u, \mathcal{T}_s(Y \wedge r)u \rangle ds \leq \langle u, Xu \rangle \tag{12}$$

for all $t, r \geq 0$ and all $u \in D(X)$. Then the QMS $\mathcal{T}$ has a normal stationary state.

Proof. Let $-b$ (with $b > 0$) be a lower bound for Y. Note that, for each $r \geq 0$ we have

$$Y \wedge r \geq -b E_r + r E_r^\perp = -(b+r)E_r + r\mathbf{1}$$

so that (12) yields

$$-(b+r)\int_0^t \langle u, \mathcal{T}_s(E_r)u \rangle ds + rt\|u\|^2 \leq \langle u, Xu \rangle$$

for all $u \in D(X)$. Normalize u and denote by $|u\rangle\langle u|$ the pure state with unit vector u. Dividing by $t(b+r)$, for all $t, r > 0$ we have then

$$\frac{1}{t}\int_0^t \mathrm{tr}(\mathcal{T}_{*s}(|u\rangle\langle u|)E_r)ds \geq \frac{r}{b+r} - \frac{\langle u, Xu \rangle}{t(b+r)}.$$

It follows that the family of states

$$\frac{1}{t}\int_0^t \mathcal{T}_{*s}(|u\rangle\langle u|)\,ds, \qquad t > 0$$

is tight. The conclusion follows then from Theorem 2.1 and Corollary 2.2.

Remark. It is worth noticing that we wrote the inequality (12) truncated (integral on $[0,t]$, and $Y \wedge r$) to cope with two difficulties: the divergence of the integral and the unboundedness of Y. Defining appropriately the supremum of a family of self-adjoint operators and then the potential $\mathcal{U}$ for positive self-adjoint operators, the formula (12) can be written as

$$\mathcal{U}(Y) \leq X.$$

This also throws light on the classical potential-theoretic meaning of our condition which is currently under investigation.

In the applications, however, the inequality (12) is hard to verify since very frequently the QMS is not explicitly given. Therefore we shall look for conditions involving the infinitesimal generator. To this end we introduce now the class of QMS with possibly unbounded generators that concerns our research. This is sufficiently general to cover a wide class of applications.

4.2 Conditions on the Generator

Here we use the notations and hypotheses of the previous chapter yielding to the construction of the minimal quantum dynamical semigroup associated to a given form-generator.

Definition 4.2. *Given two selfadjoint operators X, Y, with X positive and Y bounded from below, we write $\pounds(X) \leq -Y$ on D, whenever*

$$\langle Gu, Xu \rangle + \sum_{\ell=1}^{\infty} \langle X^{1/2}L_\ell u, X^{1/2}L_\ell u \rangle + \langle Xu, Gu \rangle \leq -\langle u, Yu \rangle, \qquad (13)$$

for all u in a linear manifold D dense in $\mathfrak{h}$, contained in the domains of G, X and Y, which is a core for X and G, such that $L_\ell(D) \subseteq D(X^{1/2})$, $(\ell \geq 1)$.

Theorem 4.2. *Assume that the hypothesis (**H-min**) of the previous chapter holds and that the minimal QDS associated with G, $(L_\ell)_{\ell \geq 1}$ is Markov. Suppose that there exist two self-adjoint operators X and Y, with X positive and Y bounded from below and with finite dimensional spectral projections associated with bounded intervals, such that*

(i) $\pounds(X) \leq -Y$ on D;
(ii) G is relatively bounded with respect to X;
(iii)$L_\ell(n + X)^{-1}(D) \subseteq D(X^{1/2})$, $(n, \ell \geq 1)$.

Then the minimal quantum dynamical semigroup associated with G, $(L_\ell)_{\ell \geq 1}$ has a stationary state.

It is worth noticing that the above sufficient conditions always hold for a finite dimensional space $\mathfrak{h}$. Indeed, by the hypothesis (**H-min**), it suffices to take $X = 1$, $Y = 0$ and $D = h$.

We begin the proof by building up approximations $\mathcal{T}^{(n)}$ of $\mathcal{T}^{(\min)}$.

Lemma 4.1. *Under the hypotheses of Theorem 4.2, for all integer $n \geq 1$ the operators $G^{(n)}$ and $L_\ell^{(n)}$ with domain D defined by*

$$G^{(n)} = nG(n + X)^{-1}, \quad L_\ell^{(n)} = nL_\ell(n + X)^{-1},$$

admit a unique bounded extension. The operator on $\mathfrak{L}(\mathfrak{h})$ defined by

$$\mathcal{L}^{(n)}(x) = G^{(n)*}x + \sum_{\ell} L_{\ell}^{(n)*}xL_{\ell}^{(n)} + xG^{(n)} \tag{14}$$

($n \geq 1$) generates a uniformly continuous quantum dynamical semigroup $\mathcal{T}^{(n)}$.

Proof. First notice that $G^{(n)}$ and the $L_{\ell}^{(n)}$'s are bounded. Indeed, by the hypothesis (ii), the resolvent $(n + X)^{-1}$ maps $\mathfrak{h}$ into the domain of the operators G and L_{ℓ}, therefore, $G^{(n)}$ and $L_{\ell}^{(n)}$ are everywhere defined. Moreover, since G is relatively bounded with respect to X, there exist two constants $c_1,\ c_2 > 0$ such that, for each $u \in \mathfrak{h}$ we have

$$\|nG(n + X)^{-1}u\| \leq c_1\|nX(n + X)^{-1}u\| + c_2\|n(n + X)^{-1}u\|.$$

By well known properties of the Yosida approximation the right hand side is bounded by $(nc_1 + c_2)\|u\|$.

On the other hand, by (**H-min**), for each $u \in \mathfrak{h}$ we also have

$$\sum_{\ell=1}^{\infty} \|nL_{\ell}(n + X)^{-1}u\|^2 = -2\mathrm{Re}\langle n(n + X)^{-1}u, G^{(n)}u\rangle \leq 2(nc_1 + c_2)\|u\|^2.$$

Thus the $L_{\ell}^{(n)}$'s are bounded. Moreover, replacing u, v in condition (**H-min**) by $n(n + X)^{-1}u,\ u \in \mathfrak{h}$, leads to

$$\langle u, \mathcal{L}^{(n)}(\mathbf{1})u\rangle = 2\mathrm{Re}\langle u, G^{(n)}u\rangle + \sum_{\ell=1}^{\infty} \|L_{\ell}^{(n)}u\|^2 = 0.$$

It follows that the sum $\sum_{\ell=1}^{\infty} L_{\ell}^{(n)*}L_{\ell}^{(n)}$ converges strongly. Therefore by Lindblad's theorem, the equation (14) defines the generator of a uniformly continuous quantum Markov semigroup.

We recall the following well-known result on the convergence of semigroups

Proposition 4.1. *Let A, $A^{(n)}$ ($n \geq 1$) be infinitesimal generators of strongly continuous contraction semigroups $(T_t)_{t \geq 0}$, $(T_t^{(n)})_{t \geq 0}$ on a Banach space and let D_0 be a core for A. Suppose that each element x of D_0 belongs to the domain of $A^{(n)}$ for n big enough and the sequence $(A^{(n)}x)_{n \geq 1}$ converges strongly to Ax. Then the operators $T_t^{(n)}$ converge strongly to T_t uniformly for t in bounded intervals.*

We refer to [41] Th. 1.5 p.429, Th. 2.16 p. 504 for the proof.
We shall need also the following elementary lemma.

Lemma 4.2. *Let $(r_\ell)_{\ell \geq 1}$, $(s_\ell^{(n)})_{\ell \geq 1}$ ($n \geq 1$) be square-summable sequences of positive real numbers. Suppose that, for every $\ell \geq 1$, $s_\ell^{(n)}$ is an infinitesimum as n tends to infinity and that there exists a positive constant c such that*

$$\sum_{\ell \geq 1} \left(s_\ell^{(n)} \right)^2 \leq c$$

for every $n \geq 1$. Then

$$\lim_{n \to \infty} \sum_{\ell \geq 1} r_\ell s_\ell^{(n)} = 0.$$

Proof. Suppose that our conclusion is false. Then, by extracting a subsequence (in n) if necessary, we find an $\varepsilon > 0$ such that, for every n,

$$\sum_{\ell \geq 1} r_\ell s_\ell^{(n)} > \varepsilon. \tag{15}$$

The sequences $s^{(n)}$ may be understood as vectors in $l^2(\mathbb{N})$ uniformly bounded in norm by c. Therefore we can extract a subsequence $(n_m)_{m \geq 1}$ such that $(s^{(n_m)})_{m \geq 1}$ converges weakly as m tends to infinity. Since $s_\ell^{(n)}$ is an infinitesimum as n tends to infinity for each $\ell \geq 1$, it follows that the weak limit must be the vector 0. This contradicts (15).

Lemma 4.3. *Let $G^{(n)}$, $L_\ell^{(n)}$ the operators on $\mathfrak{h}$ defined in Lemma 4.1. Then, under the hypotheses of Theorem 4.2, for all $u \in D(X)$, we have*

$$\lim_{n \to \infty} G^{(n)} u = Gu, \qquad \lim_{n \to \infty} L_\ell^{(n)} u = L_\ell u.$$

Moreover the operators $\mathcal{T}_{*t}^{(n)}$ on $\mathcal{I}_1(h)$ converge strongly, as n tends to infinity, to $\mathcal{T}_{*t}$, uniformly for t in bounded intervals.

Proof. For all $u \in D(X)$, we have

$$\left\| G^{(n)} u - Gu \right\| = \left\| G(n(n+X)^{-1} - 1)u \right\|$$
$$\leq c_1 \left\| (n(n+X)^{-1} - 1)Xu \right\| + c_2 \left\| (n(n+X)^{-1} - 1)u \right\|.$$

Therefore the sequence $\left(G^{(n)} u \right)_{n \geq 1}$ converges strongly to Gu as n tends to infinity by well-known properties of Yosida approximations. Moreover, by (**H-min**), for $u \in D(X)$, we have also

$$\sum_{\ell=1}^{\infty} \left\| L_\ell^{(n)} u - L_\ell u \right\|^2 = -2\mathrm{Re} \left\langle (n(n+X)^{-1} - 1)u, G(n(n+X)^{-1} - 1)u \right\rangle. \tag{16}$$

This shows the convergence of sequences $\left(L_\ell^{(n)} u \right)_{n \geq 1}$ to $L_\ell u$ for all $\ell \geq 1$.

For all $u, v \in D(X)$ we have

$$\mathcal{L}_*^{(n)}(|u\rangle\langle v|) - \mathcal{L}_*^{(\min)}(|u\rangle\langle v|) = |(G^{(n)} - G)u\rangle\langle v| + |u\rangle\langle(G^{(n)} - G)v|$$
$$+ \sum_{\ell=1}^{\infty} |(L_\ell^{(n)} - L_\ell)u\rangle\langle v|$$
$$+ \sum_{\ell=1}^{\infty} |u\rangle\langle(L_\ell^{(n)} - L_\ell)v|.$$

Therefore the trace norm of $\mathcal{L}_*^{(n)}(|u\rangle\langle v|) - \mathcal{L}_*^{(\min)}(|u\rangle\langle v|)$ can be estimated by

$$\|v\| \cdot \|(G^{(n)} - G)u\| + \|u\| \cdot \|(G^{(n)} - G)v\|$$

$$+ \sum_{\ell=1}^{\infty} \|L_\ell u\| \cdot \|(L_\ell^{(n)} - L_\ell)v\| + \sum_{\ell=1}^{\infty} \|L_\ell v\| \cdot \|(L_\ell^{(n)} - L_\ell)u\|.$$

Clearly the first two terms vanish as n tends to infinity. Moreover, by the inequality (16), since the operators $X(n + X)^{-1}$ are contractive, we have

$$\sum_{\ell=1}^{\infty} \left\| L_\ell^{(n)} u - L_\ell u \right\|^2 \leq 2\|(n + X)^{-1} X u\| \cdot \|(G^{(n)} - G)u\|$$

$$\leq 2\|u\| \left(c_1 \left\| (n(n + X)^{-1} - 1)Xu \right\| + c_2 \left\| (n(n + X)^{-1} - 1)u \right\| \right)$$
$$= 2\|u\| \left(c_1 \left\| (X(n + X)^{-1})Xu \right\| + c_2 \left\| (X(n + X)^{-1})u \right\| \right)$$
$$\leq 2\|u\| \left(c_1 \|Xu\| + c_2 \|u\| \right).$$

An application of Lemma 4.2 shows then that the trace norm of $\mathcal{L}_*^{(n)}(|u\rangle\langle v|) - \mathcal{L}_*^{(\min)}(|u\rangle\langle v|)$ converges to 0 as n tends to infinity.

Since the minimal QDS associated with G, $(L_\ell)_{\ell \geq 1}$ is Markov and $D(X)$ is a core for G (it contains D), the linear manifold generated by $|u\rangle\langle v|$ with $u \in D(X)$ is a core for $\mathcal{L}_*^{(\min)}$. The conclusion follows then from Proposition 4.1.

Lemma 4.4. *Let* $Y \wedge r$ *the operator defined by (11) and let* $X^{(n)} = nX(n + X)^{-1}$, *$(n \geq 1)$. Define* $Y_r^{(n)} = n^2(n+X)^{-1}(Y \wedge r)(n+X)^{-1}$. *Then, under the hypotheses of Theorem 4.2, the operator*

$$X^{(n)} - \int_0^t \mathcal{T}_s^{(n)}(Y_r^{(n)})ds,$$

is positive for each $t \geq 0$.

Proof. Notice that $Y_r \leq Y$. Therefore, by the hypothesis (i) of Theorem 4.2, we have the inequality

$$\langle Gu, Xu \rangle + \sum_{\ell=1}^{\infty} \langle X^{1/2} L_\ell u, X^{1/2} L_\ell u \rangle + \langle Xu, Gu \rangle \leq -\langle u, (Y \wedge r)u \rangle, \quad (17)$$

for all $u \in D$.

The domain D being a core for X and G being relatively bounded with respect to X, for every $u \in D(X)$ we can find a sequence $(u_n)_{n \geq 1}$ in D such that $(Xu_n)_{n \geq 1}$ converges to Xu and $(Gu_n)_{n \geq 1}$ converges to Gu. Then the convergence of $(L_\ell u_n)_{n \geq 1}$ to $L_\ell u$ (for all $\ell \geq 1$) follows readily from the hypothesis (**H-min**). Moreover, for every $n, m \geq 1$, the inequality (17) yields

$$\sum_{\ell=1}^{\infty} \|X^{1/2} L_\ell (u_n - u_m)\|^2 \leq -2\operatorname{Re}\langle G(u_n - u_m), X(u_n - u_m)\rangle$$

$$- \langle (u_n - u_m), (Y \wedge r)(u_n - u_m)\rangle.$$

Therefore, replacing u by u_n, and letting n tend to infinity we show that (17) holds for all $u \in D(X)$.

Since $n(n + X)^{-1}$ is a contraction and $Y_r \le Y$, under the hypotheses of Theorem 4.2, for all $u \in \mathfrak{h}$ we have

$$\langle u, \mathcal{L}^{(n)}(X^{(n)})u \rangle \le 2\text{Re}\langle G^{(n)}u, X^{(n)}u \rangle + \sum_{\ell=1}^{\infty} \langle X^{1/2}L_\ell^{(n)}u, X^{1/2}L_\ell^{(n)}u \rangle$$
$$\le -\langle n(n+X)^{-1}u, Y_r n(n+X)^{-1}u \rangle$$
$$= -\langle u, Y_r^{(n)}u \rangle.$$

It follows then $\mathcal{L}^{(n)}(X^{(n)}) \le -Y_r^{(n)}$.

Now, notice that

$$\frac{d}{dt}\left(X^{(n)} - T_t^{(n)}(X^{(n)}) - \int_0^t T_s^{(n)}(Y_r^{(n)})ds \right)$$
$$= -T_t^{(n)}\left(\mathcal{L}^{(n)}(X^{(n)}) + Y_r^{(n)} \right)$$
$$\ge 0.$$

Therefore,

$$X^{(n)} - \int_0^t T_s^{(n)}(Y_r^{(n)})ds \ge T_t^{(n)}(X^{(n)}) \ge 0$$

for all $t \ge 0$.

Proof. *(of Theorem 4.2).* By Lemma 4.4 for each $u \in D(X), t, r \ge 0$ and $n \ge 1$, we have

$$\int_0^t \text{tr}(T_{*s}^{(n)}(|u\rangle\langle u|)Y_r^{(n)})ds \le \langle u, X^{(n)}u \rangle.$$

The sequence $(Y_r^{(n)})_{n\ge 1}$ converges strongly to $Y \wedge r$. Thus, by Lemma 4.3, we can let n tend to infinity to obtain

$$\int_0^t \text{tr}(T_{*s}(|u\rangle\langle u|)(Y \wedge r))ds \le \langle u, Xu \rangle.$$

This inequality coincides with (12). Therefore Theorem 4.2 follows from Theorem 4.1.

4.3 Examples

4.4 A Multimode Dicke Laser Model

We follow Alli and Sewell [4] where a model is proposed for a Dicke laser or maser. We begin by establishing the corresponding notations.

The system consists of N identical two-level atoms coupled with a radiation field corresponding to n modes. Therefore, one can choose the Hilbert space $\mathfrak{h}$ which consists of the tensor product of N copies of $\mathbb{C}^2$ and n copies of $l^2(\mathbb{N})$. To simplify notations we simply identify any operator acting on a factor of the above tensor product with its canonical extension to $\mathfrak{h}$.

Let σ_1, σ_2, σ_3 be the Pauli matrices and define the spin raising and lowering operators $\sigma_\pm = (\sigma_1 \pm i\sigma_2)/2$. The atoms are located on the sites $r = 1, \ldots, N$ of a one dimensional lattice, so that we denote by $\sigma_{\epsilon,r}$ ($\epsilon = 1, 2, 3, +, -$) the spin component of the atom at the site r. The free evolution of the atoms is described by a generator $\mathcal{L}_{\mathrm{mat}}$ which is bounded and given in Lindblad form as

$$\mathcal{L}_{\mathrm{mat}}(x) = i[H, x] - \frac{1}{2} \sum_j (V_j^* V_j x - 2V_j^* x V_j + x V_j^* V_j), \tag{18}$$

where the sum contains a finite number of elements, H is bounded self-adjoint and the V_j's are bounded operators.

Moreover, we denote by a_j^*, a_j, the creation and annihilation operators corresponding to the j-th mode of the radiation, ($j = 1, \ldots, n$). These operators satisfy the canonical commutation relations:

$$[a_j, a_k^*] = \delta_{jk}\mathbf{1}, \quad [a_j, a_k] = 0.$$

The free evolution of the radiation is given by the formal generator

$$\mathcal{L}_{\mathrm{rad}}(x) = \sum_{\ell=1}^{n} \left(\kappa_\ell(-a_\ell^* a_\ell x + 2a_\ell^* x a_\ell - x a_\ell^* a_\ell) + i\omega_\ell[a_\ell^* a_\ell, x]\right), \tag{19}$$

where $\kappa_\ell > 0$ are the damping and $\omega_\ell \in \mathbb{R}$ are the frequencies corresponding to the ℓ-th mode of the radiation.

The coupling between the matter and the radiation corresponds to a Hamiltonian interaction of the form:

$$H_{\mathrm{int}} = \frac{i}{N^{1/2}} \sum_{r=1}^{N} \sum_{\ell=1}^{n} \lambda_\ell(\sigma_{-,r} a_\ell^* e^{-2\pi i k_\ell r} - \sigma_{+,r} a_\ell e^{2\pi i k_\ell r}), \tag{20}$$

where k_ℓ is the wave number of the ℓ-th mode and the λ's are real valued, N independent coupling constants.

With the above notations, the formal generator of the whole dynamics is given by

$$\mathcal{L}(x) = \mathcal{L}_{\mathrm{mat}}(x) + \mathcal{L}_{\mathrm{rad}}(x) + i[H_{\mathrm{int}}, x] \tag{21}$$

To identify L_ℓ and G in our notations, we use in force the convention on the abridged version of tensor products with the identity. That is, here we find

$$L_\ell = \sqrt{\kappa_\ell}\, a_\ell, \quad (\ell = 1, \ldots, n) \tag{22}$$

All the remaining L_ℓ's are bounded operators. Among them a finite number (indeed at most $3N$) coincides with some of the V_j's appearing in (18) and the other vanish.

So that the operator G becomes formally:

$$G = -\frac{1}{2}\sum_\ell L_\ell^* L_\ell - i\sum_\ell \omega_\ell a_\ell^\dagger a_\ell - iH - iH_{\text{int}}, \tag{23}$$

where the sum contains only a finite number of non zero terms.

To make the above expression rigorous some preliminary work is needed. Call $(f_m)_{m\geq 0}$ the canonical orthonormal basis on the space $l^2(\mathbb{N})$. In the radiation space, which consists of the tensor product of n copies of $l^2(\mathbb{N})$, we denote

$$f_\alpha = f_{\alpha_1}^{(1)} \otimes \ldots \otimes f_{\alpha_n}^{(n)},$$

where $\alpha = (\alpha_1,\ldots,\alpha_n)$ and $f_{\alpha_\ell}^{(\ell)}$ is an element of the canonical basis of the ℓ copy of $l^2(\mathbb{N})$. Thus, $(f_\alpha)_{\alpha\in\mathbb{N}^n}$ is the canonical orthonormal basis of the radiation space.

With these notations we have

$$a_\ell^* f_\alpha = \sqrt{\alpha_\ell+1} f_{\alpha+1_\ell}, \quad a_\ell f_\alpha = \begin{cases} \sqrt{\alpha_\ell} f_{\alpha-1_\ell} & \text{if } \alpha_\ell > 0 \\ 0 & \text{if } \alpha_\ell = 0 \end{cases}, \tag{24}$$

where 1_ℓ is the vector with a 1 at the ℓth coordinate and zero elsewhere.

Thus, the operator G is well defined over vectors of the form uf_α where $u \in \mathbb{C}^{2N}$ and the symbol of tensor product is dropped.

It is well known (see [41], Thm. 2.7 p.499) that a perturbation of a negative selfadjoint operator, relatively bounded with relative bound less than 1, is the infinitesimal generator of a contraction semigroup. Therefore, we choose X formally given by $X = \sum_{\ell=1}^n a_\ell^* a_\ell$. That is, $Xuf_\alpha = |\alpha|uf_\alpha$, where, $|\alpha| = \alpha_1 + \ldots + \alpha_n$.

Since $X^k uf_\alpha = |\alpha|^k uf_\alpha$ it follows that the linear span of vectors of the form uf_α is a dense subset of the analytic vectors for X. Therefore, by a theorem of Nelson (see e.g. [53]), X is essentially self-adjoint on the referred domain. From now on we identify X with its closure which is selfadjoint.

We show now that H_{int} is relatively bounded with respect to X. Let ξ be a finite linear combination of elements of the form uf_α. By Schwarz' inequality, and elementary inequalities like $\sqrt{t+s} \leq \sqrt{t} + \sqrt{s} \leq \sqrt{2(t+s)}, 2\sqrt{ts} \leq \epsilon t + \epsilon^{-1}s$, we obtain

$$\|H_{\text{int}}\xi\| \leq \frac{1}{N^{1/2}}\sum_{r=1}^N\sum_{\ell=1}^n |\lambda_\ell|(\|a_\ell^*\xi\| + \|a_\ell\xi\|)$$

$$\leq \frac{1}{N^{1/2}}\sum_{r=1}^N\sum_{\ell=1}^n |\lambda_\ell|\left(4\langle\xi, a_\ell^* a_\ell\xi\rangle + 2\|\xi\|^2\right)^{1/2}$$

$$\leq \frac{1}{N^{1/2}}\left[\sum_{r=1}^N\sum_{\ell=1}^n 2(|\lambda_\ell|^2\langle\xi, a_\ell^* a_\ell\xi\rangle)^{1/2} + \sqrt{2}\sum_{r=1}^N\sum_{\ell=1}^n |\lambda_\ell|\|\xi\|\right]$$

$$\leq \frac{1}{N^{1/2}}\left[\sum_{r=1}^N\sum_{\ell=1}^n (\epsilon\|a_\ell^* a_\ell\xi\| + \epsilon^{-1}|\lambda_\ell|^2\|\xi\|) + \sqrt{2}\sum_{r=1}^N\sum_{\ell=1}^n |\lambda_\ell|\|\xi\|\right]$$

Finally, by the elementary inequality $\sum_{\ell=1}^{n} \|s_\ell\| \leq \sqrt{n} \|\sum_{\ell=1}^{n} s_\ell\|$, it follows

$$\|H_{\text{int}}\xi\| \leq \frac{n^{1/2}\epsilon}{N^{1/2}} \sum_{r=1}^{N} \|\sum_{\ell=1}^{n} a_\ell^* a_\ell \xi\| + \frac{1}{N^{1/2}} \sum_{r=1}^{N} \sum_{\ell=1}^{n} (\sqrt{2} + \epsilon^{-1}|\lambda_\ell|)|\lambda_\ell|\|\xi\|$$

$$\leq \epsilon N^{1/2} n^{1/2} \|X\xi\| + \frac{1}{N^{1/2}} \sum_{r=1}^{N} \sum_{\ell=1}^{n} (\sqrt{2} + \epsilon^{-1}|\lambda_\ell|)|\lambda_\ell|\|\xi\|,$$

thus, choosing $\epsilon < (Nn)^{-1/2}$, the above inequality yields the required relative boundedness of H_{int} with respect to X.

As a result, the operator G appears as a dissipative perturbation of $-\frac{1}{2}X$, relatively bounded with respect to X, with bound strictly less than 1. Therefore, G is the generator of a contraction semigroup. Moreover, the domain of G coincides with that of X and hypothesis (**H-min**) easily checked.

To apply our main result, we fix the domain D as the space of vectors ξ which are finite linear combinations of the form uf_α. Notice that this is an invariant for X, G, and all the L_ℓ's. To identify an appropriate operator Y to have $\mathfrak{L}(X) \leq -Y$, we first perform the computation of $\mathfrak{L}(X)$. For the sake of clarity, we avoid handling forms in the computations below. However, the reader may easily notice that all the expressions are well defined since the domain D is invariant under the action of the operators X, G and L_ℓ.

Firstly, it holds $\mathcal{L}_{\text{mat}}(X) = 0$, since the V_j's act on the tensor product of N copies of $\mathbb{C}^2$ and leave the domain D invariant. Secondly, a straightforward computation using the canonical commutation relations, yields

$$\mathcal{L}_{\text{rad}}(X) = -2 \sum_{\ell=1}^{n} \kappa_\ell a_\ell^\dagger a_\ell.$$

Another easy computation yields

$$i[H_{\text{int}}, X] = -iN^{-1/2} \sum_{r=1}^{N} \sum_{\ell=1}^{n} \lambda_\ell (\sigma_{-,r} a_\ell^\dagger e^{-2\pi i k_\ell r} + \sigma_{+,r} a_\ell e^{2\pi i k_\ell r}).$$

Summing up,

$$\mathfrak{L}(X) = -2 \sum_{\ell=1}^{n} \kappa_\ell a_\ell^\dagger a_\ell - \frac{i}{N^{1/2}} \sum_{r=1}^{N} \sum_{\ell=1}^{n} \lambda_\ell (\sigma_{-,r} a_\ell^\dagger e^{-2\pi i k_\ell r} + \sigma_{+,r} a_\ell e^{2\pi i k_\ell r}).$$

To identify Y it suffices to control the term $i[H_{\text{int}}, X]$. For each $\xi \in D$, it follows

$$|\langle \xi, i[H_{\text{int}}, X]\xi\rangle| = \frac{1}{N^{1/2}} \left| \sum_{r=1}^{N} \sum_{\ell=1}^{n} \lambda_\ell \langle \xi, (\sigma_{-,r}a_\ell e^{-2\pi i k_\ell r} + \sigma_{+,r}a_\ell^* e^{2\pi i k_\ell r}\xi)\rangle \right|$$

$$\leq \frac{1}{2N^{1/2}} \sum_{r=1}^{N} \sum_{\ell=1}^{n} 2|\lambda_\ell|\|\xi\|(\|a_\ell\xi\| + \|a_\ell^*\xi\|)$$

$$\leq \frac{\epsilon}{2N^{1/2}} \sum_{r=1}^{N} \sum_{\ell=1}^{n} (\langle \xi, a_\ell^* a_\ell \xi\rangle + \langle \xi, a_\ell a_\ell^* \xi\rangle)$$

$$+ \frac{1}{2\epsilon N^{1/2}} \sum_{r=1}^{N} \sum_{\ell=1}^{n} |\lambda_\ell|^2\|\xi\|^2$$

$$\leq \epsilon N^{1/2}\langle \xi, X\xi\rangle + \frac{\epsilon N^{1/2}n}{2}\|\xi\|^2 + \frac{1}{2\epsilon N^{1/2}} \sum_{r=1}^{N} \sum_{\ell=1}^{n} |\lambda_\ell|^2\|\xi\|^2.$$

So that choosing $0 < \epsilon < 2N^{-1/2} \min \kappa_\ell$ the required operator Y may be taken as

$$Y = (2\min \kappa_\ell - \epsilon N^{1/2})(X + c),$$

where $c > 0$ is a suitable constant.

The spectrum of X coincides with $\mathbb{N}$. For each $m \in \mathbb{N}$, the corresponding eigenspace is generated by the f_α with $|\alpha| = m$. Therefore, it follows that all spectral projections of X and Y associated with bounded intervals are finite dimensional.

Similar arguments allow to check the hypotheses of Proposition 3.2 (with $C = X$) to show that the minimal QDS associated to the operators G and L_ℓ, $(1 \leq \ell \leq n)$, is Markov (this was also proved in [4] by another method).

To summarize, our main theorem implies the following

Corollary 4.1. *There exists an invariant state for the multimode Dicke model.*

4.5 A Quantum Model of Absorption and Stimulated Emission

This example corresponds to a family of models introduced by Gisin and Percival in [37]. The framework is given by the Hilbert space $\mathfrak{h} = l^2(\mathbb{N})$ where, as usual, we call $(e_n)_{n\in\mathbb{N}}$ the canonical orthonormal basis. The operators defining the form-generator $\mathfrak{L}(\cdot)$ are

$$L_1 = \nu a^\dagger a, \ L_2 = \mu a, \ H = \xi(a^\dagger + a),$$

where $\mu, \nu > 0$ and $\xi \in \mathbb{R}$.

Thus,

$$G = -i\xi(a^\dagger + a) - \frac{1}{2}\left(\nu^2(a^\dagger a)^2 + \mu^2 a^\dagger a\right),$$

Let us check the existence of an invariant state by means of Theorem 4.2. Take, for instance $X = (N+1)^2$, where $N = a^\dagger a$ is the number operator. A straightforward computation yields

$$\mathfrak{L}(X) = i\xi\left((a - a^\dagger)(N+1) + (N+1)(a - a^\dagger)\right) + \mu^2 N(1 - 2N). \quad (25)$$

We first study the term $i\xi\left((a - a^\dagger)(N + 1) + (N + 1)(a - a^\dagger)\right)$, which corresponds to $i[H, X]$.

Call D the linear manifold generated by $(e_n)_{n \in \mathbb{N}}$. For any $u \in D$, we have

$$|\langle u, i[H, X] \rangle| = |\xi| \langle (a - a^\dagger)u, (N + 1)u \rangle$$

$$\leq \frac{1}{\epsilon} \left\| (a - a^\dagger)u \right\| \epsilon \left\| (N + 1)u \right\|$$

$$\leq \frac{2}{\epsilon} \left\| N^{1/2}u \right\| \epsilon \left\| (N + 1)u \right\|$$

$$\leq \frac{2}{\epsilon^2} \langle u, Nu \rangle + \frac{\epsilon^2}{2} \langle u, (N + 1)^2 \rangle.$$

Thus,

$$\mathcal{L}(X) \leq -2\mu^2 N^2 + \mu^2 N + \frac{\epsilon^2}{2}(N + 1)^2 + \frac{2}{\epsilon^2} N.$$

Finally, if we call

$$Y = \left(2\mu^2 - \frac{\epsilon^2}{2}\right) N^2 - \left(\mu^2 + \epsilon^2 + \frac{2}{\epsilon^2}\right) N - \frac{\epsilon^2}{2}\mathbf{1},$$

we obtain

$$\mathcal{L}(X) \leq -Y.$$

In a similar way one may verify that the quantum dynamical semigroup is Markov too. To summarize,

Corollary 4.2. *The quantum semigroup which corresponds to the model of absorption and stimulated emission here before is Markov and has a stationary state.*

4.6 The Jaynes-Cummings Model

We follow our article [27] to introduce the Jaynes-Cummings model in Quantum Optics. To this aim we use the same space $\mathfrak{h} = l^2(\mathbb{N})$, since here $n = 1$, we drop the index ℓ from the notations of creation and annihilation operators and S denotes the right-shift operator.

In this framework the formal generator is given by

$$\mathcal{L}(x) = -\frac{\mu^2}{2}(a^*ax - 2a^*xa + xa^*a) - \frac{\lambda^2}{2}(aa^*x - 2axa^* + xaa^*)$$
$$+ R^2(\cos(\phi\sqrt{aa^*})x\cos(\phi\sqrt{aa^*}) - x) + R^2\sin(\phi\sqrt{aa^*})S^*xS\sin(\phi\sqrt{aa^*}),$$

where λ, μ, R and ϕ are positive constants. In [27] the rigorous construction of the minimal QDS was done showing also that it is identity preserving.

The above Jaynes-Cummings generator has a faithful invariant state if and only if $\mu^2 > \lambda^2$. This state can be computed explicitly. The interested reader is referred to [27].

To check conditions of Theorem 4.2, one can take $X = a^*a$, the value of $\mathcal{L}(X)$ becoming

$$\mathcal{L}(X) = -(\mu^2 - \lambda^2)a^*a + R^2\sin^2(\phi\sqrt{aa^*}).$$

Thus, it suffices to take $Y = (\mu^2 - \lambda^2)a^*a - R^2$ to prove the existence of a stationary state via our main result.

To prove the necessity, one can follow the argument explained in [27] inspired from classical probability.

5 Faithful Stationary States and Irreducibility

Consider a probability space $(\Omega, \mathcal{F}, \mathbf{P})$, a measurable space $(E, \mathcal{E})$ and a Markov process

$$(\Omega, \mathcal{F}, \mathbf{P}, (\mathcal{F}_t)_{t\in\mathbb{R}^+}, E, \mathcal{E}, (X_t)_{t\in}),$$

defined on $(\Omega, \mathcal{F}, \mathbf{P})$ with states in E. The Markov semigroup generated by this process is given by $T_t f(x) = \mathbb{E}(f(X_t)|X_0 = x)$, for all $t \geq 0$, $x \in E$ and any bounded and measurable function f defined on E.

A function f of the above class is *subharmonic* (resp. *superharmonic*, resp. *harmonic*) for the given semigroup if $T_t f \geq f$ (resp. $T_t f \leq f$, $T_t f = f$) for all $t \geq 0$.

Now take a positive measure μ on $(E, \mathcal{E})$ and consider the von Neumann algebra $\mathfrak{M} = L^\infty(E, \mathcal{E}, \mu)$. A state ν over this von Neumann algebra is given by a probability measure ν absolutely continuous with respect to μ. The state ν is faithful whenever $\nu(f) = 0$, for a positive $f \in \mathfrak{M}$, implies that $f = 0$, (i.e. $f(x) = 0$, μ-almost surely).

One important question then arises: when does $(T_t)_{t\in\mathbb{R}^+}$ has a faithful stationary state in $\mathfrak{M}$?

This question appears in the non-commutative theory of Markov semigroups as well. Indeed, the existence of a faithful stationary state is a crucial hypothesis in most of the ergodic studies developed in this field (see for instance section 2 before and [35], [36], [28], [29]). In a previous paper (see [31] and section 4) we obtained sufficient conditions for the existence of a stationary state of a given quantum Markov semigroup.

5.1 The Support of an Invariant State

Definition 5.1. *Given a semifinite von Neumann algebra* $\mathfrak{M}$ *of operators over a complex separable Hilbert space* $\mathfrak{h}$, *endowed with a trace* $\mathrm{tr}(\cdot)$, *and a Quantum Markov Semigroup* $(T_t)_{t\in\mathbb{R}^+}$ *on* $\mathfrak{M}$, *a positive operator* $a \in \mathfrak{M}$ *is subharmonic (resp. superharmonic, resp. harmonic) for the semigroup if* $T_t(a) \geq a$, *(resp.* $T_t(a) \leq a$, *resp.* $T_t(a) = a$), *for all* $t \geq 0$.

Subharmonic events play a fundamental role in the Potential Theory of classical Markov semigroups. They are related to stationarity, recurrence, supermartingale properties. In our framework, we will start by showing a relation between invariant states and subharmonics projections.

Lemma 5.1. *Let p be a projection of the von Neumann algebra $\mathfrak{M}$, and $x \in \mathfrak{M}$ a positive element. If $pxp = 0$, then $p^{\perp}xp = pxp^{\perp} = 0$*

Proof.   Suppose $\mathfrak{M} \subseteq \mathcal{B}(\mathfrak{h})$ and let $u, v \in \mathfrak{h}$ with $pu = u$, $pv = 0$. Since x is positive, $\langle zu + v, x(zu + v) \rangle$ is positive for every $z \in \mathbb{C}$. Then, since $pxp = 0$, for every $z \in \mathbb{C}$, we have also

$$2\mathrm{Re}\,(z\,\langle v, xu \rangle) + \langle v, xv \rangle \geq 0.$$

Therefore $\langle v, xu \rangle$ must vanish and the conclusion readily follows.

Theorem 5.1. *The support projection of a stationary state for a quantum Markov semigroup is subharmonic.*

Proof.   Let p be the support projection of a given stationary state ρ of $(\mathcal{T}_t)_{t \in \mathbb{R}^+}$. That is, p is the projection on the closure of the rank of ρ, thus $\rho p = p\rho = \rho$, and $\mathcal{T}_{*t}(\rho) = \rho$, (for all $t \geq 0$).

Let be given an arbitrary $t \geq 0$. We first notice that $p\mathcal{T}_t(p)p \leq p$, since $p \leq 1$. Therefore,

$$\mathrm{tr}(\rho(p - p\mathcal{T}_t(p)p)) = \mathrm{tr}(\rho(p - \mathcal{T}_t(p))) = 0,$$

and, since ρ is faithful on the subalgebra $p\mathfrak{M}p$, it follows

$$p\mathcal{T}_t(p)p = p.$$

On the other hand,

$$p\mathcal{T}_t(p^{\perp})p = p\mathcal{T}_t(1)p - p\mathcal{T}_t(p)p = p - p = 0.$$

Moreover, since $\mathcal{T}_t(p^{\perp})$ is positive, the previous lemma yields

$$p\mathcal{T}_t(p^{\perp})p^{\perp} = 0 = p^{\perp}\mathcal{T}_t(p^{\perp})p.$$

To summarize,

$$\mathcal{T}_t(p) = p + p^{\perp}\mathcal{T}_t(p)p^{\perp},$$

and the projection p is subharmonic.

Proposition 5.1. *For any Quantum Markov semigroup $\mathcal{T}$ the following propositions are equivalent:*

a) *p is subharmonic for $\mathcal{T}$.*
b) *The subalgebra $p^{\perp}\mathfrak{M}p^{\perp}$ is invariant for $\mathcal{T}$.*
c) *For any normal state ρ such that $p\rho p = \rho$, it holds $\mathrm{tr}(\rho\mathcal{T}_t(p^{\perp})) = 0$, for all $t \geq 0$.*

Proof.   Assume that condition a) holds. By hypothesis, we have that $p\mathcal{T}_t(p)p = p$, thus

$$p\mathcal{T}_t(p^{\perp})p = p\mathcal{T}_t(1)p - p\mathcal{T}_t(p)p = 0.$$

Therefore, for any positive $x \in p^\perp \mathfrak{M} p^\perp$ it follows $pT_t(x)p = 0$ since

$$0 \leq pT_t(x)p \leq \|x\| \, pT_t(p^\perp)p = 0.$$

From Lemma 5.1, $pT_t(x)p^\perp = p^\perp T_t(x)p = 0$, since $T_t(x)$ is positive.

Thus $T_t(x) = p^\perp T_t(x)p^\perp \in p^\perp \mathfrak{M} p^\perp$. The same conclusion holds for any arbitrary $x \in p^\perp \mathfrak{M} p^\perp$ since all those elements may be decomposed as a linear combination of four positive elements of $p^\perp \mathfrak{M} p^\perp$.

Now we prove that b) implies c). By hypothesis, for any $x \in p^\perp \mathfrak{M} p^\perp$, it holds $T_t(x) = p^\perp y p^\perp$ for some $y \in \mathfrak{M}$. Clearly, $y = T_t(x)$ since $\langle v, T_t(x)u \rangle = \langle v, yu \rangle$ for all vectors u and v for which $u = p^\perp u$, $v = p^\perp v$.

Therefore, $T_t(x) = p^\perp T_t(x)p^\perp$. In particular,

$$T_t(p^\perp) = p^\perp T_t(p^\perp)p^\perp.$$

From this it follows that, given a state ρ such that $p\rho = \rho p = \rho$, we have $p^\perp \rho = \rho p^\perp = 0$ which yields $\mathrm{tr}(\rho T_t(p^\perp)) = 0$.

We finally prove that c) implies a). From c) we obtain that

$$\mathrm{tr}(\rho p T_t(p^\perp)p) = 0,$$

thus $pT_t(p^\perp)p = 0$. As a result, by Lemma 5.1,

$$T_t(p^\perp) = p^\perp T_t(p^\perp)p^\perp \leq p^\perp,$$

which gives $T_t(p) = T_t(1 - p^\perp) \geq p$.

Definition 5.2. *We say that a quantum Markov semigroup is* irreducible *if there is no non-trivial subharmonic projection.*

5.2 Subharmonic Projections. The Case $\mathfrak{M} = \mathfrak{L}(\mathfrak{h})$

Now we concentrate on the case of $\mathfrak{M} = \mathfrak{L}(\mathfrak{h})$. The quantum Markov semigroup is the minimal obtained from an unbounded generator given as a sesquilinear form

$$\mathfrak{L}(x)(v, u) = \langle Gv, xu \rangle + \sum_{\ell \geq 1}\langle L_\ell v, xL_\ell u \rangle + \langle v, xGu \rangle,$$

under the hypotheses of section 3, that is, (**H-min**) is supposed satisfied and the associated minimal semigroup is assumed to be Markov.

Thus, for all $x \in \mathfrak{L}(\mathfrak{h})$, the quadratic form $\mathfrak{L}(x)$ is defined over the domain $D(G) \times D(G)$.

We refer the reader to section 3 for the notations and concepts connected with the minimal quantum Markov semigroup associated to $\mathfrak{L}(\cdot)$.

In which follows, we use the same notation p for both, a closed subspace and the projection determined by this subspace.

Lemma 5.2. *Let $(P_t)_{t\in\mathbb{R}^+}$ be the semigroup generated by G. Then a closed subspace p is invariant for $(P_t)_{t\in\mathbb{R}^+}$ if and only if for any $u \in D(G) \cap Rk(p)$ it holds $Gpu = pGpu$.*

Proof. If p is invariant for the semigroup, $P_t p = pP_t p$, then it is also invariant for the resolvent:

$$R(\lambda; G)p = pR(\lambda; G)p.$$

As a result for any $u \in D(G)$ such that $pu = u$, if we define $u_\lambda = \lambda R(\lambda; G)u$, then we obtain $pu_\lambda = u_\lambda$ too. Therefore $D(G) \cap Rk(p)$ is dense in $Rk(p)$.

Moreover, for any $u \in D(G) \cap Rk(p)$ we have

$$p\frac{1}{t}\left(P_t u - u\right) = \frac{1}{t}\left(P_t pu - pu\right) = \frac{1}{t}\left(P_t u - u\right).$$

Therefore, letting $t \to \infty$, we obtain $Gpu = pGpu$.

Conversely, if $Gpu = pGpu$ for all $u \in D(G) \cap Rk(p)$, then $p^\perp Gp$ is zero on $D(G)$ and

$$\frac{d}{dt}p^\perp P_t pu = p^\perp GP_t pu = (p^\perp Gp)P_t u = 0.$$

It follows that $p^\perp P_t pu = 0$ for all $u \in D(G) \cap Rk(p)$ and $t \geq 0$, hence $p^\perp P_t p = 0$ since $D(G) \cap Rk(p)$ is dense in $Rk(p)$.

Theorem 5.2. *Under the previous hypotheses, a projection p is subharmonic for $\mathcal{T}$ if and only if the following conditions are satisfied:*

$$Gpu = pGpu, \quad L_\ell pu = pL_\ell pu, \tag{26}$$

for all $u \in D(G) \cap Rk(p)$, $\ell \geq 1$.

Proof. We start assuming that p is subharmonic, thus $\mathcal{T}_t(p) \geq p$ for all $t \geq 0$. From equation (7) we obtain

$$p^\perp \geq \mathcal{T}_t(p^\perp) \geq P_t^* p^\perp P_t.$$

Therefore, for all $u \in Rk(p)$,

$$\langle u, P_t^* p^\perp P_t u \rangle = \left\| p^\perp P_t u \right\|^2 = 0,$$

that is $p^\perp P_t p = 0$. Thus $P_t p = pP_t p$, for all $t \geq 0$. Then Lemma 5.2 implies that $Gpu = pGu$ for all $u \in D(G) \cap Rk(p)$.

In addition, the equation satisfied by the minimal quantum semigroup yields

$$\int_0^t \left(\langle Gu, \mathcal{T}_s(p^\perp)u \rangle + \sum_{\ell \geq 1}\langle L_\ell u, \mathcal{T}_s(p^\perp)L_\ell u \rangle + \langle u, \mathcal{T}_s(p^\perp)Gu \rangle \right) ds \leq 0,$$

for all $t \geq 0$ and all $u \in D(G)$. As a result, computing the derivative in 0 of the above equation, we obtain:

$$\langle Gu, p^\perp u \rangle + \sum_{\ell \geq 1} \langle L_\ell u, p^\perp L_\ell u \rangle + \langle u, p^\perp Gu \rangle \leq 0.$$

Now, if $u \in D(G) \cap Rk(p)$ the above inequality gives

$$\sum_{\ell \geq 1} \left\| p^\perp L_\ell p u \right\|^2 \leq 0,$$

that is $p^\perp L_\ell p u = 0$ or, equivalently,

$$p L_\ell p u = p L_\ell u,$$

for all $\ell \geq 1$ and $u \in D(G) \cap Rk(p)$.

Conversely, we assume condition (26). We will prove that p is subharmonic by an induction argument which relays on the sequence $(\mathcal{T}^{(n)})_{n \in \mathbb{N}}$ used in the construction of $\mathcal{T}$.

Firstly, p is subharmonic for $\mathcal{T}^{(0)}$ since

$$\mathcal{T}_t^{(0)}(p^\perp) = P_t^* p^\perp P_t = p^\perp P_t^* p^\perp P_t p^\perp \leq p^\perp.$$

Secondly, assume that p is subharmonic for $\mathcal{T}^{(n)}$, we prove that it is subharmonic for $\mathcal{T}^{(n+1)}$ too. Indeed, for all $u \in D(G) \cap Rk(p)$, the definition of $\mathcal{T}^{(n+1)}$ and the induction hypothesis yield

$$\langle u, \mathcal{T}_t^{(n+1)}(p^\perp) u \rangle \leq \langle u, P_t^* p^\perp P_t u \rangle + \sum_{\ell \geq 1} \int_0^t \langle L_\ell P_{t-s} u, p^\perp L_\ell P_{t-s} u \rangle ds = 0,$$

for any $t \geq 0$.

It follows that $p \mathcal{T}_t^{(n+1)}(p^\perp) p = 0$ and Lemma 5.1 implies $p^\perp \mathcal{T}_t^{(n+1)}(p^\perp) p = p \mathcal{T}_t^{(n+1)}(p^\perp) p^\perp = 0$. Therefore,

$$\mathcal{T}_t^{(n+1)}(p^\perp) \leq p^\perp,$$

for all $t \geq 0$ and p is subharmonic for $\mathcal{T}^{(n)}$.

Hence, p is subharmonic for the minimal semigroup $\mathcal{T}$ and the proof is complete.

5.3 Examples

Gisin-Percival Model of Absorption and Stimulated Emission, Continued

We continue the analysis of the model introduced in 4.5, and refer to the notations therein. We want to characterize *all* common invariant closed subspaces for G, L_1, L_2. Since L_1 is a multiple of the number operator N, its invariant subspaces $\mathfrak{I}_K$ are spanned by $\{e_k : k \in K\}$, where $K \subseteq \mathbb{N}$. However, notice that if K is finite and $k_0 = \max K$, to have $\mathfrak{I}_K$ invariant also for L_2, one needs to have *all* of the e_k for $k \leq k_0$ inside $\mathfrak{I}_K$, since L_2 is a multiple of the annihilation operator. Define

$\mathfrak{h}_0 = \{0\}$; $\mathfrak{h}_k$, the space generated by $\{e_0, \ldots, e_{k-1}\}$ for any $k \geq 1$ and $\mathfrak{h}_\infty = \mathfrak{h}$. The unique collection of invariant subspaces for L_1 and L_2 is $(\mathfrak{h}_k)_{k \in \mathbb{N} \cup \{\infty\}}$.

As a result, if $\xi = 0$, there is a full collection of non trivial invariant subspaces.

Suppose $\xi \neq 0$. In this case there is no $\mathfrak{h}_k$ invariant under G for $k > 0$. Therefore, the only common invariant spaces for G, L_1, L_2 are trivial: $\mathfrak{h}_0$ and $\mathfrak{h}$. As a result, the QMS has a faithful stationary state, say ρ_∞.

Moreover, in the next section we will see that given any other state ρ, the semigroup $\mathcal{T}$ associated to (G, L_1, L_2) satisfy that

$$\mathrm{tr}(\mathcal{T}_{*t}(\rho) X) \to \mathrm{tr}(\rho_\infty X),$$

for all $X \in \mathfrak{L}(\mathfrak{h})$. This means that any other w^*-limit of $\mathcal{T}_{*t}(\rho)$, ρ' say, has to satisfy $\mathrm{tr}(\rho' X) = \mathrm{tr}(\rho_\infty X)$ for any $X \in \mathfrak{L}(\mathfrak{h})$, so that $\rho' = \rho_\infty$ due to the faithfulness of ρ_∞.

6 The Convergence Towards the Equilibrium

To illustrate the results of this section we will restrict the proofs to the case of a norm-continuous semigroup. This is the framework of the seminal paper of Frigerio and Verri (see [36]. The case of a semigroup with unbounded generators has been treated by us in [29], (see also [33], [34]).

Throughout this section we assume that the QMS *has a faithful normal stationary state* ρ and we want to derive conditions under which $\mathcal{T}_{*t}(\sigma)$ converges in the w^* topology towards ρ as $t \to \infty$, for any initial state σ.

Under the above basic assumtion, there exists a conditional expectation $x \mapsto \mathbb{E}^{\mathfrak{F}(T)}(x)$ in the sense of Umegaki, defined over the von Neumann algebra $\mathfrak{F}(T)$ of invariant elements under the action of $\mathcal{T}$. We recall an early result of Frigerio and Verri ([36], Theorem 3.3, p.281). Denote $\mathfrak{N}(\mathcal{T})$ the set of elements $x \in \mathcal{B}(\mathfrak{h})$ for which $\mathcal{T}_t(x^*x) = \mathcal{T}_t(x^*)\mathcal{T}_t(x)$ and $\mathcal{T}_t(xx^*) = \mathcal{T}_t(x)\mathcal{T}_t(x^*)$, for all $t \geq 0$

Theorem 6.1 (Frigerio–Verri). *If the semigroup $\mathcal{T}$ has a faithful stationary state ρ and the set of fixed points of $\mathcal{T}$ coincides with $\mathfrak{N}(\mathcal{T})$, then*

$$w^* - \lim_{t \to \infty} \mathcal{T}_t(x) = \mathbb{E}^{\mathfrak{F}(T)}(x), \tag{27}$$

for all $x \in \mathcal{B}(h)$.

The basic idea of the proof consists in associating with $\mathcal{T}$ a strongly continuous contraction semigroup on the Hilbert space of the GNS representation based on the state ρ. Alternatively, one can prove that $\mathfrak{M}$ decomposes as a direct sum of the von Neumann algebra $\mathfrak{N}(\mathcal{T})$ and a Banach space $\mathfrak{D}(\mathcal{T})$ which corresponds to the space of observables x for which

$$\lim_{t \to \infty} \omega(\mathcal{T}_t(x)) = 0, \tag{28}$$

for any state ω. Thus, given any $x \in \mathfrak{L}(\mathfrak{h})$, we can write $x = \mathbb{E}^{\mathfrak{N}(T)}(x) + (x - \mathbb{E}^{\mathfrak{N}(T)}(x))$. The element $(x - \mathbb{E}^{\mathfrak{N}(T)}(x))$ belongs to $\mathfrak{D}(T)$, so that given any state ω, $\lim_{t\to\infty}\left|\omega(T_t(x)) - \omega(T_t(\mathbb{E}^{\mathfrak{N}(T)}(x)))\right| = 0$. If $\mathfrak{N}(T) = \mathfrak{F}(T)$, then $T_t(\mathbb{E}^{\mathfrak{N}(T)}(x)) = T_t(\mathbb{E}^{\mathfrak{F}(T)}(x)) = \mathbb{E}^{\mathfrak{F}(T)}(x)$ and

$$\lim_{t\to\infty}\left|\omega(T_t(x)) - \omega(\mathbb{E}^{\mathfrak{F}(T)}(x))\right| = 0.$$

6.1 Main Results

We begin by establishing a property of the space $\mathfrak{N}(T)$.

Proposition 6.1. *Under the assumption of this section about the existence of a faithful normal stationary state for the norm continuous quantum Markov semigroup T, the space $\mathfrak{N}(T)$ is a von Neumann algebra which coincides with the generalized commutator algebra of $\mathbb{L} = (L_k, L_k^*; k \geq 1)$, denoted by $\mathbb{L}'$.*

Proof. The property of $\mathfrak{N}(T)$ being a von Neumann algebra is obtained in [36] through an application of Tomita-Takesaki theory (the conditional expectation exists since $\mathfrak{N}(T)$ is invariant under the modular automorphism associated to the faithful normal stationary state). Define

$$\Gamma(x,y) := \mathcal{L}(xy) - x\mathcal{L}(y) - \mathcal{L}(x)y,$$

for all $x, y \in \mathfrak{L}(\mathfrak{h})$. As it was established in the previous chapter [52] of this volume, $\Gamma(x^*, x) \geq 0$ for any $x \in \mathfrak{L}(\mathfrak{h})$. Notice that $x \in \mathfrak{N}(T)$ if and only if $\Gamma(x^*, x) = 0 = \Gamma(x, x^*)$. A straightforward computation shows that

$$\Gamma(x^*, x) = \sum_{k\geq 1}(L_k^* x^* x L_k - x^* L_k^* x L_k + x^* L_k^* L_k x - L_k^* x^* L_k x)$$
$$= \sum_{k\geq 1}[L_k, x]^*[L_k, x].$$

Similarly, $\Gamma(x, x^*) = \sum_{k\geq 1}[L_k^*, x]^*[L_k^*, x]$. Since each term $[L_k, x]^*[L_k, x]$ (respectively $[L_k^*, x]^*[L_k^*, x]$) is positive, we have $\Gamma(x^*, x) = 0 = \Gamma(x, x^*)$ if and only if $[L_k, x]^*[L_k, x] = 0$ for all $k \geq 1$, which means that x is an element of $\mathbb{L}'$.

Theorem 6.2. *A norm continuous quantum Markov semigroup converges in the sense that*

$$w^* - \lim_{t\to\infty} T_t(X) = T_\infty(X), \tag{29}$$

for all $X \in \mathcal{B}(h)$, whenever the generalized commutator $\mathbb{L}'$ is reduced to the trivial algebra $\mathbb{C}I$.

Proof. This result follows straightforward from the Theorem of Frigerio and Verri, since $\mathfrak{N}(T) = \mathbb{L}'$.

Proposition 6.2. *Under the above hypotheses the set* $\mathfrak{F}(T)$ *of fixed elements for the semigroup is given by*

$$\mathfrak{F}(T) = \{L_k, L_k^*, H; \ k \geq 1\}' \tag{30}$$

Proof. Since $\mathfrak{F}(T) \subset \mathfrak{N}(T)$, and $\mathfrak{N}(T) = \mathbb{L}'$, it follows that $\mathfrak{F}(T) \subset \mathbb{L}'$. So that, for all $x \in \mathfrak{F}(T)$ and any $u, v \in \mathfrak{h}$ it holds:

$$0 = \langle v, \mathcal{L}(x)u \rangle$$

$$= -\frac{1}{2}\sum_{k=1}^{\infty} L_k^* L_k v, xu \rangle + \frac{1}{2}\sum_{k=1}^{\infty}\langle L_k v, xL_k u \rangle$$

$$+ \frac{1}{2}\sum_{k=1}^{\infty}\langle L_k v, xL_k u \rangle - \frac{1}{2}\sum_{k=1}^{\infty}\langle v, xL_k^* L_k u \rangle$$

$$- \langle iHv, xu \rangle - \langle v, ixHu \rangle.$$

We now study the right hand side of the above equation. Since $xL_k = L_k x$, the first two terms cancel and the computation

$$\langle L_k v, xL_k u \rangle = \langle x^* L_k v, L_k u \rangle = \langle L_k x^* v, L_k u \rangle = \langle x^* v, L_k^* L_k u \rangle,$$

shows that the third and fourth terms cancel as well.

From the above we deduce $xH = Hx$. Therefore, x belongs to

$$\{L_k, L_k^*, H; \ k \geq 1\}'.$$

Reciprocally, if $X \in \{L_k, L_k^*, H; \ k \geq 1\}'$, the equation for $\langle v, \mathcal{L}(x)u \rangle$ gives 0 and we obtain that x is a fixed point of T.

The corollary which follows is easily derived from the propositions and theorem before.

Corollary 6.1. *For any norm continuous quantum Markov semigroup, the convergence towards the equilibrium holds if* $\{L_k, L_k^*, H; k \geq 1\}' = \{L_k, L_k^*; k \geq 1\}'$.

The sufficient condition obtained for proving the convergence towards the equilibrium is necessary, at least for a wide class of operators H, as we state in the following theorem.

Theorem 6.3. *Let be given a norm continuous quantum Markov semigroup for which H is a bounded self–adjoint operator with pure point spectrum.*
Then $T_t(\cdot)$ converges towards the equilibrium if and only if

$$\{L_k, L_k^*, H; k \geq 1\}' = \{L_k, L_k^*; k \geq 1\}'. \tag{31}$$

Proof. From the corollary before, (31) is a sufficient condition for the convergence towards the equilibrium. We will prove below that it is a necessary condition as well.

Indeed, the hypotheses assumed imply that

$$T_t(x) = e^{itH} x e^{-itH},$$

for all $x \in \mathfrak{N}(\mathcal{T})$.

For any two different eigenvalues λ and μ of H, choose corresponding eigenvector $v, u \in \mathfrak{h}$. Then,

$$\langle e^{-itH} v, x e^{-itH} u \rangle = e^{it(\mu - \lambda)} \langle v, Xu \rangle,$$

converges when $t \to \infty$.

Therefore, $\langle v, xu \rangle = 0$ and x commute with H. Consequently,

$$\{L_k, L_k^*, H; k \geq 1\}' = \{L_k, L_k^*; k \geq 1\}'.$$

Remark 6.1. All the results of this section have been extended by the authors in [29] (see also [33], [34]) to QMS defined on $\mathcal{L}(\mathfrak{h})$ with *unbounded generators* given as a form, under some suitable additional hypotheses. These hypotheses are the following:

1. The traditional hypothesis **(H-min)** used in the construction of the minimal semigroup.
2. The assumption that the minimal semigroup is Markov.
3. The existence of a domain D which is a core for both G and G^*.
4. For all $u \in D$, the image $R(n; G)u$ by the resolvent of G, belongs to $D(G^*)$ and the sequence $(nG^* R(n; G)u)_{n \geq 1}$ strongly converges. This hypothesis, together with the previous one, allow to construct a quantum cocycle which is a dilation of the semigroup $\mathcal{T}$. There exists also a *dual cocycle* and a corresponding semigroup $\widetilde{\mathcal{T}}$. The last technical hypothesis is
5. The semigroup associated to the dual cocycle preserves the identity.

We call **natural** a QMS which satisfies the above set of hypotheses. All the previous results still hold replacing **norm continuous** QMS by **natural** QMS with some minor changes which we precise below:

In Prop.6.2 Suppose that the closure of H (which is unbounded) is self–adjoint.
In Prop.6.1, 6.1,6.3 Suppose either that
   (a) H is bounded;
      or
   (b) H is selfadjoint and $e^{itH}(D) \subset D(G)$,

6.2 Examples

The Asymptotic Behavior of the Jaynes–Cummings Model in Quantum Optics

Here we consider again the model introduced in subsection 4.6, (see [27] and [28]), which is the quantum Markov semigroup associated to master equations in Quantum

Optics. The initial space is $\mathfrak{h} = \ell^2(\mathbb{N})$ endowed with the creation (resp. annihilation) operator $a^\dagger$ (resp. a), and the number operator denoted N. In addition, the coefficients G and L_k ($k = 1, \ldots, 4$) are given by the expressions

$$L_1 = \mu a, \quad L_2 = \lambda a^\dagger, \quad L_3 = R \cos(\phi\sqrt{aa^\dagger}),$$

$$L_4 = Ra^\dagger \frac{\sin(\phi\sqrt{aa^\dagger})}{\sqrt{aa^\dagger}}, \quad G = -\frac{1}{2} \sum_{k=1}^{4} L_k^* L_k,$$

where the parameters $\phi, R \geq 0, \lambda < \mu$, specify the physical model. In this case the natural set D to choose is the domain of the number operator.

Moreover, the existence of a stationary state for $\mathcal{T}$ has been proved in [27]. Indeed if $\lambda < \mu$, then $\mathcal{T}$ has a stationary state given by

$$\rho_\infty = \sum_{n=0}^{\infty} \pi_n |e_n\rangle\langle e_n|$$

where $(\pi_n)_{n\geq 0}$ is the sequence defined by

$$\pi_0 = c, \qquad \pi_n = c \prod_{k=1}^{n} \frac{\lambda^2 k + R^2 \sin^2(\phi\sqrt{k})}{\mu^2 k} \quad (n \geq 1).$$

where c is a suitable normalization constant.

The remaining hypotheses showing that the semigroup is natural have been checked in [27] as well.

Now, to verify the hypotheses of Theorem 6.2, it suffices to study the action of operators on the canonical basis $(e_m; m \geq 0)$ of $\mathfrak{h}$. In particular, it brings about a recurrence relationship among the elements of the basis from which it follows that $\langle e_r, X e_m \rangle = 0$ for all element X of the generalized commutator algebra of L_k, L_k^*, ($k = 1, \ldots, 4$).

Corollary 6.2. *The quantum Markov semigroup introduced before approaches the equilibrium in the sense of the w^* topology, as $t \to \infty$.*

As a trivial consequence of the above corollary, the Cesàro mean of the semigroup converge in the w^* topology. This result had been stated in [28] with a different direct proof.

A Class of Examples with a Non-Trivial Fixed Point Algebra

Keeping the notations on spaces and operators of the above example, consider a quantum Markov semigroup with generator given by

$$L_1 = \alpha(N), \quad L_k = 0, \ (k \neq 1), \quad H = \beta(N),$$

with α and β given functions, α assumed to be injective and β real–valued. So that, any faithful state which is a function of the number operator is an invariant state.

In addition the algebras $\{L_k, L_k^*, H; \ k \geq 1\}'$ and $\{L_k, L_k^*; \ k \geq 1\}'$ coincide with $\{N\}'$ if and only if the support of β is included in that of α. Therefore, the hypotheses of the Corollary are satisfied, whenever the support of β is included in that of α and the semigroup converges towards the equilibrium.

Simple Absorption and Stimulated Emission

Now we complete the example 4.5. In this case, the reader can easily verify that

$$\{H, L_k, L_k^*; \ k = 1, 2\}' = \{L_k, L_k^*; \ k = 1, 2\}'.$$

Therefore,

Corollary 6.3. *If $\xi \neq 0$, the quantum model of simple absorption and stimulated emission introduced in 4.5 has a unique faithful stationary state and the quantum Markov semigroup converges towards the equilibrium, that is*

$$\mathcal{T}_{*t}(\rho) \xrightarrow{w^*} \rho_\infty,$$

for any state ρ where ρ_∞ denotes the stationary state.

7 Recurrence and Transience of Quantum Markov Semigroups

Within this section we explore the probabilistic notion of recurrence (and transience) for non commutative Markov semigroups. These notions are closely related to the concept of potential as we will see below.

7.1 Potential

Let $\mathcal{T}$ be a Quantum Markov Semigroup (QMS) on a von Neumann algebra $\mathfrak{M}$ of operators on a complex Hilbert space $\mathfrak{h}$.

Inspired by the classical theory of Markov processes [22], this section introduces the non commutative version of *potential* and discusses its main properties. This is the main tool in the study of recurrence and transience.

Throughout this paper, the use of quadratic forms settings will follow the book of Kato (see [41]).

Definition 7.1. *Given a positive operator $x \in \mathfrak{M}$ we define the* form-potential of x *as a quadratic form $\mathfrak{U}(x)$ on the domain*

$$D(\mathfrak{U}(x)) = \left\{ u \in \mathfrak{h} : \int_0^\infty \langle u, \mathcal{T}_s(x)u \rangle ds < \infty \right\},$$

by

$$\mathfrak{U}(x)[u] = \int_0^\infty \langle u, \mathcal{T}_s(x)u \rangle ds, \ (u \in D(\mathfrak{U}(x))).$$

This is clearly a symmetric and positive form and by Thm. 3.13a and Lemma 3.14a p.461 of [41] it is also closed. Therefore, when it is densely defined, it is represented by a self-adjoint operator (see Th.2.1, p.322, Th. 2.6, p.323 and Th. 2.23 p.331 of [41]). This motivates the following definition.

Definition 7.2. *A positive $x \in \mathfrak{M}$ such that $D(\mathfrak{U}(x))$ is dense is called $\mathcal{T}$–integrable or simply* integrable. *We denote $\mathfrak{M}^+_{\mathrm{int}}$ the cone of positive integrable elements of $\mathfrak{M}$. For any $x \in \mathfrak{M}^+_{\mathrm{int}}$, we call* potential *of x the self-adjoint operator $\mathcal{U}(x)$ which represents $\mathfrak{U}(x)$.*

Note that $D(\mathcal{U}(x)^{1/2}) = D(\mathfrak{U}(x))$ (see Th. 2.23, p.331 in [41]).

We recall that a closed operator A is affiliated with a von Neumann algebra $\mathfrak{M}$ if $a'D(A) \subseteq D(A)$ and $a'A \subseteq Aa'$ for all $a' \in \mathfrak{M}'$.

Proposition 7.1. *For all $x \in \mathfrak{M}^+_{\mathrm{int}}$, the operator $\mathcal{U}(x)$ is affiliated with $\mathfrak{M}$.*

Proof. Fix $y \in \mathfrak{M}'$ and define $X_t = \int_0^t T_s(x)ds$, for all $t \geq 0$. Clearly, both X_t and $X_t^{1/2}$ belong to $\mathfrak{M}$. Given any $u \in \mathfrak{h}$,

$$\int_0^t \langle yu, T_s(x)yu \rangle ds = \langle yX_t^{1/2}u, yX_t^{1/2}u \rangle \leq \|y\|^2 \langle u, X_t u \rangle.$$

Thus, if $u \in D(\mathfrak{U}(x))$, then

$$\sup_{t \geq 0} \int_0^t \langle yu, T_s(x)yu \rangle ds \leq \|y\|^2 \int_0^\infty \langle u, T_s(x)u \rangle ds = \|y\|^2 \, \mathfrak{U}(x)[u].$$

It follows that, if $u \in D(\mathfrak{U}(x)) = D(\mathcal{U}(x)^{1/2})$, then $yu \in D(\mathfrak{U}(x))$.

Now, if $v, u \in D(\mathcal{U}(x))$, then $y^*v, \, yu \in D(\mathfrak{U}(x))$ and

$$\int_0^t \langle y^*v, T_s(x)u \rangle ds = \int_0^t \langle T_s(x)v, yu \rangle ds,$$

so that letting $t \to \infty$ and using complex polarization, we get

$$\langle y^*v, \mathcal{U}(x)u \rangle = \langle \mathcal{U}(x)v, yu \rangle.$$

That is, $\langle v, y\mathcal{U}(x)u \rangle = \langle \mathcal{U}(x)v, yu \rangle$ it follows that $yu \in D(\mathcal{U}(x))$ and $\mathcal{U}(x)yu = y\mathcal{U}(x)u$, hence $y\mathcal{U}(x) \subseteq \mathcal{U}(x)y$.

Proposition 7.2. *Let T be a Quantum Markov Semigroup and let $x \in \mathfrak{M}$ positive. Then the orthogonal projection p onto the closure of $D(\mathfrak{U}(x))$ is subharmonic.*

In particular, if T is irreducible, then $D(\mathfrak{U}(x))$ is either dense or $\{0\}$.

Proof. We first notice that $p \in \mathfrak{M}$. Indeed, arguing as in the proof before, we can show that for every $u \in D(\mathfrak{U}(x))$ and $y \in \mathfrak{M}'$, $yu \in D(\mathfrak{U}(x))$. Hence,

$$pypu = pyu = yu = ypu.$$

In other words, since $D(\mathfrak{U}(x))$ is dense in the range of p, we obtain $pyp = yp$.

On the other hand, $y^* \in \mathfrak{M}'$, so that $py^*p = y^*p$. Therefore, $pyp = py$. Hence $yp = py$, so that $p \in \mathfrak{M}'' = \mathfrak{M}$.

We now show that $\mathcal{T}_t(p) \geq p$ for any $t \geq 0$. Let ρ be a density matrix ρ such that

$$\rho = \sum_k \lambda_k |u_k\rangle\langle u_k|, \ \lambda_k \geq 0, \ \sum_k \lambda_k = 1, \ u_k \in D(\mathfrak{U}(x)).$$

Note that ρ defines a normal linear functional on $\mathfrak{M}$. Therefore $\rho \in \mathfrak{M}_*$. Moreover, denote $\varphi \in \mathfrak{M}_*$ the state given by $\varphi(a) = \operatorname{tr}(\rho a)$, for all $a \in \mathfrak{M}$. For any $t \geq 0$ there exists (see [23] Th.1, p.57) a density matrix ρ_t such that $\mathcal{T}_{*t}(\varphi)(a) = \operatorname{tr}(\rho_t a)$ for all $a \in \mathfrak{M}$. Notice that for all $s \geq 0$, $\operatorname{tr}(\rho_t \mathcal{T}_s(a)) = \mathcal{T}_{*t}(\varphi)(\mathcal{T}_s(a)) = \operatorname{tr}(\rho \mathcal{T}_{t+s}(a))$. Hence,

$$\int_0^\infty \operatorname{tr}(\rho_t \mathcal{T}_s(a))ds = \int_0^\infty \operatorname{tr}(\rho \mathcal{T}_{t+s}(a))ds = \int_t^\infty \operatorname{tr}(\rho \mathcal{T}_s(a))ds < \infty.$$

It follows that

$$\rho_t = \sum_k \lambda_k(t)|u_k(t)\rangle\langle u_k(t)|,$$

with $u_k(t) \in D(\mathfrak{U}(x))$, for all $k \geq 1$ and $t \geq 0$ such that $\lambda_k(t) > 0$.

As a result, the range of ρ_t is included in $D(\mathfrak{U}(x))$, i.e. $p\rho_t = \rho_t p = p\rho_t p = \rho_t$. Thus, $\operatorname{tr}(\rho \mathcal{T}_t(p)) = \operatorname{tr}(\rho_t p) = \operatorname{tr}(\rho_t) = 1$, and

$$0 = \operatorname{tr}(\rho(p - \mathcal{T}_t(p))) = \operatorname{tr}(\rho(p - p\mathcal{T}_t(p)p)).$$

However, we also have $p\mathcal{T}_t(p)p \leq p\mathcal{T}_t(1)p \leq p$. Therefore, $p\mathcal{T}_t(p)p = p$, i.e. $p\mathcal{T}_t(p^\perp)p = 0$, (see Lemma II.1 in [32]) so that $\mathcal{T}_t(p) \geq p$.

The second part is a trivial consequence of the above.

Potentials are a natural source of superharmonic (or excessive) operators. Indeed, heuristically,

$$\mathcal{T}_t(\mathcal{U}(x)) = \mathcal{T}_t(\int_0^\infty \mathcal{T}_s(x)ds) = \int_t^\infty \mathcal{T}_s(x)ds \leq \mathcal{U}(x),$$

however $\mathcal{U}(x)$ is possibly unbounded. Further on, bounded potentials will be associated with our concept of transience (see Theorems 7.2 and 7.4).

Theorem 7.1. *For any $x \in \mathfrak{M}_{\text{int}}^+$, the contraction*

$$y = \mathcal{U}(x)(1 + \mathcal{U}(x))^{-1}, \tag{32}$$

is superharmonic and $\mathcal{T}_t(y)$ converges strongly to 0 as $t \to \infty$.

Proof. Fix $x \in \mathfrak{M}_{\text{int}}^+$ and define $\mathcal{U}_t(x) = \int_0^t \mathcal{T}_s(x)ds$ $(t \geq 0)$. For any $s, t \geq 0$,

$$\mathcal{T}_t(\mathcal{U}_s(x)) = \int_t^{t+s} \mathcal{T}_r(x)dr = \mathcal{U}_{t+s}(x) - \mathcal{U}_t(x). \tag{33}$$

It follows:

$$T_t(\mathcal{U}_s(x)) \leq \mathcal{U}_{t+s}(x). \tag{34}$$

Since $T_t(\cdot)$ is in particular 2-positive, identity preserving and the function $x \mapsto (1+x)^{-1}$ is operator monotone (see e.g. [13]), we have

$$(1 + T_t(\mathcal{U}_s(x)))^{-1} \leq T_t((1 + \mathcal{U}_s(x))^{-1}).$$

From (34),

$$(1 + \mathcal{U}_{t+s}(x))^{-1} \leq T_t((1 + \mathcal{U}_s(x))^{-1}).$$

It follows:

$$\begin{aligned}
T_t(\mathcal{U}_s(x)(1 + \mathcal{U}_s(x))^{-1}) &= 1 - T_t((1 + \mathcal{U}_s(x))^{-1}) \\
&\leq 1 - (1 + \mathcal{U}_{t+s}(x))^{-1} \\
&= \mathcal{U}_{t+s}(x)(1 + \mathcal{U}_{t+s}(x))^{-1}.
\end{aligned}$$

The map $T_t(\cdot)$ is normal and $\mathcal{U}_{t+s}(x)(1 + \mathcal{U}_{t+s}(x))^{-1}$ strongly converges to y as $s \to \infty$. Therefore, letting $s \to \infty$ yields $T_t(y) \leq y$.

Finally, (33) implies

$$T_t(\mathcal{U}_s(x)(1 + \mathcal{U}_s(x))^{-1}) \leq T_t(\mathcal{U}_s(x)) = \mathcal{U}_{t+s}(x) - \mathcal{U}_t(x),$$

so that for all $u \in D(\mathfrak{U}(x))$,

$$\langle u, T_t(\mathcal{U}_s(x)(1 + \mathcal{U}_s(x))^{-1})u \rangle \leq \int_t^{t+s} \langle u, T_r(x)u \rangle dr.$$

Letting $s \to \infty$ again,

$$\langle u, T_t(y)u \rangle \leq \int_t^\infty \langle u, T_r(x)u \rangle dr,$$

thus, $\langle u, T_t(y)u \rangle$ vanishes, as t goes to infinity. Since $D(\mathfrak{U}(x))$ is dense and the operators $T_t(y)$ are uniformly bounded in norm by $\|y\| \leq 1$, the last statement of the theorem follows.

Proposition 7.3. *For any $x \in \mathfrak{M}^+$, let $\mathcal{K}(x) = \{u \in D(\mathfrak{U}(x)): \mathfrak{U}(x)[u] = 0\}$. Then the projection p on $\mathcal{K}(x)$ is subharmonic.*

Proof. We use here the notations of the previous proof. Note that for $x \in \mathfrak{M}^+$, $\mathfrak{U}(x)[u] = 0$ if and only if $\mathcal{U}_s(x)u = 0$ for each $s \geq 0$. Fix $s > 0$ and let $q_n(s)$ denote the spectral projection of $\mathcal{U}_s(x)$ associated with the interval $]1/n, \|\mathcal{U}_s(x)\|]$, $(n \geq 1)$.

It is worth noticing that $q(s) = \text{l.u.b.} q_n(s)$ is the projection onto the closure of the range of $\mathcal{U}_s(x)$. Equation (34) yields

$$T_t(q_n(s)) \leq n T_t(\mathcal{U}_s(x)) \leq n \mathcal{U}_{t+s}(x).$$

Since $\mathcal{T}_t(q_n(s)) \leq 1$ we obtain,

$$\mathcal{T}_t(q_n(s))^n \leq n\mathcal{U}_{t+s}(x),$$

that is

$$\mathcal{T}_t(q_n(s)) \leq n^{1/n}\mathcal{U}_{t+s}(x)^{1/n}.$$

Therefore, letting $n \to \infty$,

$$\mathcal{T}_t(q(s)) \leq q(t+s).$$

Now, notice that the family $q(s)$ is increasing with s and $q = \mathrm{l.u.b.}q(s)$ is equal to $1-p$, the projection onto the orthogonal of $\mathcal{K}(x)$. The conclusion follows from the previous inequality letting $s \to \infty$.

7.2 Defining Recurrence and Transience

Let $\mathcal{T}$ be a QMS on a von Neumann algebra $\mathfrak{M}$.

We say that a self-adjoint operator X is *strictly positive* if $\langle u, Xu \rangle > 0$ for any $u \in D(X)$, $u \neq 0$ (we will write simply $X > 0$).

Theorem 7.2. *The following statements are equivalent:*

1. *There exists a positive $x \in \mathfrak{M}$ with $\mathcal{U}(x)$ bounded and $\mathcal{U}(x) > 0$.*
2. *There exists a strictly positive $x \in \mathfrak{M}$ with $\mathcal{U}(x)$ bounded.*
3. *There exists a positive $x \in \mathfrak{M}$ with $\mathcal{U}(x) > 0$.*
4. *There exists an increasing sequence of projections $(p_n; n \geq 1)$, with $\mathrm{l.u.b.}p_n = 1$ and $\mathcal{U}(p_n)$ bounded for all n.*

Proof. $1{\Rightarrow}2$: Let $x_\lambda = \mathcal{R}_\lambda(x)$ $(\lambda > 0)$, where $\mathcal{R}_\lambda(\cdot)$ is the resolvent of the semigroup $\mathcal{T}$. Since $\mathcal{U}(x) > 0$, then $x_\lambda > 0$. Moreover, the resolvent identity implies

$$\mathcal{U}(x_\lambda) = \mathcal{U}(\mathcal{R}_\lambda(x)) = \lambda^{-1}\left(\mathcal{U}(x) - \mathcal{R}_\lambda(x)\right) \leq \lambda^{-1}\mathcal{U}(x).$$

Thus, $\mathcal{U}(x_\lambda)$ is bounded.

$2{\Rightarrow}1$: Clearly if $x > 0$ then $\mathcal{U}(x) > 0$.

$1{\Rightarrow}3$ is self-evident.

$3{\Rightarrow}1$: Let $x \in \mathfrak{M}$ positive with $\mathcal{U}(x) > 0$ and set y as in Theorem 7.1. Clearly $0 < y < 1$, and is a superharmonic operator. We may assume y in the domain of the generator $\mathcal{L}$ of $\mathcal{T}$ (otherwise replace y by $\mathcal{R}_\lambda(y)$; $\mathcal{T}_t(\mathcal{R}_\lambda(y))$ still vanishes as $t \to \infty$), then $\mathcal{L}(y) \leq 0$ and

$$\int_0^t \mathcal{T}_s(-\mathcal{L}(y))ds = y - \mathcal{T}_t(y), \ (t \geq 0).$$

Letting $t \to \infty$ yields $\mathcal{U}(-\mathcal{L}(y)) = y$. Thus $-\mathcal{L}(y)$ satisfies condition 1.

$4{\Rightarrow}1$: Define $c_n = 2^{-n}\|\mathcal{U}(p_n)\|^{-1}$, and $x = \sum_{n\geq 0} c_n p_n$. Then x is strictly positive, $\mathcal{U}(x)$ is bounded and $\mathcal{U}(x) > 0$.

$1{\Rightarrow}4$: By the argument of $1{\Rightarrow}2$ we can suppose that $x > 0$. It suffices then to take p_n as the spectral projection of x associated with the interval $]1/n, \|x\|]$.

Corollary 7.1. *If* $\mathfrak{M} = \mathcal{B}(\mathsf{h})$ *with* h *separable, then the statements of Theorem 7.2 are all equivalent to the following condition there exists an increasing sequence of finite dimensional projections* $(p_n; n \geq 1)$, *with* l.u.b.$p_n = 1$ *and* $\mathcal{U}(p_n)$ *bounded for all* n.

Proof. Clearly it suffices to prove that the statements of Theorem 7.2 imply the above condition on finite dimensional projections. Let $(p_m; m \geq 1)$ be an increasing sequence of projections satisfing the statement 4. For each m let $(p_{m,k}; k \geq 1)$ be an increasing sequence of finite dimensional projections on h with l.u.b.$_k p_{m,k} = p_m$. Note that $0 \leq \mathcal{U}(p_{m,k}) \leq \mathcal{U}(p_m)$ for all m, k. Therefore we have $\|\mathcal{U}(p_{m,k})\| \leq \|\mathcal{U}(p_m)\| < \infty$. Finally, since h is separable, a diagolisation argument shows the existence of a subsequence $(p_{m_n,k_n}; n \geq 1)$ with l.u.b.$_n p_{m_n,k_n} = 1$ and $\|\mathcal{U}(p_{m_n,k_n})\| < \infty$ for all n.

Theorem 7.3. *The following are equivalent:*

1. *For each positive* $x \in \mathfrak{M}$ *and* $u \in \mathfrak{h}$ *either* $u \notin D(\mathfrak{U}(x))$ *or* $u \in D(\mathfrak{U}(x))$ *and* $\mathcal{U}(x)[u] = 0$.
2. *For each projection* p *and* $u \in \mathfrak{h}$ *either* $u \notin D(\mathfrak{U}(p))$ *or* $u \in D(\mathfrak{U}(p))$ *and* $\mathcal{U}(p)[u] = 0$.

Proof. Clearly $1 \Rightarrow 2$. We prove then that $2 \Rightarrow 1$. Let $x \in \mathfrak{M}$ and $u \in \mathsf{h}$. If $u \in D(\mathfrak{U}(x))$ then, for each spectral projection p of x associated with an interval $]r, \|x\|]$, $u \in D(\mathfrak{U}(p))$. Therefore, by condition 2, we have $\mathfrak{U}(p)[u] = 0$ i.e. $\langle u, T_t(p)u \rangle = 0$ for all $t \geq 0$. It follows then that $\langle u, T_t(x)u \rangle = 0$ for all $t \geq 0$. As a consequence $\mathfrak{U}(x)[u] = 0$.

Corollary 7.2. *If* $\mathfrak{M} = \mathcal{B}(\mathsf{h})$ *with* h, *the statements of Theorem 7.3 are all equivalent to the following conditions: for each finite dimensional projection* p *and* $u \in \mathsf{h}$ *either* $u \notin D(\mathfrak{U}(p))$ *or* $u \in D(\mathfrak{U}(p))$ *and* $\mathfrak{U}(p)[u] = 0$.

Proof. Suppose that the above condition on finite dimensional projections holds and let p be any projection in $\mathfrak{M}$. Let $(p_n; n \geq 1)$ be an increasing sequence of finite dimensional such that l.u.b.$p_n = p$. If $u \in D(\mathfrak{U}(p))$ then $u \in D(\mathfrak{U}(p_n))$ and $\mathfrak{U}(p_n)[u]$ for all $n \geq 1$. This implies clearly $\langle u, T_t(p_n)u \rangle = 0$ for all $n \geq 1$ and all $t \geq 0$ and, letting n tend to infinity, $\langle u, T_t(p)u \rangle = 0$ for all $t \geq 0$. It follows that $\mathfrak{U}(p)[u] = 0$.

Definition 7.3. *A QMS is* transient *(resp.* recurrent*) if any of the equivalent conditions of Theorems 7.2 (resp. Theorem 7.3) holds.*

Proposition 7.4. *If a QMS is irreducible, then it is either recurrent or transient.*

Proof. Indeed, if a QMS is irreducible, then the domain of the form-potential $\mathfrak{U}(x)$ is either $\{0\}$ or dense by Proposition 7.2.

Corollary 7.3. *A transient semigroup in* $\mathcal{B}(\mathsf{h})$ *with* h *separable has no invariant state.*

Proof. Suppose that ρ is an invariant state and take $(p_n; n \geq 1)$ as in Corollary 7.1. Fix $m \geq 1$ such that $\mathrm{tr}(()\rho p_m) > 1/2$.

Since $\mathcal{U}(p_m)$ is bounded, using the separability of h and a diagonalisation argument we can find a sequence $(t_k; k \geq 1)$ diverging to $+\infty$ such that $\mathcal{T}_{t_k}(p_n) \to 0$ strongly as $k \to \infty$. On the other hand, using the invariance of ρ, we have

$$\mathrm{tr}(()\rho p_m) = \mathrm{tr}(\mathcal{T}_{*t_k}(\rho)p_m) = \mathrm{tr}(\rho \mathcal{T}_{t_k}(p_m)),$$

and letting $k \to \infty$ we obtain the contradiction $\mathrm{tr}(\rho p_m) = 0$.

Theorem 7.4. *An irreducible QMS $\mathcal{T}$ is transient if and only if there exists a non-trivial $\mathcal{T}$-superharmonic operator in $\mathfrak{M}$.*

Proof. If $\mathcal{T}$ is transient then, by Theorem 7.2 1, there exists a positive $x \in \mathfrak{M}$ with $\mathcal{U}(x)$ bounded and $\mathcal{U}(x) > 0$. We have then

$$\langle u, \mathcal{T}_t(\mathcal{U}(x))u \rangle = \int_t^\infty \langle u, \mathcal{T}_s(x)u \rangle ds$$

for all $u \in h$. It follows that $\mathcal{U}(x)$ is a superharmonic operator for $\mathcal{T}$ and it is not a multiple of the identity operator since $\mathcal{T}_t(\mathcal{U}(x))$ converges strongly to 0 as t goes to infinity.

Conversely if there exists a non-trivial $\mathcal{T}$-superharmonic operator y in $\mathfrak{M}$ by adding a multiple of $\mathbf{1}$ we can assume that y is also positive. Suppose first, in addition, that y is not harmonic i.e. $\mathcal{T}_t(y) < y$ for some $t > 0$. Note that $\mathcal{R}_\lambda(y)$ is also non-trivial and satisfies $\mathcal{T}_t(\mathcal{R}_\lambda(y)) \leq \mathcal{R}_\lambda(y)$, for all $t \geq 0$ and $\mathcal{T}_t(\mathcal{R}_\lambda(y)) < \mathcal{R}_\lambda(y)$ for some $t > 0$. Therefore, replacing y by $\mathcal{R}_\lambda(y)$ if necessary, we can assume also that y belongs to the domain of the generator $\mathcal{L}$. We have then $\mathcal{L}(y) < 0$ and

$$\int_0^t \langle u, \mathcal{T}_s(-\mathcal{L}(y))u \rangle ds = \langle u, (y - \mathcal{T}_t(y))u \rangle \leq \langle u, yu \rangle.$$

It follows then, letting t go to infinity, that $\mathcal{U}(-\mathcal{L}(y)) \leq y$ and $\mathcal{T}$ is transient by Theorem 7.2 2.

It remains to show that we can suppose that y is not harmonic. Indeed, if y is harmonic, then from the Schwarz inequality $\mathcal{T}_t(y^*)\mathcal{T}_t(y) \leq \mathcal{T}_t(y^*y)$, we have

$$y^2 = \mathcal{T}_t(y^*)\mathcal{T}_t(y) \leq \mathcal{T}_t(y^2).$$

Now, if y^2 is not harmonic, we can apply the above argument to $(1 + \|y^2\|)\mathbf{1} - y^2$. If y^2 is also harmonic we can try with y^4, y^8, ... until we find an n such that y^{2^n} is subharmonic but not harmonic. In case we do not find such an n then, arguing as in the proof of Theorem 7.1, we can show that the operators $y^{2^n}(s + y^{2^n})^{-1}$ ($n \geq 1$, $s > 0$) are $\mathcal{T}$-superharmonic. Then, if y is non-trivial, Lemma 7.1 shows that $\mathcal{T}$ is not irreducible. This completes the proof.

Lemma 7.1. *Let $\mathcal{T}$ be a QMS on a von Neumann algebra $\mathfrak{M}$ and let y be a strictly positive $\mathcal{T}$-harmonic operator in $\mathfrak{M}$. Suppose that, for every $n \geq 1$ and every $s > 0$ the operators $y^{2^n}(s + y^{2^n})^{-1}$ are $\mathcal{T}$-superharmonic. Then every spectral projection of y associated with an interval $]r, +\infty[$ is $\mathcal{T}$-superharmonic.*

Proof. Note that, for each $r > 0$, the operator

$$\lim_n \left(r^{-1}y\right)^{2^n} \left(s + (r^{-1}y)^{2^n}\right)^{-1} = \frac{1}{s+1}E\{r\} + E]r, +\infty[$$

where $E\{r\}$ denotes the orthogonal projection on the (possibly empty) eigenspace of y corresponding to r and $E]r, +\infty[$ the spectral projection of y associated with the interval $]r, +\infty[$ and the limit exists in the strong operator topology. It follows that

$$\mathcal{T}_t\left((s+1)^{-1}E\{r\} + E]r, +\infty[\right) \leq (s+1)^{-1}E\{r\} + E]r, +\infty[.$$

The conclusion follows letting s tend to infinity.

7.3 The Behavior of a d-Harmonic Oscillator

Let $\mathfrak{h} = L^2(\mathbb{R}^d; \mathbb{C})$ and let $\mathfrak{M} = \mathcal{L}(\mathfrak{h})$. Our framework here is the same as that of the harmonic oscillator in [44], Ch. III. By a d-harmonic oscillator, also called Quantum Brownian Motion, we mean a quantum Markov process with associated (minimal) semigroup $\mathcal{T}$ on $\mathfrak{M}$ associated with the form generator

$$\mathfrak{L}(x) = -\frac{1}{2}\sum_{j=1}^d \left(a_j a_j^* x - 2a_j x a_j^* + x a_j a_j^*\right) - \frac{1}{2}\sum_{j=1}^d \left(a_j^* a_j x - 2a_j^* x a_j + x a_j^* a_j\right),$$

where a_j^*, a_j are the creation and annihilation operators

$$a_j = (q_j + \partial_j)/\sqrt{2}, \qquad a_j^* = (q_j - \partial_j)/\sqrt{2},$$

∂_j being the partial derivative with respect to the jth coordinate q_j.

The commutative von Neumann subalgebra $\mathfrak{M}_q$ of $\mathfrak{M}$, generated by q, whose elements are multiplication operators M_f by a function $f \in L^\infty(\mathbb{R}^d; \mathbb{C})$ is $\mathcal{T}$-invariant and $\mathcal{T}_t(M_f) = M_{T_t f}$ where

$$(T_t f)(x) = \frac{1}{(2\pi t)^{d/2}} \int_{\mathbb{R}^d} f(y) e^{-|x-y|^2/2t} dy. \tag{35}$$

The same conclusion holds for the commutative algebra $\mathfrak{M}_p = F^* \mathfrak{M}_q F$, where F denotes the Fourier transform. Therefore, our process deserves the name of *quantum Brownian motion* since its contains a couple of non commuting classical Brownian motions.

Moreover, the von Neumann algebra $\mathfrak{M}_N$ generated by the number operator $N = \sum_j a_j^* a_j$ is also $\mathcal{T}$ invariant and the classical semigroup obtained by restriction of $\mathcal{T}$ to $\mathfrak{M}_N$ is a birth and death on $\mathbb{N}$ with birth rates $(n + 1)_{n \geq 0}$ and death rates $(n)_{n \geq 0}$.

Also, an application of subsection 5.2 shows that $\mathcal{T}$ is irreducible.

The unit vector $e_0(q) = \pi^{-d/4} \exp(-|q|^2/2)$ satisfies $a_j e_0 = 0$ for all j. The rank-one projection $|e_0\rangle\langle e_0|$ onto e_0 belongs to $\mathfrak{M}_N$ and satisfies

$$\mathcal{T}_t(|e_0\rangle\langle e_0|) = \frac{1}{(1+t)^d}\,(1+1/t)^{-N} \tag{36}$$

This formula can be checked as follows. Notice first that each Weyl operator $W(z)$ (see [44], III.4) belongs to the domain of $\mathcal{L}$ and $\mathcal{L}(W(z)) = -|z|^2 W(z)$ (e.g. by [29], Lemma 1.1) Therefore, we have $\mathcal{T}_t(x)(W(z)) = \exp(-t|z|^2)W(z)$ and the canonical commutation relation

$$W(-\zeta/\sqrt{2})W(z)W(\zeta/\sqrt{2}) = \exp(-i\sqrt{2}\mathrm{Im}\langle z,\zeta\rangle)W(z)$$

leads to the explicit formula (see [5]) for $x = W(z) \in \mathfrak{L}(\mathfrak{h})$

$$\mathcal{T}_t(x) = \frac{1}{(2\pi t)^d}\int_{\mathbb{R}^{2d}} W(-\zeta/\sqrt{2})xW(\zeta/\sqrt{2})\exp(-|\zeta|^2/2t)\,d\zeta \tag{37}$$

where $\zeta = r + is$ with $r, s \in \mathbb{R}^d$ and $d\zeta$ means $drds$. By normality this formula also holds for an arbitrary $x \in \mathfrak{L}(\mathfrak{h})$.

We now check (36). Indeed, for each unit vector e_α ($\alpha \in \mathbb{N}^d$) of the canonical orthonormal basis of $\mathfrak{h}$ given by d dimensional Hermite polynomials multiplied by the function e_0, we have

$$\langle e_\alpha, \mathcal{T}_t(|e_0\rangle\langle e_0|)e_\alpha\rangle = \frac{1}{(2\pi t)^d}\int_{\mathbb{R}^{2d}} \left|\langle e_\alpha, W(\zeta/\sqrt{2})e_0\rangle\right|^2 \exp(-|\zeta|^2/2t)\,d\zeta$$

$$= \frac{1}{(2\pi t)^d}\int_{\mathbb{R}^{2d}} \frac{|\zeta_1|^{2\alpha_1}\cdots|\zeta_d|^{2\alpha_d}}{2^{|\alpha|}\alpha_1!\cdots\alpha_d!}\exp(-(1+1/t)|\zeta|^2/2)\,d\zeta$$

where $|\alpha| = \alpha_1 + \cdots + \alpha_d$. By the change of variables $\zeta = \xi/(1+1/t)^{1/2}$ we find

$$\langle e_\alpha, \mathcal{T}_t(|e_0\rangle\langle e_0|)e_\alpha\rangle = c_\alpha\frac{(1+1/t)^{-|\alpha|}}{(1+t)^d}$$

where c_α is a strictly positive constant that can be evaluated by computing a Gaussian integral and shown to be equal to 1.

By means of (36), for each $d \geq 2$, we compute

$$\int_0^\infty \langle e_\alpha, \mathcal{T}_t(|e_0\rangle\langle e_0|)e_\alpha\rangle dt = \int_0^\infty \frac{(1+1/t)^{-|\alpha|}}{(1+t)^d}\,dt < +\infty.$$

Moreover, since the restriction of $\mathcal{T}$ to $\mathfrak{M}$ is also irreducible, for each β, we have $\mathcal{T}_{t_\beta}(|e_0\rangle\langle e_0|) \geq \kappa(\beta,t_\beta)|e_\beta\rangle\langle e_\beta|$ for some $t_\beta > 0$ and some constant $\kappa(\beta,t_\beta) > 0$. It follows that, for each $d \geq 2$, our QMS is transient.

On the other hand, when $d = 1$, suppose that $\mathcal{T}$ is again transient and let $(p_n)_{n\geq 1}$ be an increasing sequence of projections with l.u.b. $p_n = 1$ and $\mathcal{U}(p_n)$ bounded. We have then

$$\int_0^\infty \langle e_0, \mathcal{T}_t(p_n)e_0\rangle dt = |\langle e_0, p_n e_0\rangle|^2 \int_0^\infty \frac{dt}{1+t}$$

which diverges whenever $|\langle e_0, p_n e_0\rangle|^2$ is nonzero. This contradicts the fact that e_0 belongs to the domain of $\mathcal{U}(p_n)$ for all $n \geq 0$. Therefore, $\mathcal{T}$ being irreducible, it must be recurrent.

Corollary 7.4. *The d-harmonic oscillator is recurrent for $d = 1$ and transient for $d \geq 2$.*

References

1. Luigi Accardi and Carlo Cecchini. Conditional expectations in von Neumann algebras and a theorem of Takesaki. *J. Funct. Anal.*, 45(2):245–273, 1982.
2. L. Accardi, C. Fernández, H. Prado and R. Rebolledo, Sur les temps moyens de séjour quantiques. C. R. Acad. Sci. Paris Sér. I Math.**319**, (1994), 723–726.
3. Accardi, L. and Kozyrev, V. : On the structure of quantum Markov flows. Preprint Centro V. Volterra (1999).
4. Alli, G.; Sewell, G.L.: New methods and structures in the theory of the multimode Dicke laser model. *J. Math. Phys.* **36** (1995), no. 10, 5598–5626.
5. W. Arveson. The heat flow of the CCR algebra. *Bull. London Math. Soc.* **34**, no. 1, 73–83, 2002.
6. J. Azéma, M. Duflo and D. Revuz. Propriétés relatives des processus de Markov récurrents. *Z. Wahr. und Verw. Gebiete*, **13**, 286–314, 1969.
7. J. Bellissard and H. Schulz-Baldes. Anomalous transport in quasicrystals. In R. Mosseri C. Janot, editor, *Proc. Of the 5th. International Conference on Quasicrystals*, pages 439–443. World Scientific, 1995.
8. Bellissard, J. and Schulz-Baldes, H.: Anomalous transport: a mathematical framework, Rev. Math. Phys., **10**, 1-46 (1998)
9. Bellissard J., Rebolledo R., Spehner D. and von Waldenfels W.: The Quantum Flow of Electronic transport, in preparation.
10. Ph. Biane. Quelques propriétés du mouvement Brownien non-commutatif. Hommage à P.A. Meyer et J. Neveu, *Astérisque*, **236**, 73-102, 1996.
11. O. Bratteli and D.W. Robinson. *Operator Algebras and Quantum Statistical Mechanics*, volume 1. Springer-Verlag, 2nd. edition, 1987.
12. O. Bratteli and D.W. Robinson. *Operator Algebras and Quantum Statistical Mechanics*, volume 2. Springer-Verlag, 2nd. edition, 1996.
13. R. Carbone. *Exponential ergodicity of a class of quantum Markov semigroups*. Tesi di Dottorato. Università di Milano, 2000.
14. A.M. Chebotarev and F. Fagnola. Sufficient conditions for conservativity of minimal quantum dynamical semigroups. *J. Funct. Anal.* **153**, 382–404 (1998).
15. A.M. Chebotarev. *Lectures on Quantum Probability*. Aportaciones Matemáticas, Ser. Textos, **14**, México, 2000.
16. K. L. Chung. *Markov chains with stationary transition probabilities*, Second edition. Die Grundlehren der mathematischen Wissenschaften, Band 104, Springer-Verlag New York, Inc., New York, 1967.
17. H. Comman. *A Non-Commutative Topological Theory of Capacities and Applications*. PhD thesis, Pontificia Universidad Católica de Chile, Facultad de Matemáticas, 2000.
18. N. Dang Ngoc. Classification des systèmes dynamiques non commutatifs. *J.Funct.Anal.*, 15:188–201, 1974.
19. E.B. Davies. Quantum stochastic processes. *Commun. Math. Phys.*, 15:277–304, 1969.
20. E.B. Davies. Quantum dynamical semigroups and the neutron diffusion equation. *Rep.Math.Phys.*, 11:169–188, 1977.
21. E.B. Davies. Generators of dynamical semigroups. *J.Funct.Anal.*, 34:421–432, 1979.

22. C. Dellacherie and P.-A. Meyer. *Probabilités et potentiel. Chapitres XII–XVI*, Second edition, Hermann, Paris, 1987.
23. J. Dixmier. *Von Neumann Algebras*. North-Holland, 1981.
24. S.N. Ethier and T.G. Kurtz. *Markov processes, Characterization and Convergence*. Wiley Series in Probability and Statistics. John Wiley and Sons, New York, 1985.
25. F. Fagnola. Characterization of isometric and unitary weakly differentiable cocycles in fock space. *Quantum Probabability and Related Topics*, VIII:143–164, 1993.
26. F. Fagnola. Quantum markov semigroups and quantum flows. *Proyecciones, Journal of Math.*, 18(3):1–144, 1999.
27. Fagnola, F.; Rebolledo, R.; Saavedra, C.: Quantum flows associated to master equations in quantum optics. *J. Math. Phys.* **35** (1994), no. 1, 1–12.
28. F. Fagnola and R. Rebolledo. An ergodic theorem in quantum optics. pages 73–86, 1996. Proceedings of the Univ. of Udine Conference in honour of A. Frigerio, Editrice Universitaria Udinese.
29. F. Fagnola and R. Rebolledo. The approach to equilibrium of a class of quantum dynamical semigroups. *Inf. Dim. Anal. Q. Prob. and Rel. Topics*, 1(4):1–12, 1998.
30. Fagnola, F.; Rebolledo, R.: A view on Stochastic Differential Equations derived from Quantum Optics. Aportaciones Matemáticas, Soc.Mat.Mexicana, (1999).
31. F. Fagnola and R. Rebolledo. On the existence of invariant states for quantum dynamical semigroups. *J.Math.Phys.*, 2000.
32. F. Fagnola and R. Rebolledo. Subharmonic projections for a Quantum Markov Semigroup, *J.Math.Phys.*, **43**, 1074-1082, 2002.
33. F. Fagnola and R. Rebolledo. Lectures on the Qualitative Analysis of Quantum Markov Semigroups, Quantum Probability and White Noise Analysis, World Scientific, vol. XIV, 197-240, 2002.
34. F. Fagnola and R. Rebolledo. Quantum Markov Semigroups and their Stationary States. In *Stochastic Analysis and Mathematical Physics (ANESTOC 2000)*, Trends in Mathematics, Birkhäuser ISBN 3-7643-6997-3, 77-128, 2003.
35. A. Frigerio. Stationary states of quantum dynamical semigroups. *Comm. in Math. Phys.*, 63:269–276, 1978.
36. A.Frigerio and M.Verri. Long–time asymptotic properties of dynamical semigroups on w^*–algebras. *Math. Zeitschrift*, 1982.
37. N. Gisin and I. Percival: J.Phys.A, **25**, (1992), 5677.
38. A. Guichardet. Systèmes dynamiques non commutatifs. *Astérisque*, 13-14:1–203, 1974.
39. A. S. Holevo. On the structure of covariant dynamical semigroups. *J. Funct. Anal.*, 131(2):255–278, 1995.
40. Palle E. T. Jorgensen. Semigroups of measures in non-conmutative harmonic analysis. *Semigroup Forum.*, 43(3):263–290, 1991.
41. Tosio Kato, *Perturbation theory for linear operators. Corr. printing of the 2nd ed.* Springer–Verlag, N.Y., 1980.
42. I. .Kovacs and J. Szücs. Ergodic type theorems in von neumann algebras. *Acta Sc.Math.*, 27:233–246, 1966.
43. G. Lindblad. On the generators of quantum dynamical semigroups. *Commun. Math. Phys.*, 48:119–130, 1976.
44. P.-A. Meyer. *Quantum Probability for Probabilists*, volume 1538 of *Lect. Notes in Math.* Springer–Verlag, Berlin, Heidelberg, New York, 1993.
45. S.Ch. Moy. Characterization of conditional expectation as a transform of function spaces. *Pacific J. of Math.*, pages 47–63, 1954.
46. M.Orszag, *Quantum Optics*, Springer, Berlin, Heidelberg, New-York, (1999).
47. K.R. Parthasarathy. *An Introduction to Quantum Stochastic Calculus*, volume 85 of *Monographs in Mathematics*. Birkhaüser–Verlag, Basel-Boston-Berlin, 1992.

48. D. Petz. Conditional expectation in quantum probability. In *Quantum Proba. and Appl., Vol. III*, pages 251–260. Lecture Notes in Math. 1303, Springer-Verlag, 1988.

49. R. Rebolledo, *Entropy functionals in quantum probability*, Second Symposium on Probability Theory and Stochastic Processes. First Mexican-Chilean Meeting on Stochastic Analysis (Guanajuato, 1992), Soc. Mat. Mexicana, México City, 1992, pp. 13–36.

50. R. Rebolledo, On the recurrence of Quantum Dynamical Semigroups, (1997). Proc. ANESTOC'96, World Scientific Pub., 130–141.

51. R. Rebolledo, *Limit Problems for Quantum Dynamical Semigroups inspired from Scattering Theory*. Lecture Notes of the Summer School in Grenoble, QP Reports.

52. R. Rebolledo, Complete Positivity and the Markov structure of Open Quantum Systems. This volume.

53. Reed, M.; Simon B.: *Methods of Modern Mathematical Physics: II Fourier Analysis, Self-Adjointness*, Academic Press 1975.

54. H. Schulz-Baldes and J. Bellissard. A kinetic theory for quantum transport in aperiodic media. *J. Statist. Phys.*, 91(5-6):991–1026, 1998.

55. D. Spehner. *Contributions à la théorie du transport électronique dissipatif dans les solides apériodiques*. Thèse de Doctorat. IRSAMC, Université Paul Sabatier, Toulouse, 2000.

56. E. Størmer. Invariant states of von neumann algebras. *Math.Scand.*, 30:253–256, 1972.

57. M. Takesaki. Conditional expectations in operator algebras. *J. Funct, Anal.*, 9:306–321, 1972.

58. H. Umegaki. Conditional expectations in an operator algebra. *Tohoku J.Math.*, 6:177–181, 1954.

59. K. Yosida. *Functional Analysis*. Springer–Verlag, Berlin, Heidelberg, New York, 3rd. edition, 1971.

Continual Measurements in Quantum Mechanics and Quantum Stochastic Calculus

Alberto Barchielli

Dipartimento di Matematica, Politecnico di Milano,
Piazza Leonardo da Vinci 32, 20133 Milano, Italy
e-mail: Alberto.Barchielli@polimi.it
URL: http://www.mate.polimi.it/qp/

1 Introduction

1.1 Three Approaches to Continual Measurements

We speak of continual measurements in the case in which one ore more observables of a quantum system are followed with continuity in time. Traditional presentations of quantum mechanics consider only instantaneous measurements, but continual measurements on quantum systems are a common experimental practice; typical cases are the various forms of photon detection. The statements of a quantum theory about an observable are of probabilistic nature; so, it is natural that a quantum theory of continual measurements give rise to stochastic processes. Moreover, a continually observed system is certainly open. All these things shows that the development and the applications of a quantum theory of continual measurements need quantum measurement theory, open system theory, quantum optics, operator theory, quantum probability, quantum and classical stochastic processes... The first consistent paper treating continual measurements was published in 1969 and concerns counting of quanta [43], but some ideas on quantum counting formulae for photons had already been introduced before [66].

There are essentially three approaches to continual measurements [18, 21, 64]. These approaches have received various degrees of development, any one of them has its own merits and range of applicability, but "morally" all the three approaches are equivalent and one can go from one to the other and this feature is certainly at the bases of the flexibility and interest of the theory. The first approach is the *operational* one, which is based on positive operator valued measures or (generalized) observables and operation valued measures or *instruments* [22, 23, 44]. A variant of this approach is based on the Feynman integral [15, 22, 72, 78]. The second approach is based on *quantum stochastic calculus* and quantum stochastic differential equations [14, 24] and it is connected to quantum Langevin equations and the notion of input and output fields in quantum optics [54, 55]. The last approach is based on (classical) *stochastic differential equations* and the notion of *a posteriori states* [21, 29] and it is related to some notions appeared in quantum optics: quantum trajectories, Monte–Carlo wave function method, unravelling of master equation [33, 82]. This report is concerned mainly with the second approach, the one based on quantum stochastic calculus.

1.2 Quantum Stochastic Calculus and Quantum Optics

QSC Quantum stochastic calculus (QSC) [65, 84] was developed originally as a mathematical theory of quantum noise in open systems and its first applications in mathematical physics were the construction of unitary dilations of quantum dynamical semigroups [65, 84] and of quantum stochastic processes [51]. Soon after it was applied also to measurement theory in quantum mechanics [14, 24].

The "integrators" of QSC are Bose fields (annihilation and creation processes), which play the role of quantum analogues of independent Wiener processes, and some expressions quadratic in the field operators (conservation processes). The starting point for the applications of QSC in quantum optics is to take these Bose fields as an approximation of the electromagnetic field. The explicit introduction of QSC in quantum optics was made in Ref. 54, but the use of the related δ-correlated noise is older [70].

There are various kinds of applications of QSC to quantum optics. The Bose fields are merely considered as a source of noise and quantum stochastic differential equations (QSDE's) are used for guessing the correct master equation for the   QSDE
system of interest [53, 67, 76, 77]; the fields are used for modelling quantum input and output channels [13, 14, 17, 18, 40, 54, 69]; QSDE's are used for describing various arrangements for detecting photons [14, 16, 18, 31, 32, 80]. All these kinds of applications are related; there is not a sharp distinction [18, 55].

A central point in QSC is the 'quantum stochastic Schrödinger equation' or Hudson–Parthasarathy equation (39), giving the unitary dynamics of a quantum system interacting with the Bose fields.

1.3 Some Notations: Operator Spaces

Let A and B be two Banach spaces; then, we denote by $\mathcal{B}(A; B)$ the vector space of $\mathcal{B}(\cdot; \cdot)$
linear bounded operators from A into B and we set $\mathcal{B}(A) := \mathcal{B}(A; A)$. $\mathcal{B}(\cdot)$

Let $\mathcal{K}$ be a complex separable Hilbert space; we denote by $\mathcal{U}(\mathcal{K})$ the set of $\mathcal{U}(\cdot)$
unitary operators on it. Let us recall that in a general quantum theory the states are positive, normalized, linear functionals over a C^*- or W^*-algebra; we shall consider only the so called normal states, which are represented by trace–class operators. So, we introduce the *trace class* $\mathcal{T}(\mathcal{K}) := \{t \in \mathcal{B}(\mathcal{K}) : \mathrm{Tr}\left[\sqrt{t^*t}\right] < \infty\}$ and the set of $\mathcal{T}(\cdot)$
statistical operators, or *states*, $\mathcal{S}(\mathcal{K}) := \{\mathfrak{s} \in \mathcal{T}(\mathcal{K}) : \mathfrak{s} \geq 0, \ \mathrm{Tr}[\mathfrak{s}] = 1\}$; we denote $\mathcal{S}(\cdot)$
by $\|t\|_1 := \mathrm{Tr}\left[\sqrt{t^*t}\right]$ the norm in $\mathcal{T}(\mathcal{K})$. $\|\cdot\|_1$

Let $\mathcal{H}$ and $\mathcal{K}$ be two complex separable Hilbert spaces; the *partial trace* over $\mathcal{K}$ partial
is defined by: for $t \in \mathcal{T}(\mathcal{H} \otimes \mathcal{K})$, $\mathrm{Tr}_{\mathcal{K}}\{t\} \in \mathcal{T}(\mathcal{H})$ is the operator satisfying   trace

$$\mathrm{Tr}_{\mathcal{H}}\left(X \, \mathrm{Tr}_{\mathcal{K}}\{t\}\right) = \mathrm{Tr}_{\mathcal{H} \otimes \mathcal{K}}\{(X \otimes \mathbb{1})t\}, \qquad \forall X \in \mathcal{B}(\mathcal{H}). \tag{1}$$

Let us end by recalling the definition of core for a selfadjoint operator; see, for instance, [84] p. 64.

A closable and densely defined operator T in a Hilbert space $\mathcal{H}$ is called *symmetric* if $T \subset T^*$. The operator T is called *selfadjoint* if $T = T^*$. The operator T is called *essentially selfadjoint* if its closure is selfadjoint.

Let $D_0 \subset D(T)$ be a linear manifold and let T_0 be the restriction of T to D_0. If T is selfadjoint and the closure of T_0 is T, then D_0 is called a *core* for T.   core

2 Unitary Evolution and States

2.1 Quantum Stochastic Calculus

We are assuming that the reader is familiar with the main features of QSC, in the version based on the symmetric Fock space, and the Hudson–Parthasarathy equation [65, 84]. In the following we recall a few notions and results of QSC.

The Fock Space

Γ
$\mathcal{Z}$

We denote by Γ the symmetric (or boson) *Fock space* over the "one–particle space" $L^2(\mathbb{R}_+) \otimes \mathcal{Z}$, where $\mathcal{Z}$ is a separable complex Hilbert space ([84] p. 124); we shall see in the physical examples how to choose $\mathcal{Z}$. The space $L^2(\mathbb{R}_+) \otimes \mathcal{Z}$ is naturally identified with $L^2(\mathbb{R}_+; \mathcal{Z})$, so that a vector f in it is a square integrable function from $\mathbb{R}_+$ into $\mathcal{Z}$. So, we have

$$\Gamma := \Gamma_{\text{symm}}\left(L^2(\mathbb{R}_+) \otimes \mathcal{Z}\right), \qquad L^2(\mathbb{R}_+) \otimes \mathcal{Z} \simeq L^2(\mathbb{R}_+; \mathcal{Z}). \qquad (2)$$

$e(f)$

Let us denote by $e(f)$, $f \in L^2(\mathbb{R}_+; \mathcal{Z})$, the *exponential vectors*, whose components in the $0, 1, \ldots, n, \ldots$ particle spaces are

$$e(f) := \left(1, f, (2!)^{-1/2} f \otimes f, \ldots, (n!)^{-1/2} f^{\otimes n}, \ldots\right); \qquad (3)$$

the internal product between two exponential vectors is given by

$$\langle e(g)|e(f)\rangle = \exp\langle g|f\rangle. \qquad (4)$$

$\psi(f)$

Once normalized, the exponential vectors are called *coherent vectors*:

$$\psi(f) := \exp\left(-\frac{1}{2} \|f\|^2\right) e(f). \qquad (5)$$

$\mathcal{E}(\cdot)$
$\mathcal{E}$

In particular the vector $e(0) \equiv \psi(0)$ is the *Fock vacuum*. If $\mathcal{M}$ is a dense linear manifold in $L^2(\mathbb{R}_+; \mathcal{Z})$, then the linear span $\mathcal{E}(\mathcal{M})$ of the vectors $e(f)$, $f \in \mathcal{M}$, is dense in Γ; we call *exponential domain* the set $\mathcal{E} := \mathcal{E}\left(L^2(\mathbb{R}_+; \mathcal{Z})\right)$, i.e. the linear span of all the exponential vectors.

$\Gamma_{(s,t)}$
$\Gamma_{(t}$

An important feature of the Fock space Γ is its structure of continuous tensor product. For any choice of the times $0 \leq s < t$ let us introduce the spaces

$$\Gamma_{(s,t)} := \Gamma_{\text{symm}}\left(L^2(s,t) \otimes \mathcal{Z}\right), \qquad \Gamma_{(t} := \Gamma_{\text{symm}}\left(L^2(t,+\infty) \otimes \mathcal{Z}\right), \qquad (6)$$

and for any of such spaces its exponential vectors. Then we have the natural identification ([84] p. 179)

$$\Gamma \simeq \Gamma_{(0,s)} \otimes \Gamma_{(s,t)} \otimes \Gamma_{(t} \qquad (7)$$

(here $0 < s < t$) based on the factorization of the exponential vectors

$$e(f) \simeq e\big(f_{(0,s)}\big) \otimes e\big(f_{(s,t)}\big) \otimes e\big(f_{(t)}\big), \tag{8}$$

where

$$f_{(s,t)}(\tau) := 1_{(s,t)}(\tau) \, f(\tau), \qquad f_{(t}(\tau) := 1_{(t,+\infty)}(\tau) \, f(\tau). \tag{9}$$

$f_{(s,t)},$
$f_{(t}$

Similarly, if P is any orthogonal projection, one has the factorization

$$\Gamma = \Gamma_{\mathrm{symm}}\big(PL^2(\mathbb{R}_+; \mathcal{Z})\big) \otimes \Gamma_{\mathrm{symm}}\big((\mathbb{1} - P)L^2(\mathbb{R}_+; \mathcal{Z})\big). \tag{10}$$

The Weyl Operators and the Bose Fields

The Weyl operator $\mathcal{W}(g; U)$, $g \in L^2(\mathbb{R}_+; \mathcal{Z})$, $U \in \mathcal{U}\big(L^2(\mathbb{R}_+; \mathcal{Z})\big)$, is the unique unitary operator ([84] Section 20) defined by

$\mathcal{W}(g; U)$

$$\mathcal{W}(g; U) \, e(f) = \exp\left\{-\tfrac{1}{2}\,\|g\|^2 - \langle g|Uf\rangle\right\} e(Uf + g) \tag{11}$$

or by

$$\mathcal{W}(g; U) \, \psi(f) = \exp\left\{\mathrm{i}\,\mathrm{Im}\langle Uf|g\rangle\right\} \psi(Uf + g). \tag{12}$$

From the definition one obtains easily the inverse

$$\mathcal{W}(g; U)^{-1} = \mathcal{W}(g; U)^* = \mathcal{W}(-U^*g; U^*) \tag{13}$$

and the composition law

$$\mathcal{W}(h; V) \, \mathcal{W}(g; U) - \exp\left\{-\mathrm{i}\,\mathrm{Im}\langle h|Vg\rangle\right\} \mathcal{W}(h + Vg; VU); \tag{14}$$

moreover, by particularizing this equation to $V = U = \mathbb{1}$, one gets the Weyl form of the canonical commutation relations (CCR):

CCR

$$\mathcal{W}(h; \mathbb{1}) \, \mathcal{W}(g; \mathbb{1}) = \mathcal{W}(g; \mathbb{1}) \, \mathcal{W}(h; \mathbb{1}) \, \exp\left\{-2\mathrm{i}\,\mathrm{Im}\langle h|g\rangle\right\}. \tag{15}$$

The Weyl operators allow to introduce some important selfadjoint operators in Γ ([84] Section 20), which play the double role of being the starting point to construct the integrators of QSC and of representing the main observables used in the the theory of continual measurements. Let us collect in a unique theorem Propositions 20.4, 20.7, 20.11, 2.16 and Corollaries 20.5 and 20.6 of Ref. 84.

Theorem 2.1. *Let us take* $h \in L^2(\mathbb{R}_+; \mathcal{Z})$; *then, the map* $\kappa \mapsto \mathcal{W}(\mathrm{i}\kappa h; \mathbb{1})$ *is a strongly continuous one parameter group and we denote by* $Q(h)$ *its Stone generator:*

$Q(h)$

$$\mathcal{W}(\mathrm{i}\kappa h; \mathbb{1}) = \exp\{\mathrm{i}\kappa Q(h)\}. \tag{16}$$

Moreover, one has

(i) $Q(h)$ is essentially selfadjoint in the domain $\mathcal{E}(\mathcal{M})$, where $\mathcal{M}$ is any dense subset of $L^2(\mathbb{R}_+; \mathcal{Z})$;

(ii) $\mathcal{E}$ is a core for $Q(h)$;

(iii) the linear manifold of all finite particle vectors is a core for $Q(h)$;

(iv) $\mathcal{E}$ is included in the domain of the product $Q(h_1)Q(h_2)\cdots Q(h_n)$, $\forall n$, $\forall h_1,\ldots,h_n \in L^2(\mathbb{R}_+;\mathcal{Z})$;

(v) $[Q(h),Q(g)]\,e(f) = \{2\mathrm{i}\,\mathrm{Im}\langle h|g\rangle\}\,e(f)$, $\forall h,g,f \in L^2(\mathbb{R}_+;\mathcal{Z})$.

Let B be a selfadjoint operator in $L^2(\mathbb{R}_+;\mathcal{Z})$ with domain $D(B)$; then, the map $\kappa \mapsto \mathcal{W}(0;\exp\{\mathrm{i}\kappa B\})$ is a strongly continuous one parameter group and we denote by $\lambda(B)$ its Stone generator:

$$\mathcal{W}(0;\exp\{\mathrm{i}\kappa B\}) = \exp\{\mathrm{i}\kappa\lambda(B)\}. \tag{17}$$

Moreover, one has

(a) $\mathcal{E}(D(B))$ is included in the domain of $\lambda(B)$;
(b) $\mathcal{E}(D(B^2))$ is a core for $\lambda(B)$;
(c) if B is bounded, $\lambda(B)$ is essentially selfadjoint in the domain $\mathcal{E}$;
(d) $\mathrm{i}[\lambda(B_1),\lambda(B_2)]\,e(f) = \lambda\left(\mathrm{i}[B_1,B_2]\right)e(f)$, for any two bounded selfadjoint operators B_1, B_2 and $\forall f \in L^2(\mathbb{R}_+;\mathcal{Z})$;
(e) $\mathcal{E}$ is included in the domain of the product $T_1 T_2 \cdots T_n$, where $T_i = Q(h_i)$, $h_i \in L^2(\mathbb{R}_+;\mathcal{Z})$, or $T_i = \lambda(B_i)$, $B_i = B_i^* \in \mathcal{B}\left(L^2(\mathbb{R}_+;\mathcal{Z})\right)$.

For any $h \in L^2(\mathbb{R}_+;\mathcal{Z})$ and any selfadjoint operator B in $L^2(\mathbb{R}_+;\mathcal{Z})$ let us set

$$\lambda(B,h) := \mathcal{W}(-h;\mathbb{1})\,\lambda(B)\,\mathcal{W}(h;\mathbb{1}). \tag{18}$$

Then, the operator $\lambda(B,h)$ is the generator of the unitary group

$$\kappa \mapsto \mathcal{W}(-h;\mathbb{1})\mathcal{W}(0;\exp\{\mathrm{i}\kappa B\})\mathcal{W}(h;\mathbb{1})$$
$$\equiv \exp\left\{\mathrm{i}\,\mathrm{Im}\left\langle \mathrm{e}^{\mathrm{i}\kappa B}h|h\right\rangle\right\}\mathcal{W}\left((\mathrm{e}^{\mathrm{i}\kappa B}-\mathbb{1})h;\mathrm{e}^{\mathrm{i}\kappa B}\right) \tag{19}$$

and it is essentially selfadjoint on the linear manifold generated by $\{e(f-h):\ f \in D(B^2)\}$. When B is also bounded, $\mathcal{E}$ is a core for $\lambda(B,h)$ and, on the exponential domain $\mathcal{E}$, one has

$$\lambda(B,h) = \lambda(B) + a(Bh) + a^\dagger(Bh) + \langle h|Bh\rangle\mathbb{1}. \tag{20}$$

The operators $a(\cdot)$ and $a^\dagger(\cdot)$ are defined here below in eq. (21).

By defining

$$a(h) = \frac{1}{2}\left(Q(h)+\mathrm{i}Q(\mathrm{i}h)\right),\qquad a^\dagger(h) = \frac{1}{2}\left(Q(h)-\mathrm{i}Q(\mathrm{i}h)\right), \tag{21}$$

one obtains two mutually adjoint operators, satisfying ([84] Proposition 20.12) the eigenvalue relation

$$a(h)\,e(f) = \langle h|f\rangle\,e(f) \tag{22}$$

and, at least in the domain $\mathcal{E}$, the CCR

$$[a(h),a(g)] = [a^\dagger(h),a^\dagger(g)] = 0,\qquad [a(h),a^\dagger(g)] = \langle h|g\rangle. \tag{23}$$

So, we recognize the annihilation and creation operators and we call them, collectively, the (smeared) *Bose fields*; in quantum optics the two selfadjoint operators $Q(h)$ and $Q(ih)$ are sometimes referred to as two conjugated *field quadratures*.

If $B \in \mathcal{B}\big(L^2(\mathbb{R}_+; \mathcal{Z})\big)$ is bounded but not selfadjoint, one defines $\lambda(B) := \lambda\left(\frac{1}{2}(B + B^*)\right) + i\lambda\left(\frac{1}{2i}(B - B^*)\right)$. All matrix elements on exponential vectors and commutation relations involving $Q(h)$, $a(h)$, $a^\dagger(h)$, $\lambda(B)$ are deduced from the properties of the Weyl operators and of the exponential vectors and are given in Ref. 84 Section 20. Moreover, also the linear manifold of all finite particle vectors is contained in the domains of $Q(h)$, $a(h)$, $a^\dagger(h)$, $\lambda(B)$ ([84] Proposition 20.14).

The annihilation, creation and conservation processes

Let us fix a complete orthonormal system (c.o.n.s.) $\{z_k,\ k \geq 1\}$ in $\mathcal{Z}$ and for any $f \in L^2(\mathbb{R}_+; \mathcal{Z})$ let us set $f_k(t) := \langle z_k | f(t) \rangle$. We denote by $A_k(t)$, $A_k^\dagger(t)$, $\Lambda_{kl}(t)$ the *annihilation, creation and conservation processes* associated with such a c.o.n.s. ([65] Sect. 2):

$$A_k(t) := a\big(z_k \otimes 1_{(0,t)}\big), \qquad A_k^\dagger(t) := a^\dagger\big(z_k \otimes 1_{(0,t)}\big), \qquad (24a)$$

$$\Lambda_{kl}(t) := \lambda\left((|z_k\rangle\langle z_l|) \otimes 1_{(0,t)}\right); \qquad (24b)$$

for these processes one has

$$A_k(t)\, e(f) = \int_0^t f_k(s)\,\mathrm{d}s\, e(f), \qquad (25a)$$

$$\langle e(g) | A_k^\dagger(t) e(f) \rangle = \int_0^t \overline{g_k(s)}\,\mathrm{d}s\, \langle e(g) | e(f) \rangle, \qquad (25b)$$

$$\langle e(g) | \Lambda_{kl}(t) e(f) \rangle = \int_0^t \overline{g_k(s)}\, f_l(s)\,\mathrm{d}s\, \langle e(g) | e(f) \rangle. \qquad (25c)$$

c.o.n.s.
$\{z_k\}$,
$f_k(t)$
$A_k(t)$,
$A_k^\dagger(t)$

$\Lambda_{kl}(t)$

Let us recall that by construction these operators are defined at least on $\mathcal{E}$ and on this domain $A_k^\dagger(t)$ is the adjoint of $A_k(t)$; moreover, $A_k(t) + A_k^\dagger(t)$, $i\big(A_k^\dagger(t) - A_k(t)\big)$, $\Lambda_{kk}(t)$ are essentially selfadjoint on $\mathcal{E}$. Another form of the CCR follows from (24a) and (23): on the exponential domain, and on the finite particle vectors, one has

$$[A_k(t), A_l(s)] = [A_k^\dagger(t), A_l^\dagger(s)] = 0, \qquad [A_k(t), A_l^\dagger(s)] = \delta_{kl}\min\{t, s\}. \quad (26)$$

In theoretical physics it is usual to write formally

$$A_k(t) = \int_0^t a_k(s)\,\mathrm{d}s, \qquad A_k^\dagger(t) = \int_0^t a_k^\dagger(s)\,\mathrm{d}s, \qquad \Lambda_{kl}(t) = \int_0^t a_k^\dagger(s) a_l(s)\,\mathrm{d}s, \tag{27}$$

where the "Bose fields" $a_k(t)$, $a_k^\dagger(t)$ satisfies the (heuristic) CCR

$$[a_k(t), a_l^\dagger(s)] = \delta_{kl}\,\delta(t - s), \qquad [a_k(t), a_l(s)] = [a_k^\dagger(t), a_l^\dagger(s)] = 0. \tag{28}$$

Quantum Stochastic Integrals

The System Space

We are interested in a quantum system interacting with the Bose fields we have introduced; this quantum system is described in a complex separable Hilbert space $\mathcal{H}$, called the *system space* or the *initial space* ([84] p. 179); let us call this quantum system "system $S_{\mathcal{H}}$" or simply "the system". Operators acting in $\mathcal{H}$ are extended to $\mathcal{H} \otimes \Gamma$ by the convention that they act as the identity on Γ; the tensor product with the identity is not always indicated. A similar extension is understood for operators acting in Γ. So, $K \in \mathcal{B}(\mathcal{H})$ or $K \otimes \mathbb{1}$, $A_k(t)$ or $\mathbb{1} \otimes A_k(t)$ are the same.

Adapted processes

An adapted process is a time dependent family of operators $\{L(t),\, t \geq 0\}$, such that $L(t)$ acts trivially as the identity on $\Gamma_{(t}$ and possibly non trivially in $\mathcal{H} \otimes \Gamma_{(0,t)}$; an adapted process is something containing the fields only up to time t. In the case of a bounded adapted process, this simply means $L(t) \in \mathcal{B}(\mathcal{H} \otimes \Gamma_{(0,t)})$; in the general case the definition is the following.

Definition 2.1 ([84] p. 180). *Let D and $\mathcal{M}$ be dense linear manifolds in $\mathcal{H}$ and $L^2(\mathbb{R}_+; \mathcal{Z})$, respectively, such that $1_{(s,t)}f \in \mathcal{M}$ whenever $f \in \mathcal{M}$ for all $0 \leq s < t < \infty$. Denote by $D \underline{\otimes} \mathcal{E}(\mathcal{M})$ the linear manifold generated by all the vectors of the form $u \otimes e(f)$, $u \in D$, $f \in \mathcal{M}$. A family $\{L(t),\, t \geq 0\}$ of operators in $\mathcal{H} \otimes \Gamma$ is an* adapted process *with respect to $(D, \mathcal{M})$ if*

(i) for any t, the domain of $L(t)$ contains $D \underline{\otimes} \mathcal{E}(\mathcal{M})$;
(ii) $L(t)u \otimes e(f_{(0,t)}) \in \mathcal{H} \otimes \Gamma_{(0,t)}$ and $L(t)u \otimes e(f) = \{L(t)u \otimes e(f_{(0,t)})\} \otimes e(f_{(t})$ for all $t \geq 0$, $u \in D$, $f \in \mathcal{M}$.

It is said to be regular *if, in addition, the map $t \mapsto L(t)u \otimes e(f)$ from $\mathbb{R}_+$ into $\mathcal{H} \otimes \Gamma$ is continuous for every $u \in D$, $f \in \mathcal{M}$.*

It is convenient to fix $\mathcal{M}$ once for all; we follow the choice of Ref. 71:

$$\mathcal{M} := \{f \in L^2(\mathbb{R}_+; \mathcal{Z}) \cap L^\infty(\mathbb{R}_+; \mathcal{Z}) : f_k(t) \equiv 0$$
$$\text{for all but a finite number of indices } k\}. \quad (29)$$

For $f \in \mathcal{M}$ let us set $\dim f := \max\{k \mid f_k \text{ is a non-zero vector in } L^2(\mathbb{R}_+)\}$. Let us note that the definitions of $\mathcal{M}$ and $\dim f$ depend on the initial choice of the c.o.n.s. $\{z_k\}$ in $\mathcal{Z}$ and that, being $\mathcal{M}$ dense in $L^2(\mathbb{R}_+; \mathcal{Z})$, then $\mathcal{E}(\mathcal{M})$ is total in Γ.

Quantum stochastic integrals and Ito table

In QSC integrals of "Ito type" with respect to $dA_k(t)$, $dA_k^\dagger(t)$, $d\Lambda_{kl}(t)$ are defined ([84] pp. 188–190, 224–225); the integrands are adapted processes with some conditions to assure the existence of the quantum stochastic integrals. We shall use the class of integrands given in the definition below. It is the one used in Ref. 71 and it allows to give meaning to all the integrals we need; it is a bit larger of the one introduced by Parthasarathy [84], but all the results of Ref. 84 continue to hold.

Definition 2.2 ([65] Proposition 3.2; [84] pp. 189, 221–222, 224; [71]). *A family* $\{L_l^k, \; k,l \geq 0\}$ *of* $(\mathcal{H}, \mathcal{M})$ *adapted processes is said to be* stochastically integrable *if, for all* $t \geq 0, l \geq 0, u \in \mathcal{H}, f \in \mathcal{M}$,

$$\int_0^t \sum_{k=0}^\infty \|L_l^k(s) u \otimes e(f)\|^2 (1 + \|f(s)\|^2) \mathrm{d}s < +\infty.$$

We denote by $\mathbb{L}(\mathcal{M})$ *the class of stochastically integrable families of* $(\mathcal{H}, \mathcal{M})$ *adapted processes.*

Note that $\mathcal{M}$ and, so, the class of stochastically integrable processes $\mathbb{L}(\mathcal{M})$ depend on the initial choice of the c.o.n.s. $\{z_k\}$ in $\mathcal{Z}$.

The definition of quantum stochastic integral goes trough a suitable limit on the integral of sequences of "simple processes" ([84] Section 27); then, the following result holds.

Proposition 2.1 ([84] Proposition 27.1). *Let* $\{L_l^k, \; k,l \geq 0\} \in \mathbb{L}(\mathcal{M})$, $X(0) \in \mathcal{B}(\mathcal{H})$; *then*

$$X(t) := X(0) + \int_0^t \left\{ L_0^0(s)\mathrm{d}s + \sum_{k=1}^\infty L_0^k(s)\mathrm{d}A_k^\dagger(s) \right.$$
$$\left. + \sum_{l=1}^\infty I_l^0(s)\mathrm{d}A_l(s) + \sum_{k,l=1}^\infty L_l^k(s)\mathrm{d}\Lambda_{kl}(s) \right\} \quad (30)$$

is a regular $(\mathcal{H}, \mathcal{M})$ *adapted process and,* $\forall u \in \mathcal{H}, \forall f \in \mathcal{M}, \forall t \geq 0$,

$$\|[X(t) - X(0)] u \otimes e(f)\|^2 \leq 2 \exp\left\{ \int_0^t (1 + \|f(s)\|^2)\mathrm{d}s \right\}$$
$$\times \sum_{l=0}^{\dim f} \int_0^t \sum_{k=0}^\infty \|L_l^k(s) u \otimes e(f)\|^2 (1 + \|f(s)\|^2)\mathrm{d}s. \quad (31)$$

The main practical rules to manipulate the quantum stochastic integrals and their products are the facts that

1. $\mathrm{d}A_k(t), \mathrm{d}A_k^\dagger(t), \mathrm{d}\Lambda_{kl}(t)$ commute with adapted processes at time t, so that they can be shifted towards the right or the left, according to the convenience;
2. the products of the fundamental differentials satisfy the Ito table

$$\mathrm{d}A_k(t)\,\mathrm{d}A_l^\dagger(t) = \delta_{kl}\,\mathrm{d}t\,, \qquad\qquad \mathrm{d}A_k(t)\,\mathrm{d}\Lambda_{rl}(t) = \delta_{kr}\,\mathrm{d}A_l(t)\,,$$
$$\mathrm{d}\Lambda_{kr}(t)\,\mathrm{d}A_l^\dagger(t) = \delta_{rl}\,\mathrm{d}A_k^\dagger(t)\,, \qquad \mathrm{d}\Lambda_{kr}(t)\,\mathrm{d}\Lambda_{sl}(t) = \delta_{rs}\,\mathrm{d}\Lambda_{kl}(t)\,; \quad (32)$$

all the other products and the products involving $\mathrm{d}t$ vanish;
3. $\mathrm{d}A_k(t)\,e(f) = \mathrm{d}t\,f_k(t)\,e(f), \qquad \langle e(f)|\,\mathrm{d}A_k^\dagger(t) = \overline{f_k(t)}\,\mathrm{d}t\,\langle e(f)|,$
   $\mathrm{d}\Lambda_{kl}(t)\,e(f) = \mathrm{d}A_k^\dagger(t)\,f_l(t)\,e(f).$

By means of these rules the matrix elements on exponential vectors of quantum stochastic integrals can be computed and products of quantum stochastic integrals can
be differentiated (quantum Ito's formula); see [84] pp. 221–224. To be more precise,
the following results hold.

Proposition 2.2 ([84] Corollary 27.2). *Let* $\{L_l^k,\ k,l \geq 0\}$, $\{M_l^k,\ k,l \geq 0\} \in$
$\mathbb{L}(\mathcal{M})$, $X(0)$, $Y(0) \in \mathcal{B}(\mathcal{H})$; *let* $X(t)$ *be defined by eq.* (30) *and* $Y(t)$ *be defined
in a similar way in terms of the processes* M_l^k. *Then, for all* $u,\ v \in \mathcal{H}$, $f,\ g \in \mathcal{M}$
the matrix elements $\langle v \otimes e(g)|X(t)u \otimes e(f)\rangle$, $\langle v \otimes e(g)|Y(t)u \otimes e(f)\rangle$, $\langle Y(t)v \otimes$
$e(g)|X(t)u\otimes e(f)\rangle$ *are just the ones that one can compute by means of the practical
rules given above, i.e.*

$$\langle v \otimes e(g)|\,[X(t) - X(0)]\,u \otimes e(f)\rangle = \int_0^t \mathrm{d}s \Big\langle v \otimes e(g)\Big| \Big\{ L_0^0(s) + \sum_{k=1}^{\infty} \overline{g_k(s)}$$

$$\times L_0^k(s) + \sum_{l=1}^{\infty} L_l^0(s)f_l(s) + \sum_{k,l=1}^{\infty} \overline{g_k(s)}L_l^k(s)f_l(s) \Big\} u \otimes e(f)\Big\rangle, \quad (33)$$

$$\langle Y(t)\,v \otimes e(g)|X(t)\,u \otimes e(f)\rangle - \langle Y(0)\,v \otimes e(g)|X(0)\,u \otimes e(f)\rangle$$

$$= \int_0^t \mathrm{d}s \Big\{ \langle Y(s)\,v \otimes e(g)|L_0^0(s)\,u \otimes e(f)\rangle$$

$$+ \langle M_0^0(s)\,v \otimes e(g)|X(s)\,u \otimes e(f)\rangle + \sum_{k=1}^{\infty} \langle M_0^k(s)\,v \otimes e(g)|L_0^k(s)\,u \otimes e(f)\rangle$$

$$+ \sum_{k=1}^{\infty} \overline{g_k(s)} \Big[\langle Y(s)\,v \otimes e(g)|L_0^k(s)\,u \otimes e(f)\rangle$$

$$+ \langle M_k^0(s)\,v \otimes e(g)|X(s)\,u \otimes e(f)\rangle + \sum_{l=1}^{\infty} \langle M_k^l(s)\,v \otimes e(g)|L_0^l(s)\,u \otimes e(f)\rangle \Big]$$

$$+ \sum_{l=1}^{\infty} \Big[\langle Y(s)\,v \otimes e(g)|L_l^0(s)\,u \otimes e(f)\rangle + \langle M_0^l(s)\,v \otimes e(g)|X(s)\,u \otimes e(f)\rangle$$

$$+ \sum_{k=1}^{\infty} \langle M_0^k(s)\,v \otimes e(g)|L_l^k(s)\,u \otimes e(f)\rangle \Big] f_l(s)$$

$$+ \sum_{k,l=1}^{\infty} \overline{g_k(s)} \Big[\langle Y(s)\,v \otimes e(g)|L_l^k(s)\,u \otimes e(f)\rangle + \langle M_k^l(s)\,v \otimes e(g)|X(s)\,u \otimes e(f)\rangle$$

$$+ \sum_{r=1}^{\infty} \langle M_k^r(s)\,v \otimes e(g)|L_l^r(s)\,u \otimes e(f)\rangle \Big] f_l(s) \Big\}. \quad (34)$$

Moreover, the quantum stochastic integral (30) *is uniquely determined by the matrix
elements* (33).

To handle a product $Y(t)X(t)$ is a more delicate problem; in general we do not even know if $\mathcal{H} \otimes \mathcal{E}(\mathcal{M})$ is included in the domain of this product.

Definition 2.3. *Let* $\{L_l^k, \ k,l \geq 0\}$, $\{L_l^{\dagger k}, \ k,l \geq 0\} \in \mathbb{L}(\mathcal{M})$ *be such that* $\langle v \otimes e(g)|L_k^l(t)u \otimes e(f)\rangle = \langle L_l^{\dagger k}(t)v \otimes e(g)|u \otimes e(f)\rangle$, $\forall k,l,t, \forall u,v \in \mathcal{H}, \forall f,g \in \mathcal{M}$; *then*, $\{L_l^k, \ k,l \geq 0\}$, $\{L_l^{\dagger k}, \ k,l \geq 0\}$ *is called an* adjoint pair *in* $\mathbb{L}(\mathcal{M})$.

adjoint
pair

Proposition 2.3 ([84] Proposition 25.12, [71] p. 518). *Let* $\{L_l^k, \ k,l \geq 0\}$, $\{L_l^{\dagger k}, \ k,l \geq 0\}$ *be an adjoint pair in* $\mathbb{L}(\mathcal{M})$, $X(0) \in \mathcal{B}(\mathcal{H})$, $X^\dagger(0) = X(0)^*$; *let* $X(t)$ *be defined by eq. (30) and* $X^\dagger(t)$ *be defined in a similar way in terms of* $L_l^{\dagger k}$, $X^\dagger(0)$. *Then*, $X(t)^* = X^\dagger(t)$, $\forall t \geq 0$, *on* $\mathcal{H} \otimes \mathcal{E}(\mathcal{M})$.

Proposition 2.4 ([84] Proposition 25.26). *Let* $\{L_l^k, \ k,l \geq 0\}$, $\{M_l^k, \ k,l \geq 0\}$, $\{M_l^{\dagger k}, \ k,l \geq 0\} \in \mathbb{L}(\mathcal{M})$, *with* $\{M_l^k, \ k,l \geq 0\}$, $\{M_l^{\dagger k}, \ k,l \geq 0\}$ *an adjoint pair*, $X(0), Y(0) \in \mathcal{B}(\mathcal{H})$; *let* $X(t)$ *be defined by eq. (30) and* $Y(t)$ *be defined in a similar way in terms of* M_l^k, $Y(0)$. *Let us define*, $\forall k, l \geq 0, \forall t \geq 0$,

$$F_l^k(t) := M_l^k(t)X(t) + Y(t)L_l^k(t) + \sum_{j=1}^{\infty} M_j^k(t)L_l^j(t). \tag{35}$$

Suppose that

i) $Y(t)X(t)$, $t \geq 0$, *in an* (H, M) *adapted process*,
ii) $\{F_l^k, \ k,l \geq 0\} \in \mathbb{L}(\mathcal{M})$.

Then, on $\mathcal{H} \otimes \mathcal{E}(\mathcal{M})$, *one has*

$$Y(t)X(t) = Y(0)X(0) + \int_0^t \left\{ F_0^0(s)\mathrm{d}s + \sum_{k=1}^{\infty} F_0^k(s)\mathrm{d}A_k^\dagger(s) \right.$$
$$\left. + \sum_{l=1}^{\infty} F_l^0(s)\mathrm{d}A_l(s) + \sum_{k,l=1}^{\infty} F_l^k(s)\mathrm{d}A_{kl}(s) \right\}. \tag{36}$$

Parthasarathy gives the proof of this proposition only in the case of finitely many integrators, but nothing changes in the case of infinitely many integrators and of integrands in $\mathbb{L}(\mathcal{M})$; the existence of the adjoint pair is needed in the proof, but it does not appear in the final statement. The meaning of Proposition 2.4 is that, under the hypotheses given, $Y(t)X(t)$ can be differentiated according the practical rules of QSC.

2.2 The Unitary System–Field Evolution

The Hudson–Parthasarathy Equation

The ingredients of the Hudson–Parthasarathy equation are system operators and fields. We consider the simplest version of this equation: only bounded system operators are involved. This is not enough for all physical applications, but includes significant cases and allows to develop a general theory which gives an idea of what could be done even in other cases.

The system operators

H
R_k
S_{kl}

Let H, R_k, $k \geq 1$ and S_{kl}, $k, l \geq 1$, be bounded operators in $\mathcal{H}$ such that $H^* = H$ and $\sum_j S_{jk}^* S_{jl} = \sum_j S_{kj} S_{lj}^* = \delta_{kl}$; if $\mathcal{Z}$ is infinite dimensional, the set of indices is infinite and the previous series and $\sum_k R_k^* R_k$ are assumed to be strongly convergent to bounded operators. From these conditions, we have that also the series $\sum_k R_k^* S_{kl}$ is strongly convergent to a bounded operator. It is useful to construct from H, R_k, S_{kl} three operators

$$S \in \mathcal{U}(\mathcal{H} \otimes \mathcal{Z}), \qquad R \in \mathcal{B}(\mathcal{H}; \mathcal{H} \otimes \mathcal{Z}), \qquad K \in \mathcal{B}(\mathcal{H}) \qquad (37\text{a})$$

R, S, K by

$$Ru = \sum_k (R_k u) \otimes z_k, \quad \forall u \in \mathcal{H}, \qquad (37\text{b})$$

or, equivalently,

$$\langle v \otimes z_k | Ru \rangle = \langle v | R_k u \rangle, \quad \forall u, v \in \mathcal{H}, \qquad (37\text{c})$$

$$S := \sum_{kl} S_{kl} \otimes |z_k\rangle\langle z_l|, \qquad (37\text{d})$$

$$K := -\mathrm{i}H - \frac{1}{2} \sum_k R_k^* R_k \equiv -\mathrm{i}H - \frac{1}{2} R^* R. \qquad (37\text{e})$$

$C(f)$ Another useful operator, for $f \in \mathcal{Z}$, is

$$C(f) : \mathcal{H} \to \mathcal{H} \otimes \mathcal{Z}, \qquad C(f)u = u \otimes f, \quad \forall u \in \mathcal{H}. \qquad (37\text{f})$$

By the positions (37) we have also the useful relations

$$\sum_k \overline{g_k(s)} R_k = C(g(s))^* R, \qquad (38\text{a})$$

$$\sum_{kl} \overline{g_k(s)} (S_{kl} - \delta_{kl}) f_l(s) = C(g(s))^* (S - \mathbb{1}) C(f(s)), \qquad (38\text{b})$$

$$\sum_{kl} R_k^* S_{kl} f_l(s) = R^* S C(f(s)). \qquad (38\text{c})$$

The evolution equation

Theorem 2.2 ([84] Proposition 27.5 p. 225, Theorem 27.8 p. 228). *In the hypotheses above, there exists a unique $(\mathcal{H}, \mathcal{M})$ adapted process $U(t)$ satisfying the initial condition $U(0) = \mathbb{1}$ and the QSDE*

$U(t)$

$$\mathrm{d}U(t) = \left\{ \sum_k R_k \, \mathrm{d}A_k^\dagger(t) + \sum_{kl} (S_{kl} - \delta_{kl}) \, \mathrm{d}\Lambda_{kl}(t) \right.$$

$$\left. - \sum_{kl} R_k^* S_{kl} \, \mathrm{d}A_l(t) + K \, \mathrm{d}t \right\} U(t). \qquad (39)$$

Moreover, $U(t)$ is uniquely extended to a unitary process, which turns out to be strongly continuous in t.

The adjoint process $U(t)^*$ is strongly continuous and it is the unique unitary adapted process satisfying $U(0)^* = \mathbb{1}$ and the adjoint equation

$$dU(t)^* = U(t)^* \left\{ \sum_k R_k^* \, dA_k(t) + \sum_{kl} (S_{kl}^* - \delta_{kl}) \, d\Lambda_{lk}(t) \right.$$
$$\left. - \sum_{kl} S_{kl}^* R_k \, dA_l^\dagger(t) + K^* \, dt \right\}. \quad (40)$$

By the practical rules of QSC (Proposition 2.2) we get from (39): $\forall u, v \in \mathcal{H}$, $\forall f, g \in \mathcal{M}$

$$\langle v \otimes e(g) | U(t) \, u \otimes e(f) \rangle - \langle v | u \rangle \, \exp\{\langle g | f \rangle\} = \int_0^t ds \, \Big\langle v \otimes e(g) \Big| \Big\{ \sum_k \overline{g_k(s)} R_k$$
$$+ \sum_{kl} \overline{g_k(s)} \, (S_{kl} - \delta_{kl}) \, f_l(s) - \sum_{kl} R_k^* S_{kl} f_l(s) + K \Big\} U(s) \, u \otimes e(f) \Big\rangle; \quad (41)$$

it is also true that this equation uniquely determines U.

Corollary 2.1 ([56] Corollary 3.2 p. 26). *The solution U of (39) is the unique unitary adapted process satisfying (41) for every $t \geq 0$, $u, v \in \mathcal{H}$, $f, g \in \mathcal{M}$.*

Proof. Let U' be a unitary adapted process satisfying (41) and let us define

$$\widetilde{U}'(t) := \mathbb{1} + \int_0^t \left\{ \sum_k R_k \, dA_k^\dagger(s) + \sum_{kl} (S_{kl} - \delta_{kl}) \, d\Lambda_{kl}(s) \right.$$
$$\left. - \sum_{kl} R_k^* S_{kl} \, dA_l(s) + K \, ds \right\} U'(s).$$

We see that, by the hypotheses on the system operators and the unitarity of $U'(s)$, the coefficients $R_k U'(s), \ldots$ satisfy the conditions in Definition 2.2; then, $\widetilde{U}'(t)$ is well defined by Proposition 2.1 and its matrix elements on exponential vectors are given by Proposition 2.2. So, for every $t \geq 0$, $u, v \in \mathcal{H}$, $f, g \in \mathcal{M}$, we have $\langle v \otimes e(g) | \widetilde{U}'(t) \, u \otimes e(f) \rangle = \langle v \otimes e(g) | U'(t) \, u \otimes e(f) \rangle$; having equal matrix elements on a total set of vectors, $\widetilde{U}'(t)$ and $U'(t)$ are equal and this implies that $\widetilde{U}'(t)$ satisfies the QSDE (39). The uniqueness of the solution of (39) implies $\widetilde{U}'(t) = U'(t) = U(t)$. $\qquad\square$

The Hamiltonian Evolution

As soon as Hudson and Parthasarathy introduced their equation, Frigerio and Maassen independently realized that each unitary solution U is naturally associated to a strongly continuous one–parameter unitary group V [50, 51, 73, 74]. This is important for physical applications: a strongly continuous one–parameter unitary group is what we want in quantum mechanics to give the dynamics of an isolated system, here system $S_{\mathcal{H}}$ and fields. To obtain this result we need to enlarge the Fock space Γ; it is convenient to consider this ampliation of Γ only in this subsection, because it has no effect in the rest of the paper.

With the notations of eq. (6), we have $\Gamma \equiv \Gamma_{(0}$; now, let us introduce the spaces

$\Gamma_{t)}$, $\widetilde{\Gamma}$ $\qquad \Gamma_{t)} := \Gamma_{\text{symm}}\left(L^2(-\infty, t) \otimes \mathcal{Z}\right)$ and $\widetilde{\Gamma} := \Gamma_{\text{symm}}\left(L^2(\mathbb{R}) \otimes \mathcal{Z}\right) \equiv \Gamma_{0)} \otimes \Gamma_{(0}$.
With the usual convention of not to write the tensor products with the identity, the solution $U(t)$ of eq. (39) can be understood as a unitary operator on $\mathcal{H} \otimes \widetilde{\Gamma}$.

Then, we introduce the strongly continuous one–parameter unitary group θ of the shift operators on $L^2(\mathbb{R}; \mathcal{Z})$ and its second quantization Θ on $\widetilde{\Gamma}$: for every $t \in \mathbb{R}$

θ_t, Θ_t

$$(\theta_t f)(r) = f(r + t), \qquad \forall f \in L^2(\mathbb{R}; \mathcal{Z}), \qquad (42\text{a})$$

$$\Theta_t\, e(f) = e(\theta_t f), \qquad \forall f \in L^2(\mathbb{R}; \mathcal{Z}); \qquad (42\text{b})$$

we extend Θ_t to the space $\mathcal{H} \otimes \widetilde{\Gamma}$. Now one can prove the *cocycle* property (43) for U, which Accardi already showed to give rise to groups [1,7].

Theorem 2.3 ([50, 51, 73, 74]). *Let Θ be the strongly continuous one–parameter unitary group defined by (42) and let U be the solution of the QSDE (39) with system operators satisfying the conditions of Theorem 2.2. Then, they are related by the cocycle property (of U with respect to Θ)*

$$U(s + t) = \Theta_s^* \, U(t) \, \Theta_s \, U(s), \qquad \forall s, t \geq 0, \qquad (43)$$

V_t $\qquad$ *and the family of unitary operators $V = \{V_t\}_{t \in \mathbb{R}}$, defined by*

$$V_t = \begin{cases} \Theta_t \, U(t), & \text{if } t \geq 0, \\ U(|t|)^* \, \Theta_t, & \text{if } t \leq 0, \end{cases} \qquad (44)$$

is a strongly continuous one–parameter unitary group.

$U(t, s)$ $\qquad$ *Moreover, the two–parameter family of unitary operators*

$$U(t, s) := \Theta_t^* \, V_{t-s} \, \Theta_s \equiv \Theta_s^* \, U(t - s) \, \Theta_s, \qquad s \leq t, \qquad (45)$$

is strongly continuous in t and s and satisfies the composition law

$$U(t, s)\, U(s, r) = U(t, r), \qquad r \leq s \leq t. \qquad (46)$$

The operator $U(t, s)$ is adapted to $\mathcal{H} \otimes \Gamma_{(s,t)}$, i.e. it acts as the identity on $\Gamma_{s)} \otimes \Gamma_{(t}$ and with respect to t it satisfies the Hudson–Parthasarathy equation (39) with initial condition $U(s, s) = \mathbb{1}$.

Proof. Let us note that eq. (39) gives also

$$U(t+s) - U(s) = \int_s^{t+s} \left\{ \sum_k R_k \, dA_k^\dagger(r) + \sum_{kl} (S_{kl} - \delta_{kl}) \, d\Lambda_{kl}(r) \right.$$

$$\left. - \sum_{kl} R_k^* S_{kl} \, dA_l(r) + K \, dr \right\} U(r). \quad (47)$$

Let us consider $X_s(t) := \Theta_s U(s+t) U(s)^* \Theta_{-s}$; note that $X_s(0) = \mathbb{1}$. By Proposition 2.2, the definition of Θ (42), the previous equation and a change of integration variable we get

$$\left\langle v \otimes e(g) \middle| (X_s(t) - \mathbb{1}) u \otimes e(f) \right\rangle = \left\langle v \otimes e(\theta_{-s}g) \middle| \right.$$

$$\int_s^{t+s} dr \left\{ \sum_k \overline{g_k(r-s)} \, R_k + \sum_{kl} \overline{g_k(r-s)} \, (S_{kl} - \delta_{kl}) \, f_l(r-s) \right.$$

$$\left. - \sum_{kl} R_k^* S_{kl} f_l(r-s) + K \right\} U(r) U(s)^* \, u \otimes e(\theta_{-s}f) \right\rangle$$

$$= \left\langle v \otimes e(g) \middle| \int_0^t d\tau \left\{ \sum_k \overline{g_k(\tau)} \, R_k + \sum_{kl} \overline{g_k(\tau)} \, (S_{kl} - \delta_{kl}) \, f_l(\tau) \right.\right.$$

$$\left.\left. - \sum_{kl} R_k^* S_{kl} f_l(\tau) + K \right\} X_s(\tau) \, u \otimes e(f) \right\rangle.$$

Therefore $X_s(t)$ satisfies eq. (41) as $U(t)$ and, by Corollary 2.1, $U(t) = X_s(t)$, which is equivalent to the cocycle property (43).

Let us prove now the group property for V; from the definition (44) one has the unitarity of V_t and

$$V_t^* = V_{-t}, \qquad \forall t \in \mathbb{R}. \quad (48)$$

From the cocycle property (43) and the fact that Θ is a group one gets, $\forall t, s \geq 0$,

$$V_t V_s = \Theta_t U(t) \Theta_s U(s) = \Theta_{t+s} \Theta_s^* U(t) \Theta_s U(s) = \Theta_{t+s} U(t+s) = V_{t+s}. \quad (49)$$

All the other combinations of positive and negative times can be examined and give the same result. For instance, for $s \leq 0$, $t+s \geq 0$, one has from (49) $V_{t+s} V_{-s} = V_t$, which is equivalent to the group property $V_{t+s} = V_t V_s$ due to (48).

Being V a unitary group, it is enough to prove its strong continuity in 0, which follows from the unitarity and the strong continuity of U and Θ. For $t \geq 0$ and $\Upsilon \in \mathcal{H} \otimes \widetilde{\Gamma}$ we have

$$\left\| (V_t - \mathbb{1}) \Upsilon \right\| = \left\| (\Theta_t U(t) - \mathbb{1}) \Upsilon \right\| \leq \left\| \Theta_t (U(t) - \mathbb{1}) \Upsilon \right\| + \left\| (\Theta_t - \mathbb{1}) \Upsilon \right\|$$

$$= \left\| (U(t) - \mathbb{1}) \Upsilon \right\| + \left\| (\Theta_t - \mathbb{1}) \Upsilon \right\| \to 0 \qquad \text{as } t \downarrow 0.$$

For $t \leq 0$, we have

$$\left\| (V_t - \mathbb{1}) \Upsilon \right\| = \left\| (U(|t|)^* \Theta_t - \mathbb{1}) \Upsilon \right\|$$

$$\leq \left\| U(|t|)^* (\Theta_t - \mathbb{1}) \Upsilon \right\| + \left\| (U(|t|)^* - \mathbb{1}) \Upsilon \right\|$$

$$= \left\| (\Theta_t - \mathbb{1}) \Upsilon \right\| + \left\| (U(|t|) - \mathbb{1}) \Upsilon \right\| \to 0 \qquad \text{as } t \uparrow 0.$$

From $U(t,s) = \Theta_s^* U(t-s)\Theta_s$ and Corollary 2.1 one has immediately that $U(t,s)$ is adapted to $\mathcal{H} \otimes \Gamma_{(s,t)}$ and satisfies the QSDE (39). From $U(t,s) = \Theta_t^* V_{t-s}\Theta_s$ and the fact that Θ and V_t are strongly continuous unitary groups one has immediately the composition law (46) and, by using also $U(t+r,s) - U(t,s) = \Theta_t^*(\Theta_r^* V_r - \mathbb{1})V_{t-s}\Theta_s$ and $U(t,s+r) - U(t,s) = \Theta_t^* V_{t-s}(V_r^* \Theta_r - \mathbb{1})\Theta_s$, the strong continuity in t and s.   □

The interaction picture

Theorem 2.3 is essential for the interpretation of the whole construction. The group V (44) is the reversible evolution of the isolated system $S_\mathcal{H}$ plus Bose fields. The free evolution of the fields is supposed to be given by Θ, and hence, physically, the degree of freedom described by $L^2(\mathbb{R})$ is thought to be the conjugate moment of the one–particle energy. Then, $U(t) = U(t,0) = \Theta_t^* V_t$ is the evolution operator giving the state dynamics from time 0 to time t of the whole system in the *interaction picture* with respect to the free field dynamics.

In quantum mechanics probabilities, mean values, . . . can be reduced to expressions like as

$$\langle B, \tilde{s} \rangle_t := \mathrm{Tr}_{\mathcal{H} \otimes \widetilde{\Gamma}} \left\{ B V_t \tilde{s} V_t^* \right\}, \tag{50}$$

where $\tilde{s} \in \mathcal{S}(\mathcal{H} \otimes \widetilde{\Gamma})$ is the initial state at time zero and $B \in \mathcal{B}(\mathcal{H} \otimes \widetilde{\Gamma})$. To pass to the interaction picture means to write (50) as

$$\langle B, \tilde{s} \rangle_t = \mathrm{Tr}_{\mathcal{H} \otimes \widetilde{\Gamma}} \left\{ B(t) U(t) \tilde{s} U(t)^* \right\}, \qquad B(t) := \Theta_t^* B \Theta_t. \tag{51}$$

If $B(t) \in \mathcal{B}(\mathcal{H} \otimes \Gamma)$ at a certain time t (i.e. it acts as the identity on $\Gamma_{0)}$) and $\tilde{s} = s_{0)} \otimes s$ with $s_{0)} \in \mathcal{S}(\Gamma_{0)})$ and $s \in \mathcal{S}(\mathcal{H} \otimes \Gamma)$, we can get ride of $\Gamma_{0)}$ and we write

$$\langle B, \tilde{s} \rangle_t = \mathrm{Tr}_{\mathcal{H} \otimes \Gamma} \left\{ B(t) U(t) s U(t)^* \right\}. \tag{52}$$

This is the case when $B \in \mathcal{B}(\mathcal{H})$ (one gets $B(t) = B, \forall t$); but this is not the only possible case and also field observables can be involved.

By (25a) and (42) one obtains

$$\Theta_s^* A_k(t)\Theta_s = A_k(t+s) - A_k(s), \tag{53}$$

which shows that the fields $A_k(t)$, and therefore also $A_k^\dagger(t)$, $\Lambda_{kl}(t)$, are already in the interaction picture. This conclusion is clearer if we use the heuristic notation (27) and the "energy representation" of the fields:

$$\hat{a}_k(\omega) = \frac{1}{\sqrt{2\pi}} \int_{-\infty}^{+\infty} e^{i\omega t} a_k(t) \mathrm{d}t. \tag{54}$$

Indeed we get

$$\Theta_s^* A_k(t)\Theta_s = \int_0^t \mathrm{d}\tau \frac{1}{\sqrt{2\pi}} \int_{-\infty}^{+\infty} \mathrm{d}\omega \, e^{-i\omega(\tau+s)} \hat{a}_k(\omega), \tag{55}$$

$$\Theta_s^* \hat{a}_k(\omega) \Theta_s = e^{-i\omega s} \hat{a}_k(\omega) . \tag{56}$$

From the next subsection on, we shall go back to use Γ and not $\widetilde{\Gamma}$, because automatically we work in the interaction picture and consider only states and observables adapted to $(0, +\infty)$ in the sense above.

The Hamiltonian operator

Being a strongly continuous unitary group, V is differentiable on the domain of its generator, it gives rise to a Schrödinger equation of usual type and a selfadjoint operator with the role of Hamiltonian exists. The characterization of this Hamiltonian required some effort; the results are in Refs. 34–38,57,58. Also before these works, it was clear that this Hamiltonian must be unbounded either from above, either from below; in this sense, the dynamics V, or, equivalently, U, must necessarily be an approximation to a "true" dynamics with a physical Hamiltonian bounded from below. Let us also stress that the unitary groups Θ and V are strongly differentiable on the dense domains of their Hamiltonians; being the two Hamiltonians unbounded, there is no surprise that the product U of the two groups be not differentiable. From this point of view the surprise is that U satisfies a closed equation, the QSDE (39).

2.3 The System–Field State

Once the dynamical equation (39) is constructed, one needs an *initial state*

$$\mathfrak{s} \in \mathcal{S}(\mathcal{H} \otimes \Gamma') .$$

From a mathematical point of view any statistical operator is a possible initial state; from a physical point of view the whole theory is an approximation and whether this approximation is good or not can depend also on the initial state. So, the interpretation of the theory depends not only on the dynamics (37)–(39), but also on the initial state; moreover, such a state must be sufficiently simple to allow computations.

The usual choice for the initial state is a factorized state

$$\mathfrak{s} = \rho_0 \otimes \sigma , \qquad \rho_0 \in \mathcal{S}(\mathcal{H}) , \quad \sigma \in \mathcal{S}(\Gamma) . \tag{57}$$

Let us use the notation

$$\eta(f) := |\psi(f)\rangle\langle\psi(f)| , \qquad f \in L^2(\mathbb{R}_+; \mathcal{Z}) , \tag{58}$$

for a pure coherent state. The coherent states are the only statistical operators in $\mathcal{S}(\Gamma)$ which enjoy the factorization property $\sigma = \sigma_{(0,s)} \otimes \sigma_{(s}, \sigma_{(0,s)} \in \mathcal{S}(\Gamma_{(0,s)})$, $\sigma_{(s} \in \mathcal{S}(\Gamma_{(s})), \forall s \geq 0$; indeed, from (8) one has

$$\eta(f) = \eta\left(f_{(0,s)}\right) \otimes \eta\left(f_{(s}\right) , \qquad \forall s \geq 0 , \tag{59}$$

$$\text{with} \quad \eta\left(f_{(0,s)}\right) \in \mathcal{S}(\Gamma_{(0,s)}) , \quad \eta\left(f_{(s}\right) \in \mathcal{S}(\Gamma_{(s}) .$$

In the following, we list and comment possible choices for σ.

S1. $\sigma = \eta(0)$,   the vacuum [65].

This is the case when the fields act essentially as a reservoir. The reduced dynamics of system $S_\mathcal{H}$ turns out to be time homogeneous and to be given by a quantum dynamical semigroup.

S2. $\sigma = \eta(f)$,   a coherent state for the field [18].

Now the fields provide also a coherent source which stimulates the system $S_\mathcal{H}$. This is the typical choice when a "perfectly" coherent laser is present; a coherent and monochromatic laser could be represented by $\eta(f)$ with

$$f(t) = 1_{[0,T]}(t)\,\mathrm{e}^{-i\omega t}\,\lambda\,, \qquad \lambda \in \mathcal{Z}\,, \quad \omega > 0, \tag{60}$$

where T is a very large time; eventually, one takes $T \to +\infty$ at the end in the physical quantities. Equation (59) implies a related factorization property for the state at time t in the interaction picture, because U is an adapted process:

$$U(t)\,[\rho_0 \otimes \eta(f)]\,U(t)^* = \{U(t)\,[\rho_0 \otimes \eta\,(f_{(0,t)})]\,U(t)^*\} \otimes \eta\,(f_{(t}), \tag{61}$$
$$\text{with} \qquad \{U(t)\,[\rho_0 \otimes \eta\,(f_{(0,t)})]\,U(t)^*\} \in \mathcal{S}(\Gamma_{(0,t)})\,, \quad \eta\,(f_{(t}) \in \mathcal{S}(\Gamma_{(t})\,.$$

S3. $\sigma = \mathbb{E}^c\,[\eta(f)]$ a mixture of coherent states [28].

$\mathbb{E}^c$ Here we mean that there is a random variable f in a probability space $(\Omega^c, \mathcal{F}^c, P^c)$ with values in $L^2(\mathbb{R}_+; \mathcal{Z})$. Then, the expectation is given by

$$\mathbb{E}^c\,[\eta(f)] = \int_{\Omega^c} \eta\big(f(\cdot; \mathfrak{w})\big)\,P^c(\mathrm{d}\mathfrak{w}), \tag{62}$$

where the integral can be understood in the topology induced in $\mathcal{T}(\Gamma)$ by the duality with $\mathcal{B}(\Gamma)$; being $\eta(f)$ a pure state, this means: $\forall A \in \mathcal{B}(\Gamma)$,

$$\mathrm{Tr}_\Gamma\{A\sigma\} = \int_{\Omega^c} \langle\psi\big(f(\cdot; \mathfrak{w})\big)|A\,\psi\big(f(\cdot; \mathfrak{w})\big)\rangle P^c(\mathrm{d}\mathfrak{w}). \tag{63}$$

It is always possible to think f as a stochastic process defined for times in $\mathbb{R}_+$ with values in $\mathcal{Z}$ and with trajectories in $L^2(\mathbb{R}_+; \mathcal{Z})$; moreover, by taking a filtration containing its natural one, it is always possible to take as f an adapted, or non anticipating, process. So we complete **S3** with

S3'. $(\Omega^c, \mathcal{F}^c, P^c)$ is a probability space with a filtration $\{\mathcal{F}^c_t, \ t \geq 0\}$, i.e. $\mathcal{F}^c_t$ is a σ-algebra with $\mathcal{F}^c_s \subset \mathcal{F}^c_t \subset \mathcal{F}^c$ for all times $0 \leq s \leq t$; $f(t), \ t \geq 0$, is a progressively measurable process, and hence adapted, with $f(t; \mathfrak{w}) \in \mathcal{Z}$, $f(\cdot; \mathfrak{w}) \in L^2(\mathbb{R}_+; \mathcal{Z})$, for $\mathfrak{w} \in \Omega^c$.

Now the factorization property (61) does not hold, but we have only

$$U(t)\,[\rho_0 \otimes \sigma]\,U(t)^* = \mathbb{E}^c\,\big[\{U(t)\,[\rho_0 \otimes \eta\,(f_{(0,t)})]\,U(t)^*\} \otimes \mathbb{E}^c\,[\eta\,(f_{(t})\,|\mathcal{F}^c_t]\big] \tag{64}$$

As an example take the "phase diffusion model" of a laser, used in Ref. 28 in the study of the fluorescence spectrum of a two–level atom: the field state is given by **S3**, where f is the process

$$f(t) = e^{-i(\omega t + \sqrt{B} W(t))} 1_{(0,T)}(t) \lambda, \quad \lambda \in \mathcal{Z}, \quad \omega > 0, \quad B \geq 0; \qquad (65)$$

$W(t)$ is a real standard Wiener process canonically realized in the Wiener probability space $(\Omega^c, \mathcal{F}^c, P^c)$.

In the case **S3** we have a *demixture* $\{\eta(f(\cdot; \mathfrak{w})), P^c(d\mathfrak{w})\}$ of a statistical operator σ into pure states $\eta(f)$. Often in quantum mechanics the point of view is taken that in a single, individual experiment a pure state is realized; then, if there is not a perfect control of the preparation, in replicas of the experiment other pure states are realized according to some probability law: mixed states arise due to our imperfect knowledge of the initial state. This interpretation is not always justified, certainly not in the typical situations of statistical mechanics when thermal states and thermodynamical limits enter into play. Another difficulty is that, given a mixed state σ, there are infinitely many demixtures, one of them being the one determined by its eigenvalues and eigenvectors. However, in our case the demixture $\{\eta(f(\cdot; \mathfrak{w})), P^c(d\mathfrak{w})\}$ can be thought as a special one among the possible demixtures of σ, as the physical one. For instance in the example (65), we have a laser nearly monochromatic, nearly coherent, but with a fluctuating phase, fluctuating in time and from an experiment to another; $t \mapsto f(t; \mathfrak{w})$ gives the history of the laser in a single experiment, while from an experiment to another it is $\mathfrak{w}$ to change.

In quantum optics the mixtures of coherent states **S3**, and the particular cases **S2** and **S1**, are called *classical states*; all the other states are called *quantum states*. Among the quantum states are the *Scrödinger cats*, which are quantum superpositions of coherent vectors, such as $|\alpha\psi(f) + \beta\psi(g)\rangle\langle\alpha\psi(f) + \beta\psi(g)|$. To my knowledge, quantum states have never been used as initial states in the "quantum stochastic framework", when the QSC is based on Fock space; some "non classical" situations have been approached by using versions of QSC based on non Fock representations of the CCR (thermal and squeezed noise) [54, 55].

2.4 The Reduced Dynamics

System Observables in the Heisenberg Picture

Let us consider any observable $X \in \mathcal{B}(\mathcal{H})$ of the quantum system $S_{\mathcal{H}}$; X can be in particular a selfadjoint operator or even a projection operator... We already discussed the interaction picture, but we can use also the Heisenberg picture; being V_t the unitary group giving the system field dynamics, the observable X in the Heisenberg picture becomes $X(t) = V_t^*(X \otimes \mathbb{1})V_t$. Recalling that $V_t = \Theta_t U(t)$ and that Θ_t commutes with $X \otimes \mathbb{1}$, we obtain $X(t) = U(t)^*(X \otimes \mathbb{1})U(t)$. By the rules of QSC, $X(t)$ can be differentiated; the result is a "quantum stochastic" Heisenberg equation.

j_t

Proposition 2.5 ([84] Corollary 27.9). *Let us define*

$$j_t(X) := U(t)^*(X \otimes \mathbb{1})U(t), \qquad \forall X \in \mathcal{B}(\mathcal{H}) ; \qquad (66)$$

then, we have

$$dj_t(X) = j_t\big(\mathcal{L}_0'[X]\big)dt + \sum_{k=1}^{\infty} j_t\big(\mathcal{R}_k[X]\big)dA_k^{\dagger}(t)$$

$$+ \sum_{k=1}^{\infty} j_t\big(\mathcal{R}_k[X^*]^*\big)dA_k(t) + \sum_{k,l=1}^{\infty} j_t\big(\mathcal{S}_{kl}[X]\big)d\Lambda_{kl}(t), \quad (67)$$

$\mathcal{L}_0'$ *where*

$$\mathcal{L}_0'[X] := \mathrm{i}[H, X] - \frac{1}{2}\sum_{k=1}^{\infty}\big(R_k^*[R_k, X] + [X, R_k^*]R_k\big), \qquad (68a)$$

$$\mathcal{R}_k[X] := \sum_{l=1}^{\infty} S_{lk}^*[X, R_l] \qquad (68b)$$

$$\mathcal{S}_{kl}[X] := \sum_{j=1}^{\infty} S_{jk}^* X S_{jl} - \delta_{kl} X. \qquad (68c)$$

The Master Equation

$\rho(t)$ The reduced statistical operator is defined by the partial trace

$$\rho(t) := \mathrm{Tr}_\Gamma\{U(t)\mathfrak{s}U(t)^*\} \qquad (69)$$

or, equivalently, by

$$\mathrm{Tr}_\mathcal{H}\{X\rho(t)\} = \mathrm{Tr}_{\mathcal{H}\otimes\Gamma}\{j_t(X)\mathfrak{s}\}, \qquad \forall X \in \mathcal{B}(\mathcal{H}).$$

$\rho(f;t)$ If we set

$$\rho(f;t) := \mathrm{Tr}_\Gamma\{U(t)(\rho_0 \otimes \eta(f))U(t)^*\}, \qquad (70)$$

then the reduced statistical operator, for the three choices discussed in Section 2.3, turns out to be given by

case S1 $\rho(t) = \rho(0;t)$;
case S2 $\rho(t) = \rho(f;t)$;
case S3 $\rho(t) = \mathbb{E}^c[\rho(f;t)]$.

We can always write

$$\rho(t) = \mathbb{E}^c[\rho(f;t)] \qquad (71)$$

by understanding that the classical expectation $\mathbb{E}^c$ has no effect when f is not random.

By Propositions 2.5 and 2.2, $\rho(f;t)$ satisfies the integral equation

$$\rho(f;t) = \rho_0 + \int_0^t \mathrm{d}s\, \mathcal{L}\big(f(s)\big)[\rho(f;s)], \tag{72}$$

where the integral can be interpreted again in the topology induced by the duality with $\mathcal{B}(\mathcal{H})$ and $\mathcal{L}\big(f(s)\big)$ is the bounded operator on $\mathcal{T}(\mathcal{H})$ given by

$\mathcal{L}(f)$,
$R_k(f)$,
$H(f)$

$$\mathcal{L}(f)[\varrho] = -\mathrm{i}\,[H(f), \varrho] + \frac{1}{2} \sum_{k=1}^{\infty} \left([R_k(f)\varrho, R_k(f)^*] + [R_k(f), \varrho R_k(f)^*]\right), \tag{73a}$$

$$R_k(f) = C(z_k)^* \left[R + (S - \mathbb{1})\, C(f)\right], \tag{73b}$$

$$H(f) = H + \mathrm{i}\,[C(f)^*R - R^*C(f)] + \frac{\mathrm{i}}{2}\Big[C(f)^* (S - S^*)\, C(f) \\ - R^* (S - \mathbb{1})\, C(f) + C(f)^* (S^* - \mathbb{1})\, R\Big]. \tag{73c}$$

For a sufficiently regular f, eq. (72) can be differentiated and becomes the quantum *master equation*

$$\frac{\mathrm{d}}{\mathrm{d}t}\,\rho(f;t) = \mathcal{L}\big(f(t)\big)[\rho(f;t)]. \tag{74}$$

The infinitesimal generator $\mathcal{L}\big(f(t)\big)$ is known as *Liouville operator* or *Liouvillian*; the name comes from the Liouville equation in classical statistical mechanics, which became in quantum mechanics the Liouville–von Neumann equation $\mathrm{d}\rho(t)/\mathrm{d}t = -\mathrm{i}[H, \rho(t)]$ for an isolated system and the master equation (74) for an open system without memory. From (73a) the Lindblad structure ([84] Corollary 30.13 p. 268) of the Liouvillian is apparent; by (37), a more compact form is

$$\mathcal{L}(f)[\varrho] = -\left(\frac{1}{2}\,R^*R + R^*SC(f) + \mathrm{i}H\right)\varrho - \varrho\left(\frac{1}{2}\,R^*R + C(f)^*S^*R - \mathrm{i}H\right) \\ + \mathrm{Tr}_{\mathcal{Z}}\left\{(R + SC(f))\varrho(R^* + C(f)^*S^*)\right\} - \|f\|^2\varrho. \tag{75}$$

In the case **S1** one has the autonomous master equation

$$\frac{\mathrm{d}}{\mathrm{d}t}\,\rho(t) = \mathcal{L}_0[\rho(t)], \tag{76}$$

where $\mathcal{L}_0 \equiv \mathcal{L}(0)$ is given by eq. (73a) with $f = 0$ and its adjoint $\mathcal{L}_0'$ is given in eq. (68a). In the case **S2** the reduced operator $\rho(t)$ satisfies the master equation (74), with a time dependent Liouvillian. Finally, in the case **S3**, the reduced operator does not satisfy any closed equation.

2.5 Physical Basis of the Use of QSC

The Quasi–Monochromatic Paraxial Approximation of the Electromagnetic Field

In 1978 Yuen and Shapiro [89] started to develop a theory of quantum–field propagation as a boundary–value problem; in particular, they treated the case of *quasi–monochromatic paraxial fields*, i.e. fields whose significant (nonvacuum state) modes all have temporal frequencies in the vicinity of a nominal carrier frequency ω_0 and satisfy $|k^\perp| \mathbb{L}(\omega_0/c)^2$, where $k^\perp$ is the wave–vector component orthogonal to the direction of propagation. In such approximations they found, for the positive and negative frequency parts of the electric field, the CCR with a Dirac delta in time, as in (28). Moreover, they showed that, in the quasimonochromatic paraxial limit, the spatial propagation along the axial direction involves pure translation in time, as for the solutions of the one–dimensional wave equation. These ideas were used by Frigerio and Ruzzier in an attempt to develop a relativistic version of QSC [52] and by the author in developing a photon detection theory [18]. A consequence of the idea of the quasimonochromatic paraxial approximation is that the Hilbert space $\mathcal{Z}$, which appear in the definition of the Fock space Γ, must contain the field degrees of freedom linked to the *nominal carrier frequencies* involved in the problem, the *directions of propagation*, the *polarization*. The possible polarization states for the electromagnetic field are only two (two linear or two circular polarizations) and often, for sake of simplicity, polarization is even not considered. The other two things, nominal carrier frequencies and directions of propagation, depend on what is relevant in the matter–field interaction and, so, the choice of $\mathcal{Z}$ depends on the specific problem and, in particular, on the quantum system $S_\mathcal{H}$.

Approximations in the System–Field Interaction

In Ref. 54 Gardiner and Collet discuss the physical approximations needed to pass from a quasi–physical Hamiltonian to an evolution like (39), at least in the case $S = \mathbb{1}$. They work with the Heisenberg equations of motion for system operators and obtain at the end equations like (67); let us give an idea of these approximations by working with the evolution operator [18]. Let us consider our system $S_\mathcal{H}$ interacting with some Bose fields $a_j(\omega)$ (we are in the frequency domain), satisfying the CCR's

$$[a_i(\omega), a_j(\omega')] = 0, \qquad [a_i(\omega), a_j^\dagger(\omega')] = \delta_{ij}\,\delta(\omega - \omega')\,. \tag{77}$$

A generic system–field interaction, linear in the field operators and in the rotating wave approximation, can be written as

$$H_{\mathrm{I}} = \mathrm{i} \sum_j \frac{1}{\sqrt{2\pi}} \int_{\Omega_j - \theta_j}^{\Omega_j + \theta_j} k_j(\omega)\big[R_j a_j^\dagger(\omega) - R_j^* a_j(\omega)\big]\mathrm{d}\omega\,, \tag{78}$$

where the R_j are system operators and the $k_j(\omega)$ are real coupling constants. We shall consider the limiting case in which the coupling constants become independent of frequency and the field spectrum becomes flat and broad [$k_j(\omega) \to 1$, $\theta_j \to +\infty$]. In quantum optics this is often a good approximation [54].

First we pass to the interaction picture with respect to the free dynamics of the fields and take $k_j(\omega) = 1$ (flat spectrum). The interaction Hamiltonian becomes

$$H_I(t) = i \sum_j \left[R_j \tilde{a}_j^\dagger(t) - R_j^* \tilde{a}_j(t) \right] , \tag{79}$$

$$\tilde{a}_j(t) := \frac{1}{\sqrt{2\pi}} \int_{\Omega_j - \theta_j}^{\Omega_j + \theta_j} a_j(\omega) \, e^{-i\omega t} \, d\omega . \tag{80}$$

Moreover, the time evolution operator $\tilde{U}_t$ in the interaction picture can be written as

$$\tilde{U}(t) = \overleftarrow{T} \exp\left\{ -i \int_0^t [H + H_I(s)] \, ds \right\} , \tag{81}$$

where H is the system Hamiltonian and $\overleftarrow{T}$ is the time–ordering prescription, which is usual in theoretical physics; for a mathematical presentation of the time ordered exponentials in QSC see Ref. 63.

By taking the limit $\theta_j \to +\infty$ (broad–band approximation), we obtain formally that the quantities $a_j(t) = \lim_{\theta_j \to +\infty} \tilde{a}_j(t)$ satisfy the CCR (28). Moreover, we have

$$\int_0^t \tilde{a}_j(s) \, ds \overset{\theta_j \to +\infty}{\longrightarrow} A_j(t), \tag{82a}$$

$$-iH_I(s) \, ds \overset{\theta_j \to +\infty}{\longrightarrow} \sum_j \left[R_j \, dA_j^\dagger(s) - R_j^* \, dA_j(s) \right] , \tag{82b}$$

$$\tilde{U}(t) \overset{\theta_j \to +\infty}{\longrightarrow} U(t), \tag{82c}$$

$$U(t) \equiv \overleftarrow{T} \exp\left\{ -i \int_0^t \left[H \, ds + i \sum_j \left(R_j \, dA_j^\dagger(s) - R_j^* \, dA_j(s) \right) \right] \right\} . \tag{83}$$

By writing $U(t + dt)$ in the form

$$U(t + dt) = \exp\left\{ -iH \, dt + \sum_j \left[R_j \, dA_j^\dagger(t) - R_j^* \, DA_j(t) \right] \right\} U_t , \tag{84}$$

by expanding the exponential and by using the multiplication table (32), one sees that $U(t)$ satisfies the Hudson–Parthasarathy equation (39) with $S = \mathbb{1}$. Note that the term $-\frac{1}{2} R_j^\dagger R_j \, dt$ in K (37e) comes from the second order term in the expansion of the exponential.

Let us stress that this construction shows that the fields $a_j(t)$ used in QSC are the formal limit $\theta_j \to +\infty$ of the quantities (80). This explains once again the fact that a field $a_j(t)$ has to be considered as a wave packet with some carrier frequency Ω_j and bandwidth $2\theta_j$; then, the approximation of infinite bandwidth is taken.

Many other limiting schemes justifying the Hudson–Parthasarathy equation have been developed, also in mathematically rigourous forms [4–6].

3 Continual Measurements

3.1 Field Observables and Indirect Measurements on $S_{\mathcal{H}}$

When the fields represent pure noise, it is natural to consider system observables as in Section 2.4; but in other situations, as when the fields are intended to represent the electromagnetic field, the natural observables are field quantities, from which inferences are done on the system $S_{\mathcal{H}}$. We are interested in the behaviour of the system $S_{\mathcal{H}}$, but we measure field observables; this scheme is known as *indirect measurement*.

Another way to think to this situation is the following one. We cannot act directly on our system $S_{\mathcal{H}}$, but any action is mediated by some quantum input and output channel. We can think of an atom driven by a laser (input) and emitting fluorescence light (output) or of the light entering (input) and leaving (output) an optical cavity. In these examples the role of input and output channels is played by the electromagnetic field and we can think of approximating it by the Bose fields on which QSC is based.

So, we have to identify the main field observables, which eventually we want to take under measurement with continuity in time.

Counts of Quanta

Let $P \in \mathcal{B}(\mathcal{Z})$ be an orthogonal projection; for any $t \geq 0$, we introduce the operator

$N(P;t)$

$$N(P;t) := \lambda\left(P \otimes 1_{(0,t)}\right) = \sum_{kl}\langle z_k|Pz_l\rangle \Lambda_{kl}(t). \tag{85}$$

By Theorem 2.1, this operator is essentially selfadjoint on $\mathcal{E}$ and its domain includes also the finite particle number vectors. From (25c) we have

$$\langle e(g)|N(P;t)e(f)\rangle = \exp\left\{\langle g_{(0,t)}|(1-P)f_{(0,t)}\rangle + \langle g_{(t}|f_{(t)}\rangle\right\}$$
$$\times \sum_{n=0}^{\infty}\frac{n}{n!}\left(\langle g_{(0,t)}|Pf_{(0,t)}\rangle\right)^n ; \tag{86}$$

by taking into account the factorization (10), one sees that the eigenvalues of $N(P;t)$ are the integers $n = 0, 1, \ldots$ and that the eigenspace corresponding to n is the "n-particle sector of $\Gamma_{\text{symm}}\left((P\mathcal{Z}) \otimes \left(1_{(0,t)}L^2(\mathbb{R}_+)\right)\right)$" $\otimes \Gamma_{\text{symm}}\left((1-P \otimes 1_{(0,t)})(\mathcal{Z} \otimes L^2(\mathbb{R}_+))\right)$. Therefore, we can interprete $N(P;t)$ as the number operator which counts the quanta injected in the system up to time t with state in $P\mathcal{Z}$. Another way to see that $N(P;t)$ is a number operator is to use the heuristic rules of QSC; by (85), (32) and the fact that P is a projection, we have immediately

$$\left(dN(P;t)\right)^2 = dN(P;t), \tag{87}$$

which shows that an infinitesimal increment has eigenvalues 0 and 1.

By (17) and (85), we have

$$\exp\{i\kappa N(P;t)\} = \mathcal{W}\big(0; \exp\{i\kappa P \otimes 1_{(0,t)}\}\big) \tag{88}$$

and by (14) one sees that the unitary groups generated by $N(P;t)$ and $N(P;s)$ commute; therefore, $\{N(P;t), t \geq 0\}$ is a set of jointly diagonalizable selfadjoint operators, or, in physical terms, of *compatible observables*. The same is true for

$$\{N(P_\alpha;t)\ \ t \geq 0, \ \ \alpha = 1,2,\dots\} \quad \text{with } P_\alpha P_\beta = \delta_{\alpha\beta}\, P_\alpha = \delta_{\alpha\beta}\, P_\alpha^*, \tag{89}$$

i.e. P_1, P_2, ... are mutually orthogonal projections.

In the case of photons the measurement of number operators can be experimentally realized through the so called *direct detection*, which we present in Section 3.4.

Measurements of Field Quadratures

Let us consider now the field quadratures $Q(h;t)$

$$Q(h;t) := Q\big(h_{(0,t)}\big) = \sum_k \left\{ \int_0^t \overline{h_k(s)}\, \mathrm{d}A_k(s) + \int_0^t h_k(s)\, \mathrm{d}A_k^\dagger(s) \right\}, \tag{90}$$

which are essentially selfadjoint operators on $\mathcal{E}$ (Theorem 2.1). The spectrum of $Q(h;t)$ is the whole real axis because $(\sqrt{2}\,\|h\|)^{-1} Q(h;t)$ and $(\sqrt{2}\,\|h\|)^{-1}\, Q(ih;t)$ form a couple of canonically conjugated selfadjoint operators (the commutator gives i). By (15), (16), we have that

$$\{Q(h_\alpha;t), t \geq 0, \alpha = 1,2,\dots\}, \quad \text{with } \langle h_\alpha(s)|h_\beta(s)\rangle = \delta_{\alpha\beta}\, \|h_\alpha(s)\|^2, \tag{91}$$

is a family of compatible observables. Definition 2.2 and Proposition 2.1 guarantee that the quantum stochastic integral in (90) is well defined

In the case of photons the measurement of field quadratures can be experimentally realized through the so called *heterodyne detection*, which we present in Section 3.5.

Field Observables in the Heisenberg Picture

In Section 2.2 we discussed the fact that all the fields we have introduced are expressed in the interaction picture and, so, this is true also for the field observables (85) and (90). However, in order to construct a theory of continual measurements, based on the usual rules of quantum mechanics, which require the existence of joint spectral measures, we need observables commuting at different times *in the Heisenberg picture*; so, we have to show that the observables introduced above continue to commute at different times even in the Heisenberg picture.

Let us call "input fields" the fields before the interaction with the system $S_{\mathcal{H}}$, i.e. the fields $A_k(t)$, $A_k^\dagger(t)$, $\Lambda_{kl}(t)$, ... and let us call "output fields" the fields after the interaction with the system $S_{\mathcal{H}}$ or, in other words, the fields in the Heisenberg picture. We have

$A_j^{\text{out}}(t)$

$$A_j^{\text{out}}(t) := U(t)^* A_j(t) U(t) \tag{92}$$

$A_j^{\text{out}\,\dagger}(t)$, and similar definitions for $A_j^{\text{out}\,\dagger}(t)$, $\Lambda_{ij}^{\text{out}}(t)$, $Q^{\text{out}}(h;t)$, $N^{\text{out}}(P;t)$. Note that, if D
$\Lambda_{ij}^{\text{out}}(t)$, is the domain of $A_j(t)$, then $U(t)^*$D is the domain of $A_j^{\text{out}}(t)$ and similar statements
$Q^{\text{out}}(h;t)$, for the other operators; $Q^{\text{out}}(h;t)$, $N^{\text{out}}(P;t)$ remain selfadjoint operators.
$N^{\text{out}}(P;t)$ By Theorem 2.3 we have

$$U(T) = U(T,t)U(t), \qquad \forall T \geq t, \tag{93}$$

with $U(T,t)$ adapted to $\mathcal{H} \otimes \Gamma_{(t,T)}$ and, so, commuting with $A_j(t)$, $A_j^\dagger(t)$, $\Lambda_{ij}(t)$, ...
Therefore, we have

$$A_j^{\text{out}}(t) = U(T)^* A_j(t) U(T), \qquad \forall T \geq t. \tag{94}$$

This implies immediately that the output fields satisfies the same commutation rules of the input fields, for instance the CCR (26): the output fields remain Bose free fields. Moreover, we have that now $\{N^{\text{out}}(P_j;t), t \geq 0, j = 1, 2, \ldots\}$, with P_1, P_2, ... mutually othogonal projections, and $\{Q^{\text{out}}(h_j;t), t \geq 0, j = 1, 2, \ldots\}$, with $h_1(s)$, $h_2(s)$, ... mutually orthogonal vectors for any s, are two families of *compatible observables in the Heisenberg picture* as we wanted.

By applying the formal rules of QSC we can express the output fields as the quantum stochastic integrals [14]

$$A_j^{\text{out}}(t) = \int_0^t \left\{ \sum_k U(s)^* S_{jk} U(s) \mathrm{d}A_k(s) + U(s)^* R_j U(s) \, \mathrm{d}s \right\}, \tag{95a}$$

$$A_j^{\text{out}\,\dagger}(t) = \int_0^t \left\{ \sum_k U(s)^* S_{jk}^* U(s) \mathrm{d}A_k^\dagger(s) + U(s)^* R_j^* U(s) \, \mathrm{d}s \right\}, \tag{95b}$$

$$\Lambda_{ij}^{\text{out}}(t) = \int_0^t \left\{ \sum_{kl} U(s)^* S_{ik}^* S_{jl} U(s) \, \mathrm{d}\Lambda_{kl}(s) + \sum_k U(s)^* S_{ik}^* R_j U(s) \, \mathrm{d}A_k^\dagger(s) \right.$$
$$\left. + \sum_l U(s)^* R_i^* S_{jl} U(s) \, \mathrm{d}A_l(s) + U(s)^* R_i^* R_j U(s) \, \mathrm{d}s \right\}. \tag{95c}$$

From these equations one explicitly sees that the output fields carry information on system $S_{\mathcal{H}}$: the quantities R_k, S_{kl} are the system operators appearing in the system–field interaction. Moreover, these equations allow to write our observables as

$$N^{\text{out}}(P;t) = \int_0^t \left\{ \sum_{kl} U(s)^* C(z_k)^* S^* \left(\mathbb{1} \otimes P\right) SC(z_l)U(s)\, \mathrm{d}\Lambda_{kl}(s) \right.$$

$$+ \sum_k U(s)^* C(z_k)^* S^* \left(\mathbb{1} \otimes P\right) RU(s)\, \mathrm{d}A_k^\dagger(s)$$

$$+ \sum_l U(s)^* R^* \left(\mathbb{1} \otimes P\right) SC(z_l)U(s)\, \mathrm{d}A_l(s)$$

$$\left. + U(s)^* R^* \left(\mathbb{1} \otimes P\right) RU(s)\, \mathrm{d}s \right\}, \quad (96)$$

$$Q^{\text{out}}(h;t) = \int_0^t \left\{ \sum_k U(s)^* C(z_k)^* S^* C\big(h(s)\big) U(s)\, \mathrm{d}A_k^\dagger(s) \right.$$

$$+ \sum_l U(s)^* C\big(h(s)\big)^* SC(z_l)U(s)\, \mathrm{d}A_l(s)$$

$$\left. + U(s)^* \left[R^* C\big(h(s)\big) + C\big(h(s)\big)^* R\right] U(s)\, \mathrm{d}s \right\}; \quad (97)$$

these two equations give us an idea of how the output observables depend on the field operators and on the system ones. In eqs. (95)–(97) it is possible to check that the quantum stochastic integrals in the r.h.s. are all well defined, but a rigorous proof of anyone of these formulae needs to use two times Proposition 2.4 and to control the domain of the triple product. However we do not need this, because we can shortcut this difficulty by using Weyl operators.

3.2 Characteristic Functionals

Let us recall some very well known facts about selfadjoint operators (observables) and their distribution laws; see for instance Section 10 of Ref. 84. Let X be a selfadjoint operator and $\exp\{ikX\}$ the group generated by X; then, there exists a unique projection–valued measure (pvm) ξ on $\big(\mathbb{R}, \mathcal{B}(\mathbb{R})\big)$ such that   pvm

$$e^{ikX} = \int_{\mathbb{R}} e^{ikx}\, \xi(\mathrm{d}x), \qquad \forall k \in \mathbb{R}. \tag{98}$$

Let $\boldsymbol{X} \equiv (X_1, \ldots, X_d)$ be a set of d mutually commuting selfadjoint operators, in the sense that the groups generated by them commute or that the associated pvm ξ_j commute; then, there exists a unique pvm ξ on $\big(\mathbb{R}^d, \mathcal{B}(\mathbb{R}^d)\big)$ such that

$$e^{i\boldsymbol{k}\cdot\boldsymbol{X}} \equiv \prod_{j=1}^d e^{ik_j X_j} = \int_{\mathbb{R}^d} e^{i\boldsymbol{k}\cdot\boldsymbol{x}}\, \xi(\mathrm{d}\boldsymbol{x}), \qquad \forall \boldsymbol{k} \in \mathbb{R}^d. \tag{99}$$

Moreover, in the state ϱ, the characteristic function (Fourier transform) of the probability law

$$P_\varrho^X(\mathrm{d}\boldsymbol{x}) = \mathrm{Tr}\left\{\varrho\,\xi(\mathrm{d}\boldsymbol{x})\right\} \tag{100}$$

of the observable associated to X is

$$\int_{\mathbb{R}^d} \mathrm{e}^{\mathrm{i}\boldsymbol{k}\cdot\boldsymbol{x}}\, P_\varrho^X(\mathrm{d}\boldsymbol{x}) = \mathrm{Tr}\left\{\varrho\,\mathrm{e}^{\mathrm{i}\boldsymbol{k}\cdot\boldsymbol{X}}\right\}. \tag{101}$$

These results extend to "infinitely many" commuting selfadjoint operators; only uniqueness is lost.

Proposition 3.1. *Let $\{\xi_t,\ t \in T\}$ be a family of commuting pvm on $\left(\mathbb{R}, \mathcal{B}(\mathbb{R})\right)$ and let X_t be the selfadjoint operator associated with ξ_t. Then, there exist a measurable space $(\Omega, \mathcal{F})$, a pvm ξ on $(\Omega, \mathcal{F})$ and a family of real valued measurable functions $\left\{\tilde{X}_t,\ t \in T\right\}$ on Ω such that*

$$\mathrm{e}^{\mathrm{i}kX_t} \equiv \int_{\mathbb{R}} \mathrm{e}^{\mathrm{i}kx}\,\xi_t(\mathrm{d}x) = \int_{\Omega} \mathrm{e}^{\mathrm{i}k\tilde{X}_t(\omega)}\,\xi(\mathrm{d}\omega)\,, \qquad \forall k \in \mathbb{R},\ t \in T. \tag{102}$$

Proof. This proposition is given as an exercise by Parthasarathy ([84] Exercise 10.11 p. 59), with the hint: use Bochner's Theorem, Kolmogorov's Consistency Theorem and Zorn's Lemma. □

In the situation described in this proposition, if ϱ is a fixed state and we set

$$P_\varrho(\mathrm{d}\omega) := \mathrm{Tr}\{\varrho\,\xi(\mathrm{d}\omega)\}\,,$$

we have that $(\Omega, \mathcal{F}, P_\varrho)$ is a classical probability space and $\left\{\tilde{X}_t(\cdot),\ t \in T\right\}$ becomes a classical stochastic process. The characteristic functions of the finite–dimensional distributions of this process are given by

$$\int_{\Omega} \exp\left[\mathrm{i}\sum_{j=1}^{n} k_j \tilde{X}_{t_j}(\omega)\right] P_\varrho(\mathrm{d}\omega) = \mathrm{Tr}\left\{\varrho \exp\left[\mathrm{i}\sum_{j=1}^{n} k_j X_{t_j}\right]\right\}. \tag{103}$$

Let us consider now the case in which the index becomes time plus a discrete label: $T = \{(\alpha, s) : \alpha = 1, \ldots, d,\ 0 < s \le t\}$. Then, we denote the process by $\tilde{X}(\alpha, s; \omega)$, the operators by $X(\alpha, s)$ and we assume, for simplicity, $\tilde{X}(\alpha, 0; \omega) = 0$, $X(\alpha, 0) = 0$. Instead of considering the finite–dimensional distributions of the process $\tilde{X}(\alpha, s)$, it is equivalent and simpler to introduce the finite–dimensional distributions of the increments of the original process, whose characteristic functions are

$$\int_{\Omega} \exp\left[\mathrm{i}\sum_{\alpha=1}^{d}\sum_{j=1}^{n} k_\alpha(s_j)\left(\tilde{X}(\alpha, s_j; \omega) - \tilde{X}(\alpha, s_{j-1}; \omega)\right)\right] P_\varrho(\mathrm{d}\omega)$$

$$= \mathrm{Tr}\left\{\varrho \exp\left[\mathrm{i}\sum_{\alpha=1}^{d}\sum_{j=1}^{n} k_\alpha(s_j)\left(X(\alpha, s_j) - X(\alpha, s_{j-1})\right)\right]\right\},$$

$$0 = s_0 < s_1 < \cdots < s_n \le t. \tag{104}$$

The Characteristic Functional for the Counts of Quanta

Let us consider the family of compatible observables $\{N(P_\alpha;t),\ t \geq 0,\ \alpha = 1, 2, \ldots, d\}$; $P_1, \ldots, P_d$ are mutually othogonal projections on $\mathcal{Z}$. According to the discussion above and Proposition 3.1, we can handle the stochastic process associated to these operators by means of the finite–dimensional characteristic functions for the increments, which in turn can be summarized in a *characteristic functional*, which is suggested by the structure of eq. (104) and which now we construct.

Let us introduce the *test functions* $k \in L^\infty\left(\mathbb{R}_+;\mathbb{R}^d\right)$, the unitary operators $\mathsf{S}_t(k)$ on $L^2(\mathbb{R}_+;\mathcal{Z})$ by

$$\mathsf{S}_t(k) := \exp\left[i \sum_{\alpha=1}^{d} P_\alpha \otimes 1_{(0,t)} k_\alpha \right] \tag{105a}$$

or by

$$(\mathsf{S}_t(k)f)(s) = \exp\left[i 1_{(0,t)}(s) \sum_{\alpha=1}^{d} k_\alpha(s) P_\alpha \right] f(s)$$

$$\equiv 1_{(0,t)}(s) \sum_{\alpha=1}^{d} \left[e^{ik_\alpha(s)} - 1\right] P_\alpha f(s) + f(s) \tag{105b}$$

and the *characteristic operator*

$$\widehat{\varPhi}_t(k) := \mathcal{W}(0; \mathsf{S}_t(k)). \tag{106}$$

By (17) and (105a) the map $\kappa \mapsto \widehat{\varPhi}_t(\kappa k)$ is a unitary group with selfadjoint generator

$$\lambda\left(\sum_\alpha P_\alpha \otimes 1_{(0,t)} k_\alpha \right) = \sum_{kl} \int_0^t \langle z_k | \sum_\alpha k_\alpha(s) P_\alpha z_l \rangle \mathrm{d}\varLambda_{kl}(s)$$

$$\equiv \sum_\alpha \int_0^t k_\alpha(s)\, \mathrm{d}N(P_\alpha;s) \tag{107}$$

and, so, we can write

$$\widehat{\varPhi}_t(k) = \exp\left[i \sum_\alpha \int_0^t k_\alpha(s)\, \mathrm{d}N(P_\alpha;s) \right]. \tag{108}$$

Now, by recalling that our initial state is s and that $N(P_\alpha;s)$ and $\widehat{\varPhi}_t(k)$ contain the free–field dynamics, but not the system–field interaction, we can define the *characteristic functional* by

$$\varPhi_t(k) := \mathrm{Tr}\left\{\widehat{\varPhi}_t(k) U(t) \mathsf{s} U(t)^* \right\}. \tag{109}$$

$\tilde{N}_\alpha(s)$ By taking in $\Phi_t(k)$ a simple function k, we obtain a characteristic function of the type (104) for the increments of a process which we denote by $\tilde{N}_\alpha(s)$. All the finite–dimensional distributions for the increments of $\tilde{N}_\alpha(s)$ and for $\tilde{N}_\alpha(s)$ itself are contained in the characteristic functional (109). By using the output fields and in particular the property (94), it is suggestive to write

$$\Phi_t(k) = \mathrm{Tr}\left\{\exp\left[i\sum_\alpha \int_0^t k_\alpha(s)\,dN^{\mathrm{out}}(P_\alpha; s)\right]\mathfrak{s}\right\}, \tag{110}$$

where Heisenberg–picture commuting operators $N^{\mathrm{out}}(P_\alpha; s)$ explicitly appear.

The Characteristic Functional for the Field Quadratures

Let us consider the family of compatible observables $\{Q(h_\alpha; t),\ t \geq 0,\ \alpha = 1, 2, \ldots, d\}$, with $\langle h_\alpha(s)|h_\beta(s)\rangle = \delta_{\alpha\beta}\|h_\alpha(s)\|^2$; we can repeat the construction of the previous subsection. By (90) we have

$$\sum_{\alpha=1}^d \int_0^t k_\alpha(s)\,dQ(h_\alpha; s) = Q\left(\sum_\alpha k_\alpha h_\alpha; t\right) \tag{111}$$

and, by taking into account (16), we can write the characteristic operator as an adapted Weyl operator again:

$$\widehat{\Phi}_t(k) = \exp\left\{i\sum_{\alpha=1}^d \int_0^t k_\alpha(s)\,dQ(h_\alpha; s)\right\}$$
$$= \exp\left\{iQ\left(\sum_\alpha k_\alpha h_\alpha; t\right)\right\} = \mathcal{W}\left(i\sum_\alpha k_\alpha h_\alpha 1_{(0,t)}; \mathbb{1}\right). \tag{112}$$

$\tilde{Q}_\alpha(s)$ Then, the characteristic functional of the process $\tilde{Q}_\alpha(s)$ associated with the selfadjoint operators $Q(h_\alpha; s)$ is given by (109) again or by

$$\Phi_t(k) = \mathrm{Tr}\left\{\exp\left[i\sum_\alpha \int_0^t k_\alpha(s)\,dQ^{\mathrm{out}}(h_\alpha; s)\right]\mathfrak{s}\right\}. \tag{113}$$

Field Observables and Adapted Weyl Operators

By the use of adapted Weyl operators and characteristic operators we can unify the field observables (85) and (90) and generalize them by including observables of the type (18). By putting together the structures (19) and (112) and by imposing $\widehat{\Phi}_t(k)\widehat{\Phi}_t(k') = \widehat{\Phi}_t(k + k')$, which gives the commutativity of the associated observables, we arrive to the following theorem.

Theorem 3.1. *Let $B^1, B^2, \ldots, B^d$ be commuting selfadjoint operators in $\mathcal{Z}$, i.e.* $B^\alpha = B^{\alpha*}$, $\left[\mathrm{e}^{\mathrm{i}\kappa_\alpha B^\alpha}, \mathrm{e}^{\mathrm{i}\kappa_\beta B^\beta} \right] = 0$, $\forall \kappa_\alpha, \kappa_\beta \in \mathbb{R}$, $\forall \alpha, \beta = 1, \ldots, d$. *Let us take* $c \in L^1_{\mathrm{loc}}(\mathbb{R}_+; \mathbb{R}^d)$, $b \in L^2_{\mathrm{loc}}(\mathbb{R}_+; \mathcal{Z})$ *and* $h^\alpha \in L^2_{\mathrm{loc}}(\mathbb{R}_+; \mathcal{Z})$, $\alpha = 1, \ldots, d$, *such that*

$$\mathrm{Im}\langle h^\alpha(t) | h^\beta(t) \rangle = 0, \qquad \mathrm{e}^{\mathrm{i}\kappa B^\alpha} h^\beta(t) = h^\beta(t),$$
$$\forall t \geq 0, \quad \forall \kappa \in \mathbb{R}, \quad \forall \alpha, \beta = 1, \ldots, d. \quad (114)$$

For any test function $k \in L^\infty(\mathbb{R}_+; \mathbb{R}^d)$ *let us define* $\mathsf{S}_t(k) \in \mathcal{U}\big(L^2(\mathbb{R}_+; \mathcal{Z})\big)$, $\mathsf{r}_t(k) \in L^2(\mathbb{R}_+; \mathcal{Z})$ *and the* characteristic operator $\widehat{\Phi}_t(k) \in \mathcal{U}(\Gamma)$ *by*

$$\big(\mathsf{S}_t(k)f\big)(s) = 1_{(0,t)}(s) \left[\mathsf{S}(k(s)) - \mathbb{1} \right] f(s) + f(s), \quad \forall f \in L^2(\mathbb{R}_+; \mathcal{Z}), \tag{115a}$$

$$\mathsf{S}\big(k(s)\big) = \prod_{\alpha=1}^d \mathrm{e}^{\mathrm{i}k_\alpha(s)B^\alpha}, \tag{115b}$$

$$\mathsf{r}_t(k)(s) = 1_{(0,t)}(s)\,\mathsf{r}(k;s), \tag{115c}$$

$$\mathsf{r}(k;s) = \mathrm{i} \sum_{\alpha=1}^d k_\alpha(s)h^\alpha(s) + \left[\mathsf{S}(k(s)) - \mathbb{1} \right] b(s), \tag{115d}$$

$$\widehat{\Phi}_t(k) = \exp\left\{ \mathrm{i} \int_0^t \mathrm{d}s \left[\sum_\alpha k_\alpha(s)c^\alpha(s) + \mathrm{Im}\,\langle b(s) | \mathsf{S}(k(s))b(s) \rangle \right] \right\}$$
$$\times \mathcal{W}\big(\mathsf{r}_t(k); \mathsf{S}_t(k)\big). \tag{116}$$

Then, the characteristic operator has the following properties:

1. localization properties:

$$\widehat{\Phi}_t\big(1_{(t_1,t_2)}k\big) = \widehat{\Phi}_{t_2}\big(1_{(t_1,t_2)}k\big) \in \mathcal{U}\big(\Gamma_{(t_1,t_2)}\big), \quad 0 \leq t_1 < t_2 \leq t; \tag{117}$$

2. group property:

$$\widehat{\Phi}_t(k)\,\widehat{\Phi}_t(k') = \widehat{\Phi}_t(k + k'), \qquad \forall k, k' \in L^\infty(\mathbb{R}_+; \mathbb{R}^d); \tag{118}$$

3. continuity: $\widehat{\Phi}_t(k)$ *is strongly continuous in* $k \in L^\infty(\mathbb{R}_+; \mathbb{R}^d)$ *and in* $t \geq 0$;
4. matrix elements:

$$\langle e(g) | \widehat{\Phi}_t(k)e(f) \rangle = \langle e(g) | e(f) \rangle$$
$$\times \exp\left\{ \int_0^t \mathrm{d}s \left[-\frac{1}{2} \sum_{\alpha\beta} k_\alpha(s)\langle h^\alpha(s) | h^\beta(s) \rangle k_\beta(s) \right. \right.$$
$$+ \mathrm{i} \sum_\alpha k_\alpha(s)\big(c^\alpha(s) + \langle h^\alpha(s) | f(s) \rangle + \langle g(s) | h^\alpha(s) \rangle\big)$$
$$\left. \left. + \langle g(s) + b(s) | \big(\mathsf{S}(k(s)) - \mathbb{1}\big)\big(f(s) + b(s)\big) \rangle \right] \right\}; \tag{119}$$

5. given the initial condition $\widehat{\Phi}_0(k) = \mathbb{1}$, $\widehat{\Phi}_t(k)$ is the unique unitary solution of the QSDE

$$
\begin{aligned}
d\widehat{\Phi}_t(k) = \widehat{\Phi}_t(k) \Bigg\{ &\sum_{lm} \langle z_l | \left(S(k(t)) - \mathbb{1} \right) z_m \rangle \, d\Lambda_{lm}(t) \\
&+ \sum_l \langle z_l | r(k;t) \rangle dA_l^\dagger(t) - \sum_l \langle r(k;t) | S(k(t)) z_l \rangle dA_l(t) \\
&+ \Big[\langle b(t) | \left(S(k(t)) - \mathbb{1} \right) b(t) \rangle + i \sum_\alpha k_\alpha(t) c^\alpha(t) \\
&- \frac{1}{2} \sum_{\alpha\beta} k_\alpha(t) \langle h^\alpha(t) | h^\beta(t) \rangle k_\beta(t) \Big] dt \Bigg\}. \quad (120)
\end{aligned}
$$

$\xi(d\omega)$

$\tilde{X}(\alpha,t;\cdot)$

$X(\alpha,t)$

Moreover, there exists a measurable space $(\Omega, \mathcal{F})$, a pvm ξ on $(\Omega, \mathcal{F})$, a family of real valued measurable functions $\left\{ \tilde{X}(\alpha,t;\cdot), \ \alpha = 1,\dots,d, \ t \geq 0 \right\}$ on Ω, a family of commuting and adapted selfadjoint operators $\left\{ X(\alpha,t), \ \alpha = 1,\dots,d, \ t \geq 0 \right\}$ such that $\tilde{X}(\alpha,0;\omega) = 0$, $X(\alpha,0) = 0$ and, for any choice of n, $0 = t_0 < t_1 < \cdots < t_n \leq t$, $\kappa_\alpha^j \in \mathbb{R}$,

$$
\begin{aligned}
\widehat{\Phi}_t(k) &= \exp\left\{ i \sum_{j=1}^n \sum_{\alpha=1}^d \kappa_\alpha^j \left[X(\alpha,t_j) - X(\alpha,t_{j-1}) \right] \right\} \\
&= \int_\Omega \exp\left\{ i \sum_{j=1}^n \sum_{\alpha=1}^d \kappa_\alpha^j \left[\tilde{X}(\alpha,t_j;\omega) - \tilde{X}(\alpha,t_{j-1};\omega) \right] \right\} \xi(d\omega), \quad (121)
\end{aligned}
$$

where $k_\alpha(s) = \sum_{j=1}^n 1_{(t_{j-1},t_j)}(s) \, \kappa_\alpha^j$.

Proof. Equations (117) follow immediately from the definition of the characteristic operator and from the properties of the Weyl operators.

One can check that the definitions of $S_t(k)$ and $r_t(k)$ are such that

$$
S_t(k)S_t(k') = S_t(k+k'), \qquad S_t(k)^{-1} = S_t(k)^* = S_t(-k), \qquad (122)
$$

$$
r_t(k) + S_t(k)r_t(k') = r_t(k+k'). \qquad (123)
$$

Together with (14) and (114), these equations imply (118).

The matrix elements (119) follow by simple computations from the definition (11) of the Weyl operators.

By the unitarity of $\widehat{\Phi}_t(k)$ and the fact that $\mathcal{E}$ is dense, it is enough to prove the strong continuity on the exponential vectors. By the unitarity and the properties (117) and (118), the strong continuity on the exponential vectors reduces to the check of the continuity of the matrix elements (119).

By checking that the condition of Definition 2.2 is satisfied, one has that the r.h.s. of (120) is well defined. By differentiating the matrix elements (119) one gets the QSDE (120); by passing to the equation for the matrix elements, which turns out to be a closed ordinary differential equation, one obtains the uniqueness of the solution.

The last statement is an application of Proposition 3.1 to the present case.   □

Continual measurements and infinitely divisible laws

It is important to realize that in a coherent state $\psi(f)$ the process $\tilde{X}(\alpha, t)$ has independent increments; here we are not considering the interaction with system $S_{\mathcal{H}}$. Indeed, by (119), (121) and (8), one obtains

$$\langle \psi(f)|\widehat{\Phi}_t(k)\psi(f)\rangle = \prod_{j=1}^{n} \left\langle \psi(1_{(t_{j-1},t_j)}f) \left| \exp\left\{ i\sum_{\alpha=1}^{d} \kappa_\alpha^j \left[X(\alpha, t_j) \right.\right.\right.\right.$$
$$\left.\left.\left.\left. - X(\alpha, t_{j-1}) \right] \right\} \psi(1_{(t_{j-1},t_j)}f) \right\rangle; \quad (124)$$

by the localization properties (117), we can reintroduce $\psi(f)$ in every factor and we obtain, again by (121), the independence of the increments:

$$\int_\Omega \exp\left\{ i\sum_{j=1}^{n}\sum_{\alpha=1}^{d} \kappa_\alpha^j \left[\tilde{X}(\alpha, t_j; \omega) - \tilde{X}(\alpha, t_{j-1}; \omega) \right] \right\} \langle \psi(f)|\xi(\mathrm{d}\omega)\psi(f)\rangle$$

$$= \prod_{j=1}^{n} \int_\Omega \exp\left\{ i\sum_{\alpha=1}^{d} \kappa_\alpha^j \left[\tilde{X}(\alpha, t_j; \omega) - \tilde{X}(\alpha, t_{j-1}; \omega) \right] \right\} \langle \psi(f)|\xi(\mathrm{d}\omega)\psi(f)\rangle.$$
$$(125)$$

This fact implies that, in a time–homogeneous case, the increments follow an infinitely divisible law. Indeed, let us take $f = 0$, $b(s) = b$, $c(s) = c$, $h(s) = h$, $k_\alpha(s) = \kappa_\alpha$ and let us denote by $\zeta_B(\mathrm{d}x)$ the joint spectral measure of the B's; then, one has

$$\left\langle \psi(0)\middle|\widehat{\Phi}_t(\kappa)\,\psi(0)\right\rangle \equiv \left\langle \psi(0)\middle| \prod_{\alpha=1}^{d} e^{i\kappa_\alpha X(\alpha,t)}\,\psi(0) \right\rangle = \exp\left\{ it\sum_\alpha \kappa_\alpha c^\alpha \right.$$
$$\left. - \frac{t}{2}\sum_{\alpha\beta} \kappa_\alpha \langle h^\alpha|h^\beta\rangle \kappa_\beta + t\int_{\mathbb{R}^d} \left[\exp\left(i\sum_\alpha \kappa_\alpha x^\alpha \right) - 1 \right] \langle b|\zeta_B(\mathrm{d}x)b\rangle \right\}. \quad (126)$$

By comparing this result with the Lévy–Khintchin formula, one sees that (126) is the characteristic function of an infinitely divisible distribution on $\mathbb{R}^d$, but not the most general one because $\langle b|\zeta_B(\mathrm{d}x)b\rangle$ is a finite measure.

The representations of infinitely divisible laws have been studied ([84] Section 21) and from there we can take a suggestion on how we can generalize the observables of Theorem 3.1. All the properties of $\widehat{\Phi}_t(k)$ are maintained and the most general infinitely divisible distribution is obtained if everywhere in Theorem 3.1 $b(s)$ is replaced by $\sqrt{\frac{1+|B|^2}{|B|^2}}\,b(s)$ and $c_\alpha(s)$ by $c_\alpha(s) - \langle b(s)|\frac{1+|B|^2}{|B|^2}B^\alpha b(s)\rangle$; here

$|B|^2 := \sum_{\alpha=1}^{d} (B^{\alpha})^2$. This is only an indication of a possible generalization, but we do not touch any more this more general characteristic operator in this work.

The output characteristic operator

The observables in the Heisenberg description can now be introduced implicitly by defining the *output characteristic operator*

$\widehat{\Phi}_t^{\text{out}}(k)$

$$\widehat{\Phi}_t^{\text{out}}(k) = U(t)^* \widehat{\Phi}_t(k) U(t) \,. \tag{127}$$

Formally we have

$$\widehat{\Phi}_t^{\text{out}}(k) = \exp\left[i \sum_\alpha \int_0^t k_\alpha(s) \, dX^{\text{out}}(\alpha; s)\right], \tag{128}$$

but we do not need to give a rigourous meaning to the "Heisenberg" observables $X^{\text{out}}(\alpha; t)$ and to the integrals with respect to $dX^{\text{out}}(\alpha; s)$; all we need is to differentiate $\widehat{\Phi}_t^{\text{out}}(k)$, which is done in Proposition 3.2.

The Characteristic Functional and the Finite Dimensional Laws

$\Phi_t(k)$

By considering also the interaction with system $S_{\mathcal{H}}$, we have that the *characteristic functional* of the process $\tilde{X}$ is again given by (109):

$$\Phi_t(k) = \text{Tr}\left\{\widehat{\Phi}_t(k) U(t) \mathfrak{s} U(t)^*\right\} \equiv \text{Tr}\left\{\widehat{\Phi}_t^{\text{out}}(k) \mathfrak{s}\right\}. \tag{129}$$

All the probabilities describing the continual measurement of the observables $X(\alpha, t)$ are contained in $\Phi_t(k)$; let us give explicitly the construction of the joint probabilities for a finite number of increments.

The measurable functions $\left\{\tilde{X}(\alpha, t; \cdot), \ \alpha = 1, \ldots, d, \ t \geq 0\right\}$, introduced in Theorem 3.1, represent the output signal of the continual measurements. Let us

$\Delta \boldsymbol{X}(t_1, t_2)$ denote by $\Delta \boldsymbol{X}(t_1, t_2) = \left(\tilde{X}(1, t_2) - \tilde{X}(1, t_1), \ldots, \tilde{X}(d, t_2) - \tilde{X}(d, t_1)\right)$ the vec-
$\xi(d\boldsymbol{x}; t_1, t_2)$ tor of the increments of the output in the time interval (t_1, t_2) and by $\xi(d\boldsymbol{x}; t_1, t_2)$ the joint pvm on $\mathbb{R}^d$ of the increments $X(\alpha; t_2) - X(\alpha; t_1)$, $\alpha = 1, \ldots, d$. Note that, because of the properties of the characteristic operator, not only the different components of an increment are commuting, but also increments related to different time intervals; this implies that the pvm related to different time intervals commute. Moreover, the localization properties of the characteristic operator give

$$\xi(A; t_1, t_2) \in \mathcal{B}(\Gamma_{(t_1, t_2)}), \quad \text{for any Borel set } A \subset \mathbb{R}^d \,. \tag{130}$$

As in the last part of Theorem 3.1, let us consider $0 = t_0 < t_1 < \cdots < t_n \leq t$, $k_\alpha(s) = \sum_{j=1}^{n} 1_{(t_{j-1}, t_j)}(s) \kappa_\alpha^j$, $\boldsymbol{\kappa}^j = \left(\kappa_1^j, \ldots, \kappa_d^j\right)$; then we can write

$$\Phi_t(k) = \text{Tr}\left\{\exp\left(i\sum_{j=1}^n\sum_{\alpha=1}^d \kappa_\alpha^j\left[X(\alpha,t_j) - X(\alpha,t_{j-1})\right]\right)U(t)\mathfrak{s}U(t)^*\right\}$$

$$= \int_{\mathbb{R}^{nd}}\left(\prod_{j=1}^n e^{i\kappa^j\cdot x^j}\right)P_{\rho_0}\left[\Delta X(t_0,t_1)\in dx^1,\ldots,\Delta X(t_{n-1},t_n)\in dx^n\right],$$

$$(131)$$

where the physical probabilities are given by

$$P_{\rho_0}\left[\Delta X(t_0,t_1)\in A_1,\ldots,\Delta X(t_{n-1},t_n)\in A_n\right]$$

$$= \text{Tr}\left\{\left(\prod_{j=1}^n \xi(A_j;t_{j-1},t_j)\right)U(t)\mathfrak{s}U(t)^*\right\}. \quad (132)$$

Obviously, $\Phi_t(k)$ is the characteristic function of the physical probabilities $P_{\rho_0}\left[\Delta X(t_0,t_1)\in A_1,\ldots,\Delta X(t_{n-1},t_n)\in A_n\right]$ and it uniquely determines them.

3.3 The Reduced Description of the Continual Measurements

An essential step in the theory is to eliminate the degrees of freedom of the fields and to pass to a reduced description based only on system $S_\mathcal{H}$. This is similar to the passage to the reduced dynamics in Section 2.4 and, indeed, the quantities $\rho(f;t)$ (70) and $\mathcal{L}(f(t))$ (73) will be involved also here.

The Reduced Characteristic Operator

Definition 3.1. For $f \in L^2(\mathbb{R}^+;\mathcal{Z})$, $k \in L^\infty(\mathbb{R}_+;\mathbb{R}^d)$, $t \geq 0$, let us define the reduced characteristic operator $\mathcal{G}_t(f;k) \in \mathcal{B}(\mathcal{T}(\mathcal{H}))$ by: $\forall a \in \mathcal{B}(\mathcal{H})$, $\forall \varrho \in \mathcal{T}(\mathcal{H})$, $\qquad \mathcal{G}_t(f;k)$

$$\text{Tr}_\mathcal{H}\left\{a\,\mathcal{G}_t(f;k)[\varrho]\right\} = \text{Tr}_{\mathcal{H}\otimes\Gamma}\left\{a\otimes\widehat{\Phi}_t(k)\,U(t)\,\varrho\otimes\eta(f)\,U(t)^*\right\}. \quad (133)$$

Let us recall that the symbol $\eta(f)$ for a coherent state is defined in eq. (58).

By the definition of reduced characteristic functional, for the three choices of initial state proposed in Section 2.3 the characteristic functional is given by

S1. $\qquad \Phi_t(k) = \text{Tr}_\mathcal{H}\left\{\mathcal{G}_t(0;k)[\rho_0]\right\}$,
S2. $\qquad \Phi_t(k) = \text{Tr}_\mathcal{H}\left\{\mathcal{G}_t(f;k)[\rho_0]\right\}$,
S3. $\qquad \Phi_t(k) = \mathbb{E}^c\left[\text{Tr}_\mathcal{H}\left\{\mathcal{G}_t(f;k)[\rho_0]\right\}\right]$;
let us recall that in this case f is random.

In (133) only unitary operators are involved and this implies that it is easy to apply Proposition 2.4 to the triple product; we get the following results.

Proposition 3.2. Let $\widehat{\Phi}_t^{\text{out}}(k)$ and $\mathcal{G}_t(f;k)$ be defined by (127), (133), with all the hypotheses given in the previous sections.

The output characteristic operator (127) satisfies the QSDE

$$
\begin{aligned}
\mathrm{d}\widehat{\varPhi}_t^{\mathrm{out}}(k) = U(t)^* \Bigg\{ & \sum_{lm} C(z_l)^* S^* \left(\mathsf{S}\big(k(t)\big) - \mathbb{1} \right) S C(z_m)\, \mathrm{d}\varLambda_{lm}(t) \\
& + \sum_l C(z_l)^* S^* \left[C\big(\mathsf{r}(k;t)\big) + \left(\mathsf{S}\big(k(t)\big) - \mathbb{1} \right) R \right] \mathrm{d}A_l^\dagger(t) \\
& + \sum_l \left[-C\big(\mathsf{r}(k;t)\big)^* \mathsf{S}\big(k(t)\big) + R^* \left(\mathsf{S}\big(k(t)\big) - \mathbb{1} \right) \right] S C(z_l)\, \mathrm{d}A_l(t) \\
& + \bigg[\langle b(t)| \left(\mathsf{S}\big(k(t)\big) - \mathbb{1} \right) b(t) \rangle + \mathrm{i} \sum_\alpha k_\alpha(t) c^\alpha(t) \\
& - \frac{1}{2} \sum_{\alpha\beta} k_\alpha(t) \langle h^\alpha(t)|h^\beta(t)\rangle k_\beta(t) + R^* C\big(\mathsf{r}(k;t)\big) \\
& - C\big(\mathsf{r}(k;t)\big)^* \mathsf{S}\big(k(t)\big) R + R^* \left(\mathsf{S}\big(k(t)\big) - \mathbb{1} \right) R \bigg] \mathrm{d}t \Bigg\} U(t) \widehat{\varPhi}_t^{\mathrm{out}}(k). \quad (134)
\end{aligned}
$$

The reduced characteristic operator satisfies the equation

$$
\mathcal{G}_t(f;k)[\varrho] - \varrho = \int_0^t \mathcal{K}_s\big(f;k(s)\big) \circ \mathcal{G}_s(f;k)[\varrho]\, \mathrm{d}s, \quad (135)
$$

$\mathcal{K}_t(f;\kappa)$ *with generator*

$$
\begin{aligned}
\mathcal{K}_t(f;\kappa)[\varrho] = \mathcal{L}\big(f(t)\big)[\varrho] + \mathrm{Tr}_{\mathcal{Z}}\Big\{ \mathbb{1} \otimes \big(\mathsf{S}(\kappa) - \mathbb{1}\big) \mathcal{J}(f;t)[\varrho] \Big\} \\
- \frac{1}{2} \sum_{\alpha\beta} \kappa_\alpha \langle h^\alpha(t)|h^\beta(t)\rangle \kappa_\beta\, \varrho + \mathrm{i} \sum_\alpha \kappa_\alpha \big[Z^\alpha(f;t)\varrho + \varrho Z^\alpha(f;t)^* \big], \quad (136a)
\end{aligned}
$$

$$
Z^\alpha(f;t) = \frac{1}{2} c^\alpha(t) + C\big(h^\alpha(t)\big)^* \left[R + SC\big(f(t)\big) \right], \quad (136b)
$$

$$
\mathcal{J}(f;t)[\varrho] = \left[R + SC\big(f(t)\big) + C\big(b(t)\big) \right] \varrho \left[R^* + C\big(f(t)\big)^* S^* + C\big(b(t)\big)^* \right]; \quad (136c)
$$

$Z^\alpha(f;t)$, $\mathcal{J}(f;t)$ $\mathcal{L}\big(f(t)\big)$ *is given by eq. (75).*

Proof. Let us apply Proposition 2.4 to $U(t)^* a \otimes \widehat{\varPhi}_t(k)$; due to the unitarity of the adapted processes involved, there is no problem to control the domains and we get

$$
\begin{aligned}
\mathrm{d}\left(U(t)^* a \otimes \widehat{\varPhi}_t(k) \right) = U(t)^* \widehat{\varPhi}_t(k) \Bigg\{ & \sum_{lm} C(z_l)^* \left(S^* a \otimes \mathsf{S}\big(k(t)\big) - a \otimes \mathbb{1} \right) \\
& \times C(z_m)\, \mathrm{d}\varLambda_{lm}(t) - \sum_l C(z_l)^* S^* \left(R - C\big(\mathsf{r}(k;t)\big) \right) a\, \mathrm{d}A_l^\dagger(t) \\
& + \sum_l \left(R^* - C\big(\mathsf{r}(k;t)\big)^* \right) a \otimes \mathsf{S}\big(k(t)\big) C(z_l)\, \mathrm{d}A_l(t)
\end{aligned}
$$

$$+ \left[K^* a + \langle b(t)| \left(\mathsf{S}(k(t)) - \mathbb{1} \right) b(t) \rangle a + \mathrm{i} \sum_\alpha k_\alpha(t) c^\alpha(t) a \right.$$

$$\left. - \frac{1}{2} \sum_{\alpha\beta} k_\alpha(t) \langle h^\alpha(t)|h^\beta(t) \rangle k_\beta(t) a + R^* C \big(r(k;t) \big) a \right] \mathrm{d}t \Bigg\}. \quad (137)$$

By appling Proposition 2.4 again we get

$$\mathrm{d} \left(U(t)^* a \otimes \widehat{\Phi}_t(k) U(t) \right) = U(t)^* \Bigg\{ \sum_{lm} C(z_l)^* \left(\mathsf{S}^* a \otimes \mathsf{S}(k(t)) \mathsf{S} - a \otimes \mathbb{1} \right)$$

$$\times C(z_m) \, \mathrm{d}\Lambda_{lm}(t) - \sum_l C(z_l)^* \mathsf{S}^* \left[\left(R - C(r(k;t)) \right) a - a \otimes \mathsf{S}(k(t)) R \right] \mathrm{d}A_l^\dagger(t)$$

$$+ \sum_l \left[\left(R^* - C(r(k;t))^* \right) a \otimes \mathsf{S}(k(t)) - a R^* \right] \mathsf{S} C(z_l) \, \mathrm{d}A_l(t)$$

$$+ \left[\mathcal{L}_0'[a] + \langle b(t)| \left(\mathsf{S}(k(t)) - \mathbb{1} \right) b(t) \rangle a + \mathrm{i} \sum_\alpha k_\alpha(t) c^\alpha(t) a \right.$$

$$\left. - \frac{1}{2} \sum_{\alpha\beta} k_\alpha(t) \langle h^\alpha(t)|h^\beta(t) \rangle k_\beta(t) a + R^* C(r(k;t)) a \right.$$

$$\left. - C(r(k;t))^* a \otimes \mathsf{S}(k(t)) R + R^* a \otimes \left(\mathsf{S}(k(t)) - \mathbb{1} \right) R \right] \mathrm{d}t \Bigg\} \widehat{\Phi}_t(k) U(t). \quad (138)$$

By taking the matrix elements with respect to the coherent vector $|\psi(f)\rangle$ we get eqs. (135), (136a); by taking $a = \mathbb{1}$ we get (134). □

If the time dependence of the generator is sufficiently regular we have the differential equation

$$\frac{\mathrm{d}}{\mathrm{d}t} \mathcal{G}_t(f;k)[\varrho] = \mathcal{K}_t \big(f; k(t) \big) \circ \mathcal{G}_t(f;k)[\varrho], \qquad \mathcal{G}_0(f;k) = \mathbb{1}, \quad (139)$$

which is a modification of a master equation. This kind of equations was introduced in [22, 23], inside of the operational approach, and it was related to the approach based on QSDE in [24]. The problem of generators of the type (136a) was studied in [25, 61, 62].

Let us define a two–time modification of the reduced characteristic operator by $\mathcal{G}_{t_1}^{t_2}(f;\kappa)$

$$\mathcal{G}_{t_1}^{t_2}(f;\kappa)[\varrho] - \varrho = \int_{t_1}^{t_2} \mathcal{K}_s(f;\kappa) \circ \mathcal{G}_{t_1}^s(f;\kappa)[\varrho] \, \mathrm{d}s. \quad (140)$$

Then, for $0 = t_0 < t_1 < \cdots < t_n \leq t$, $k_\alpha(s) = \sum_{j=1}^n 1_{(t_{j-1}, t_j)}(s) \kappa_\alpha^j$, $\kappa^j = \left(\kappa_1^j, \ldots, \kappa_d^j \right)$, eq. (135) gives

$$\mathcal{G}_t(f;k) = \mathcal{G}_{t_n}^t(f;0) \circ \mathcal{G}_{t_{n-1}}^{t_n}(f;\kappa^n) \circ \cdots \circ \mathcal{G}_{t_0}^{t_1}(f;\kappa^1). \quad (141)$$

Moreover, we have

$$\mathcal{G}_{t_2}^{t_3}(f;\kappa) \circ \mathcal{G}_{t_1}^{t_2}(f;\kappa) = \mathcal{G}_{t_1}^{t_3}(f;\kappa). \quad (142)$$

The reduced dynamics

$\Upsilon(f;t,s)$ Let us set

$$\Upsilon(f;t,s) := \mathcal{G}_s^t(f;0);\tag{143}$$

by (141) we have the composition law

$$\Upsilon(f;t_3,t_1) = \Upsilon(f;t_3,t_2) \circ \Upsilon(f;t_2,t_1), \qquad 0 \le t_1 \le t_2 \le t_3.\tag{144}$$

Then, by comparing (72) with (135), (136a), (144) we obtain

$$\rho(f;t) = \Upsilon(f;t,0)[\rho_0] = \Upsilon(f;t,s)[\rho(f;s)].\tag{145}$$

In the case **S2** the reduced statistical operator is $\rho(t) = \rho(f;t)$ and, so, $\Upsilon(f;t,s)$ has the meaning of reduced evolution operator from s to t; eq. (144) says that this evolution is in some sense without memory. In the case **S3** we have

$$\rho(t) = \mathbb{E}^c\left[\Upsilon(f;t,0)[\rho_0]\right] = \mathbb{E}^c\left[\Upsilon(f;t,s)[\rho(f;s)]\right]\tag{146}$$

and the lack–of–memory property is lost.

From eqs. (140), (143), (136a) we get

$$\Upsilon(f;t_2,t_1)[\varrho] - \varrho = \int_{t_1}^{t_2} \mathcal{L}(f(s)) \circ \Upsilon(f;s,t_1)[\varrho]\,\mathrm{d}s,\tag{147}$$

which is another form of the master equation (74).

Instruments and Finite–Dimensional Laws

In the quantum theory of measurement an important notion is that of *instrument* and the operational approach to continual measurements, mentioned in the Introduction, is based on such a notion. Here we recall a few facts, without developing in full this side of the theory.

By using the joint pvm $\xi(\mathrm{d}x;t_1,t_2)$ of the increments $X(\alpha;t_2) - X(\alpha;t_1)$, $\alpha = 1,\ldots,d$, we define the map–valued measure $\mathcal{I}_{t_1}^{t_2}(f;\cdot)$, $0 \le t_1 < t_2$, $f \in$

$\mathcal{I}_{t_1}^{t_2}(f;A)$ $L^2(R_+;\mathcal{Z})$, by: $\forall a \in \mathcal{B}(\mathcal{H})$, $\forall \varrho \in T(\mathcal{H})$,

$$\mathrm{Tr}_{\mathcal{H}}\left\{a\,\mathcal{I}_{t_1}^{t_2}(f;A)[\varrho]\right\} = \mathrm{Tr}_{\mathcal{H}\otimes\Gamma}\left\{a \otimes \xi(A;t_1,t_2)\,U(t_2,t_1)\,\varrho \otimes \eta(f)\,U(t_2,t_1)^*\right\},\tag{148}$$

where A is a Borel set in $\mathbb{R}^d$; by the factorization properties of Γ and $\eta(f)$, only $f_{(t_1,t_2)}$, the part of f in (t_1,t_2), is relevant for the definition of $\mathcal{I}_{t_1}^{t_2}(f;A)$. The family

nstrument   of maps $\mathcal{I}_{t_1}^{t_2}(f;\cdot)$ is a completely positive *instrument* [44,64], whose characterizing properties are

1. $\mathcal{I}_{t_1}^{t_2}(f;A) \in \mathcal{B}(T(\mathcal{H}))$;
2. $\mathrm{Tr}\left\{\mathcal{I}_{t_1}^{t_2}(f;\mathbb{R}^d)[\varrho]\right\} = \mathrm{Tr}\{\varrho\}, \forall \varrho \in T(\mathcal{H})$;
3. $\displaystyle\sum_{i,j=1}^{n} \mathrm{Tr}\left\{a_i^* a_j \mathcal{I}_{t_1}^{t_2}(f;A)\left[|\psi_j\rangle\langle\psi_i|\right]\right\} \ge 0, \forall n, \forall \psi_j \in \mathcal{H}, \forall a_j \in \mathcal{B}(\mathcal{H})$;

4. for any finite or countable (Borel) partition $A_1, A_2, \ldots$ of a Borel set A one has $\sum_j \mathcal{I}_{t_1}^{t_2}(f; A_j)[\varrho] = \mathcal{I}_{t_1}^{t_2}(f; A)[\varrho], \forall \varrho \in \mathcal{T}(\mathcal{H})$.

Then, $\mathcal{G}_{t_1}^{t_2}(f; \boldsymbol{\kappa})$ is the Fourier transform of the instrument $\mathcal{I}_{t_1}^{t_2}(f; \cdot)$, i.e.

$$\mathcal{G}_{t_1}^{t_2}(f; \boldsymbol{\kappa})[\varrho] = \int_{\mathbb{R}^d} e^{i\boldsymbol{\kappa} \cdot \boldsymbol{x}} \, \mathcal{I}_{t_1}^{t_2}(f; d\boldsymbol{x})[\varrho], \qquad \forall \varrho \in \mathcal{T}(\mathcal{H}), \qquad (149)$$

and a Bochner type theorem holds which says that $\mathcal{G}$ uniquely determines $\mathcal{I}$, see [25] p. 110 Theorem 1.5.

By (131), (132), (141), (149) the probabilities (132) are given by

$$P_{\rho_0}\left[\Delta \boldsymbol{X}(t_0, t_1) \in A_1, \ldots, \Delta \boldsymbol{X}(t_{n-1}, t_n) \in A_n\right]$$
$$= \mathbb{E}^c\left[\operatorname{Tr}_{\mathcal{H}}\left\{\mathcal{I}_{t_{n-1}}^{t_n}(f; A_n) \circ \cdots \circ \mathcal{I}_{t_0}^{t_1}(f; A_1)[\rho_0]\right\}\right]; \quad (150)$$

we use the convention that the classical expectation $\mathbb{E}^c$ has no effect when f is not random.

Mean Values and Covariances

As usual, all the moments of the process $\tilde{X}(\alpha, t)$ can be obtained by derivation of the characteristic functional. If we take

$$k_\alpha(t) = \kappa_1 \delta_{\alpha\alpha_1} 1_{(0,t_1)}(t) + \kappa_2 \delta_{\alpha\alpha_2} 1_{(0,t_2)}(t), \qquad (151)$$

the mean and the second moments are given by

$$\mathbb{E}_{\rho_0}\left[\tilde{X}(\alpha_j, t_j)\right] = -i \frac{\partial}{\partial \kappa_j} \Phi_{t_1 \vee t_2}(k)\Big|_{\boldsymbol{\kappa}=0}, \qquad (152)$$

$$\mathbb{E}_{\rho_0}\left[\tilde{X}(\alpha_1, t_1)\tilde{X}(\alpha_2, t_2)\right] = -\frac{\partial^2}{\partial \kappa_1 \partial \kappa_2} \Phi_{t_1 \vee t_2}(k)\Big|_{\boldsymbol{\kappa}=0}, \qquad (153)$$

when they exist. Obviously, the covariance function is

$$\operatorname{Cov}\left[\tilde{X}(\alpha_1, t_1), \tilde{X}(\alpha_2, t_2)\right]$$
$$= \mathbb{E}_{\rho_0}\left[\tilde{X}(\alpha_1, t_1)\tilde{X}(\alpha_2, t_2)\right] - \mathbb{E}_{\rho_0}\left[\tilde{X}(\alpha_1, t_1)\right] \mathbb{E}_{\rho_0}\left[\tilde{X}(\alpha_2, t_2)\right]. \quad (154)$$

Proposition 3.3. *Let the operators B^α be bounded; then we have*

$$-i \frac{\partial}{\partial \kappa_j} \operatorname{Tr}_{\mathcal{H}}\left\{\mathcal{G}_{t_1 \vee t_2}(f; k)[\rho_0]\right\}\Big|_{\boldsymbol{\kappa}=0} = \int_0^{t_j} dt \, \operatorname{Tr}_{\mathcal{H}}\left\{Y^{\alpha_j}(f; t)\rho(f; t)\right\}, \quad (155)$$

$$-\frac{\partial^2}{\partial \kappa_1 \partial \kappa_2} \operatorname{Tr}_{\mathcal{H}}\left\{\mathcal{G}_{t_1 \vee t_2}(f; k)[\rho_0]\right\}\Big|_{\boldsymbol{\kappa}=0}$$

$$= \int_0^{t_1 \wedge t_2} dt \left[\langle h^{\alpha_1}(t) | h^{\alpha_2}(t) \rangle + \mathrm{Tr}_{\mathcal{H}} \{ J^{\alpha_1 \alpha_2}(f;t) \, \rho(f;t) \} \right]$$

$$+ \int_0^{t_1} dt \int_0^{t \wedge t_2} ds \, \mathrm{Tr}_{\mathcal{H}} \{ Y^{\alpha_1}(f;t) \Upsilon(f;t,s) \circ \mathcal{Y}^{\alpha_2}(f;s) [\rho(f;s)] \}$$

$$+ \int_0^{t_2} dt \int_0^{t \wedge t_1} ds \, \mathrm{Tr}_{\mathcal{H}} \{ Y^{\alpha_2}(f;t) \Upsilon(f;t,s) \circ \mathcal{Y}^{\alpha_1}(f;s) [\rho(f;s)] \}, \quad (156)$$

where

$$\mathcal{Y}^{\alpha}(f;t)[\varrho] = Z^{\alpha}(f;t)\varrho + \varrho Z^{\alpha}(f;t)^* + \mathrm{Tr}_{\mathcal{Z}} \{ \mathbb{1} \otimes B^{\alpha} \, \mathcal{J}(f;t)[\varrho] \}, \quad (157a)$$

$$Y^{\alpha}(f;t) = Z^{\alpha}(f;t) + Z^{\alpha}(f;t)^* + \left[R^* + C(f(t))^* S^* \right.$$
$$\left. + C(b(t))^* \right] \mathbb{1} \otimes B^{\alpha} \left[R + SC(f(t)) + C(b(t)) \right], \quad (157b)$$

$$J^{\alpha\beta}(f;t) = \left[R^* + C(f(t))^* S^* + C(b(t))^* \right]$$
$$\mathbb{1} \otimes B^{\alpha} B^{\beta} \left[R + SC(f(t)) + C(b(t)) \right], \quad (157c)$$

Proof. From the definition of $\mathcal{K}$ (136a) and eq. (151) we get

$$-\mathrm{i} \frac{\partial}{\partial \kappa_j} \mathcal{K}_s \big(f; k(s) \big) \Big|_{\kappa=0} = 1_{(0,t_j)}(s) \, \mathcal{Y}^{\alpha_j}(f;s), \quad (158)$$

$$-\frac{\partial^2}{\partial \kappa_1 \partial \kappa_2} \mathcal{K}_s \big(f; k(s) \big)[\varrho] \Big|_{\kappa=0} = 1_{(0,t_1 \wedge t_2)}(s) \Big[\langle h^{\alpha_1}(t) | h^{\alpha_2}(t) \rangle \varrho$$
$$+ \mathrm{Tr}_{\mathcal{Z}} \Big\{ \mathbb{1} \otimes B^{\alpha_1} B^{\alpha_2} \big(R + SC(f(t)) + C(b(t)) \big)$$
$$\varrho \big(R^* + C(f(t))^* S^* + C(b(t))^* \big) \Big\} \Big]. \quad (159)$$

By (135), (143) we obtain

$$-\mathrm{i} \frac{\partial}{\partial \kappa_j} \mathcal{G}_t(f;k) \Big|_{\kappa=0} = \int_0^t ds \, (-\mathrm{i}) \frac{\partial}{\partial \kappa_j} \mathcal{K}_s \big(f; k(s) \big) \Big|_{\kappa=0} \circ \Upsilon(f;s,0)$$
$$+ \int_0^t ds \, \mathcal{L}(f(s)) \circ (-\mathrm{i}) \frac{\partial}{\partial \kappa_j} \mathcal{G}_s(f;k) \Big|_{\kappa=0}, \quad (160)$$

$$-\frac{\partial^2}{\partial \kappa_1 \partial \kappa_2} \mathcal{G}_t(f;k) \Big|_{\kappa=0} = -\int_0^t ds \, \frac{\partial^2}{\partial \kappa_1 \partial \kappa_2} \mathcal{K}_s \big(f; k(s) \big) \Big|_{\kappa=0} \circ \Upsilon(f;s,0)$$
$$+ \int_0^t ds \, (-\mathrm{i}) \frac{\partial}{\partial \kappa_1} \mathcal{K}_s \big(f; k(s) \big) \Big|_{\kappa=0} \circ (-\mathrm{i}) \frac{\partial}{\partial \kappa_2} \mathcal{G}_s(f;k) \Big|_{\kappa=0}$$

$$+ \int_0^t ds \, (-\mathrm{i}) \frac{\partial}{\partial \kappa_2} \mathcal{K}_s\big(f; k(s)\big)\Big|_{\kappa=0} \circ (-\mathrm{i}) \frac{\partial}{\partial \kappa_1} \mathcal{G}_s(f; k)\Big|_{\kappa=0}$$

$$- \int_0^t ds \, \mathcal{L}\big(f(s)\big) \circ \frac{\partial^2}{\partial \kappa_1 \partial \kappa_2} \mathcal{G}_s(f; k)\Big|_{\kappa=0}. \qquad (161)$$

By using (147), one can check that the solution of eq. (160) is

$$-\mathrm{i} \frac{\partial}{\partial \kappa_j} \mathcal{G}_t(f; k)\Big|_{\kappa=0} = \int_0^{t \wedge t_j} ds \, \Upsilon(f; t, s) \circ \mathcal{Y}^{\alpha_j}(f; s) \circ \Upsilon(f; s, 0); \qquad (162)$$

then, this expression can be inserted into (161).

By applying the expressions (160), (161) to ρ_0, by taking the trace and by using the definitions (157), one gets the final result. $\qquad \square$

Note that the first and second moments turn out to be given by

$$\mathbb{E}_{\rho_0}\left[\tilde{X}(\alpha_j, t_j)\right] = \int_0^{t_j} dt \, \mathbb{E}^c\left[\mathrm{Tr}_{\mathcal{H}}\{Y^{\alpha_j}(f; t)\rho(f; t)\}\right], \qquad (163)$$

$$\mathbb{E}_{\rho_0}\left[\tilde{X}(\alpha_1, t_1)\tilde{X}(\alpha_2, t_2)\right]$$

$$= \int_0^{t_1 \wedge t_2} dt \left[\langle h^{\alpha_1}(t)|h^{\alpha_2}(t)\rangle + \mathbb{E}^c\left[\mathrm{Tr}_{\mathcal{H}}\{J^{\alpha_1\alpha_2}(f; t)\,\rho(f; t)\}\right]\right]$$

$$+ \int_0^{t_1} dt \int_0^{t \wedge t_2} ds \, \mathbb{E}^c\left[\mathrm{Tr}_{\mathcal{H}}\{Y^{\alpha_1}(f; t)\Upsilon(f; t, s) \circ \mathcal{Y}^{\alpha_2}(f; s)[\rho(f; s)]\}\right]$$

$$+ \int_0^{t_2} dt \int_0^{t \wedge t_1} ds \, \mathbb{E}^c\left[\mathrm{Tr}_{\mathcal{H}}\{Y^{\alpha_2}(f; t)\Upsilon(f; t, s) \circ \mathcal{Y}^{\alpha_1}(f; s)[\rho(f; s)]\}\right], \qquad (164)$$

3.4 Direct Detection

The Detection Scheme

The measurement of an observable of the type $N(P; t)$ can be realized according to the scheme of Fig. 1, called *direct detection*.

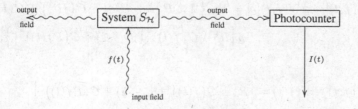

Fig. 1. Direct detection

$I(t)$

The system $S_{\mathcal{H}}$ is stimulated by some input field, say in a coherent state $\psi(f)$ or in a mixture of coherent states; then, it emits fluorescence light in various directions. A part of this output field reaches a *photoelectron counter* which produces some *output current* $I(t)$ "proportional" to the rate of arrival of the photons. By denoting by $\tilde{N}(t)$ the process which counts the photons arriving to the photocounter and by using Stieltjes integrals, the output current can be written as

$$I(t) = \int_0^t F(t-s)\, d\tilde{N}(s), \tag{165}$$

where F is a *response function* which characterizes the apparatus. A typical choice is

$F(t)$

$$F(t) = k \exp\left(-\frac{\tilde{\gamma}}{2}t\right); \tag{166}$$

$k > 0$ and $\tilde{\gamma} > 0$ are constants which depend on the apparatus. If we do not consider the time–of–flight from the system $S_{\mathcal{H}}$ to the detector and we denote by P the projection which gives the part of the field reaching the counter, we can say that the direct detection scheme described here realizes a continual measurement of the observables $N(P;t)$, $t \geq 0$. There is no conceptual difficulty in considering more counters together, but we prefer to continue to study the case of a single detector.

The general results given in the previous sections apply to the present case by taking

- $d=1$, $\quad X(\alpha,t) \equiv X(t) = N(P;t)$, $\quad \tilde{X}(\alpha,t) \equiv \tilde{X}(t) = \tilde{N}(t)$,
- $h^\alpha(t) = 0$, $c(t) = 0$, $b(t) = 0$, which gives $Z^\alpha(f;t) = 0$, $r(k;t) = 0$,
- $B^\alpha = P$, with $P^2 = P^* = P$, which gives $S(k(t)) - \mathbb{1} = \left(e^{ik(t)} - 1\right)P$.

$\tilde{\mathcal{L}}(f(t))$
$\mathcal{Y}(f;t)$
$Y(f;t)$

Then, we obtain

$$\mathcal{K}_t(f;k)[\varrho] = \tilde{\mathcal{L}}(f(t))[\varrho] + e^{ik(t)}\,\mathcal{Y}(f;t)[\varrho], \tag{167a}$$

$$\tilde{\mathcal{L}}(f)[\varrho] = -\left(\frac{1}{2}R^*R + R^*SC(f) + iH\right)\varrho - \varrho\left(\frac{1}{2}R^*R + C(f)^*S^*R - iH\right)$$
$$+ \mathrm{Tr}_{\mathcal{Z}}\left\{\mathbb{1} \otimes (\mathbb{1} - P)\left(R + SC(f)\right)\varrho\left(R^* + C(f)^*S^*\right)\right\} - \|f\|^2\varrho, \tag{167b}$$

$$\mathcal{Y}^\alpha(f;t)[\varrho] \equiv \mathcal{Y}(f;t)[\varrho] = \mathrm{Tr}_{\mathcal{Z}}\Big\{\mathbb{1} \otimes P\left[R + (S - \mathbb{1})C(f(t)) + C(Pf(t))\right]$$
$$\varrho\left[R^* + C(f(t))^*(S^* - \mathbb{1}) + C(Pf(t))^*\right]\Big\}, \tag{167c}$$

$$Y^\alpha(f;t) \equiv Y(f;t) = \left[R^* + C(f(t))^*(S^* - \mathbb{1}) + C(Pf(t))^*\right]$$
$$\mathbb{1} \otimes P\left[R + (S - \mathbb{1})C(f(t)) + C(Pf(t))\right], \tag{167d}$$

$$J^{\alpha\beta}(f;t) \equiv J(f;t) = Y(f;t) \qquad (167e)$$

Usually the detector is placed in a position in which it is not reached by the direct light of the stimulating laser; in mathematical terms we have

$$Pf(t) = 0, \qquad \forall t. \qquad (168)$$

Equations (167c) and (167d) are written in a way that puts in evidence the terms that disappear when condition (168) holds.

Moments

In the case of the counting process $\tilde{N}$, the first two moments (163), (164) become

$$\mathbb{E}_{\rho_0}\left[\tilde{N}(t)\right] = \int_0^t d\tau\, \mathbb{E}^c\left[\mathrm{Tr}_{\mathcal{H}}\left\{Y(f;\tau)\rho(f;\tau)\right\}\right], \qquad (169)$$

$$\mathbb{E}_{\rho_0}\left[\tilde{N}(t)\tilde{N}(s)\right] = \mathbb{E}_{\rho_0}\left[\tilde{N}(t \wedge s)\right]$$
$$+ \int_{t\wedge s}^{t\vee s} d\tau_1 \int_0^{t\wedge s} d\tau_2\, \mathbb{E}^c\left[\mathrm{Tr}_{\mathcal{H}}\left\{Y(f;\tau_1)\Upsilon(f;\tau_1,\tau_2)\circ\mathcal{Y}(f;\tau_2)[\rho(f;\tau_2)]\right\}\right]$$
$$+ 2\int_0^{t\wedge s} d\tau_1 \int_0^{\tau_1} d\tau_2\, \mathbb{E}^c\left[\mathrm{Tr}_{\mathcal{H}}\left\{Y(f;\tau_1)\Upsilon(f;\tau_1,\tau_2)\circ\mathcal{Y}(f;\tau_2)[\rho(f;\tau_2)]\right\}\right]. $$
$$(170)$$

Then, for the output current (165) we get

$$\mathbb{E}_{\rho_0}[I(t)] = \mathsf{k}\int_0^t d\tau\, e^{-\tilde{\gamma}(t-\tau)/2}\, \mathbb{E}^c\left[\mathrm{Tr}_{\mathcal{H}}\left\{Y(f;\tau)\rho(f;\tau)\right\}\right], \qquad (171)$$

$$\mathbb{E}_{\rho_0}[I(t)I(s)] = \mathsf{k}^2\int_0^{t\wedge s} d\tau\, e^{-\tilde{\gamma}\left(\frac{t+s}{2}-\tau\right)}\, \mathbb{E}^c\left[\mathrm{Tr}_{\mathcal{H}}\left\{Y(f;\tau)\rho(f;\tau)\right\}\right]$$
$$+ \mathsf{k}^2\left[\int_{t\wedge s}^{t\vee s} d\tau_1 \int_0^{t\wedge s} d\tau_2 + 2\int_0^{t\wedge s} d\tau_1 \int_0^{\tau_1} d\tau_2\right] e^{-\tilde{\gamma}(t+s-\tau_1-\tau_2)/2}$$
$$\times \mathbb{E}^c\left[\mathrm{Tr}_{\mathcal{H}}\left\{Y(f;\tau_1)\Upsilon(f;\tau_1,\tau_2)\circ\mathcal{Y}(f;\tau_2)[\rho(f;\tau_2)]\right\}\right]. \quad (172)$$

In photocounting problems it is usual to measure the deviations from the variance of the Poisson process by means of the *Mandel Q-parameter* [19], defined by

$$Q(t,s)$$

$$Q(t,s) = \frac{\mathrm{Var}\left[\tilde{N}(t)-\tilde{N}(s)\right] - \mathbb{E}_{\rho_0}\left[\tilde{N}(t)-\tilde{N}(s)\right]}{\mathbb{E}_{\rho_0}\left[\tilde{N}(t)-\tilde{N}(s)\right]}, \qquad t > s. \qquad (173)$$

For a Poisson process one has $Q(t, s) = 0$; the terms *sub-* and *super–Poissonian statistics* are used for the cases $Q(t, s) < 0$ and $Q(t, s) > 0$, respectively. The sub–Poissonian case $Q(t, s) < 0$ is considered as an index of "nonclassical" light. In our case we obtain the expressions

$$\mathbb{E}_{\rho_0}\left[\tilde{N}(t) - \tilde{N}(s)\right] = \int_s^t d\tau\, \mathbb{E}^c\left[\text{Tr}_{\mathcal{H}}\left\{Y(f; \tau)\rho(f; \tau)\right\}\right], \qquad (174)$$

$$\mathbb{E}_{\rho_0}\left[\left(\tilde{N}(t) - \tilde{N}(s)\right)^2\right] = \mathbb{E}_{\rho_0}\left[\tilde{N}(t) - \tilde{N}(s)\right]$$
$$+ 2\int_s^t d\tau_1 \int_s^{\tau_1} d\tau_2\, \mathbb{E}^c\left[\text{Tr}_{\mathcal{H}}\left\{Y(f; \tau_1)\Upsilon(f; \tau_1, \tau_2) \circ \mathcal{Y}(f; \tau_2)[\rho(f; \tau_2)]\right\}\right], \qquad (175)$$

$$Q(t, s) = 2\left(\int_s^t d\tau\, \mathbb{E}^c\left[\text{Tr}_{\mathcal{H}}\left\{Y(f; \tau)\rho(f; \tau)\right\}\right]\right)^{-1}$$
$$\times \int_s^t d\tau_1 \int_s^{\tau_1} d\tau_2\Big\{\mathbb{E}^c\left[\text{Tr}_{\mathcal{H}}\left\{Y(f; \tau_1)\Upsilon(f; \tau_1, \tau_2) \circ \mathcal{Y}(f; \tau_2)[\rho(f; \tau_2)]\right\}\right]$$
$$- \mathbb{E}^c\left[\text{Tr}_{\mathcal{H}}\left\{Y(f; \tau_1)\rho(f; \tau_1)\right\}\right]\mathbb{E}^c\left[\text{Tr}_{\mathcal{H}}\left\{Y(f; \tau_2)\rho(f; \tau_2)\right\}\right]\Big\}. \quad (176)$$

In reality the interesting case, for which the Mandel Q-parameter was introduced, is that of a counting process with stationary increments; so, the most useful parameter is

$$Q := \lim_{t \to +\infty} Q(t, 0), \qquad (177)$$

if this limit exists.

Probabilities for Counts

Let us note that the operator

$$\mathcal{L}(f(t)) = \mathcal{K}_t(f; 0) = \tilde{\mathcal{L}}(f(t)) + \mathcal{Y}(f; t) \qquad (178)$$

$\tilde{\Upsilon}(f; t, s)$ generates the dynamics $\Upsilon(f; t, s)$ through eq. (147). In constructing the probabilities it is useful to introduce similar maps generated by $\tilde{\mathcal{L}}(f(t))$:

$$\tilde{\Upsilon}(f; t, s)[\varrho] - \varrho = \int_s^t \tilde{\mathcal{L}}(f(\tau)) \circ \tilde{\Upsilon}(f; \tau, s)[\varrho]\, d\tau. \qquad (179)$$

Such maps turn out to be positive (even completely positive), but not trace preserving:

$$\varrho \geq 0 \quad \Rightarrow \quad \tilde{\Upsilon}(f; t, s)[\varrho] \geq 0, \quad \text{Tr}_{\mathcal{H}}\left\{\tilde{\Upsilon}(f; t, s)[\varrho]\right\} \leq \text{Tr}_{\mathcal{H}}\left\{\varrho\right\}. \qquad (180)$$

By using $\tilde{\Upsilon}(f;t,s)$ and $\mathcal{Y}(f;t)$, we define the completely positive maps on $\mathcal{T}(\mathcal{H})$

$$\mathcal{I}_s^t(f;0) = \tilde{\Upsilon}(f;t,s),$$

$$
\begin{aligned}
\mathcal{I}_s^t(f;m) = &\int_s^t dt_m \int_s^{t_m} dt_{m-1} \cdots \int_s^{t_2} dt_1 \, \tilde{\Upsilon}(f;t,t_m) \\
&\circ \mathcal{Y}(f;t_m) \circ \tilde{\Upsilon}(f;t_m,t_{m-1}) \circ \mathcal{Y}(f;t_{m-1}) \\
&\circ \cdots \circ \tilde{\Upsilon}(f;t_2,t_1) \circ \mathcal{Y}(f;t_1) \circ \tilde{\Upsilon}(f;t_1,s).
\end{aligned}
\tag{181}
$$

Then, one can check that the solution of eq. (140) with generator (167a) can be written as

$$\mathcal{G}_s^t(f;\kappa) = \sum_{m=0}^{\infty} e^{im\kappa} \mathcal{I}_s^t(f;m);\tag{182}$$

in the physical literature solutions of evolution equations written as expansions with respect to a part of the generator are called Dyson series. By the connection (149) with the instruments, we immediately identify $\{\mathcal{I}_s^t(f;m),\ m = 0,1,\ldots\}$ as the instrument giving the counts in the time interval $(s,t]$. Instruments for counts of such a type were introduced by Davies [43,44].

By combining eqs. (150) and (181), we can say that the quantities

$$P_t(0|\rho_0) = \mathbb{E}^c\left[\mathrm{Tr}_{\mathcal{H}}\left\{\tilde{\Upsilon}(f;t,0)[\rho_0]\right\}\right],\tag{183a}$$

$$
\begin{aligned}
p_t(t_m,t_{m-1},\ldots,t_1|\rho_0) = \mathbb{E}^c\Big[\mathrm{Tr}_{\mathcal{H}}\Big\{&\tilde{\Upsilon}(f;t,t_m) \circ \mathcal{Y}(f;t_m) \circ \tilde{\Upsilon}(f;t_m,t_{m-1}) \\
&\circ \mathcal{Y}(f;t_{m-1}) \circ \cdots \circ \tilde{\Upsilon}(f;t_2,t_1) \circ \mathcal{Y}(f;t_1) \circ \tilde{\Upsilon}(f;t_1,0)[\rho_0]\Big\}\Big] \quad (183b)
\end{aligned}
$$

determine all the probabilities for counts. The quantity (183a) is the probability of having no count in the time interval $(0,t]$, when the initial state of the system is ρ_0, and the probability of exactly m counts in the interval $(0,t]$ is

$$P_t(m|\rho_0) = \int_0^t dt_m \int_0^{t_m} dt_{m-1} \cdots \int_0^{t_2} dt_1 \, p_t(t_m,t_{m-1},\ldots,t_1|\rho_0).\tag{184}$$

The quantity (183b) is the probability density of a count at time t_1, a count at time $t_2,\ldots$, and no other count in the interval $(0,t]$; these quantities are called *exclusive probability densities* and from them also more complicated probabilities can be obtained. Indeed, the characteristic functional (129), from which all probabilities can be computed, turns out to be writable as

$$
\begin{aligned}
\Phi_t(k) = P_t(0|\rho_0) + \sum_{m=1}^{\infty} \int_0^t dt_m \int_0^{t_m} dt_{m-1} \\
\cdots \int_0^{t_2} dt_1 \exp\left\{i\sum_{n=1}^m k(t_n)\right\} p_t(t_m,\ldots,t_1|\rho_0).
\end{aligned}
\tag{185}
$$

252   Alberto Barchielli

3.5 Optical Heterodyne Detection

Ordinary Heterodyne Detection

Also a different kind of measurement, the so called *heterodyne detection* [87, 90], can be described by using QSC. By inserting a beam splitter (a half transparent mirror) before the photoelectron counter, the output field from $S_\mathcal{H}$ is made to beat with an intense laser field (local oscillator); only then, the intensity of the compound beam is measured by the photoelectron counter. This measurement scheme is illustrated in Fig. 2.

$A_0(t)$
$^0\Gamma$
    Let us introduce a new field $A_0(t)$ which does not interact with $S_\mathcal{H}$ and which can be used to describe the local oscillator; the initial state of this new field is taken to be a coherent vector $\psi(\ell f_0)$, where $\ell > 0$ is a parameter which we shall send to infinity in order to have a very intense laser field. The Fock space is now $^0\Gamma \otimes \Gamma$, $^0\Gamma = \Gamma(L^2(\mathbb{R}_+))$; we can also write $^0\Gamma \otimes \Gamma = \Gamma(L^2(\mathbb{R}_+) \otimes (\mathbb{C} \oplus \mathcal{Z}))$. Let us assume that the basis $\{z_k\}$ in $\mathcal{Z}$ is chosen in such a way that the index contains the direction of propagation and that only field 1 reaches the beam splitter. So, at the two input ports of the beam splitter the fields $A_1^{\text{out}}(t)$ and $A_0(t)$ arrive. Let us call $B_+(t)$ and $B_-(t)$ the fields leaving the two output ports; they are given by

$$B_+(t) = \frac{1}{\sqrt{2}} A_1^{\text{out}}(t) + \frac{1}{\sqrt{2}} A_0(t),$$
$$B_-(t) = \frac{1}{\sqrt{2}} A_0(t) - \frac{1}{\sqrt{2}} A_1^{\text{out}}(t). \tag{186}$$

In other terms the beam splitter operates the unitary transformation $(z_0, z_1) \rightarrow (z_+, z_-)$, $z_\pm = \frac{1}{\sqrt{2}}(z_0 \pm z_1)$. The phases of all the fields can always be redefined in order to have no additional phase shifts in (186). Note that the B-fields satisfy the CCR's as the input and the output A-fields.

    A photoelectron counter is placed in such a way that it collects the light coming out from one of the output ports of the beam splitter; let us say port 1. The detector

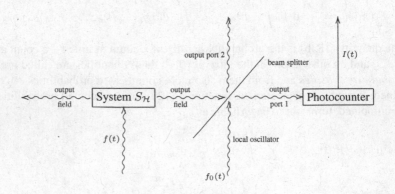

Fig. 2. Heterodyne detection

counts the photons carried by field $B_+(t)$; then, the number operator "measured" by the photodetector is

$$N_+^{\text{out}}(t) = U(t)^* N(P_+; t) U(t), \qquad P_\pm = |z_\pm\rangle\langle z_\pm|,$$
$$N(P_+; t) = \frac{1}{2}\left[\Lambda_{11}(t) + \Lambda_{10}(t) + \Lambda_{01}(t) + \Lambda_{00}(t)\right] \tag{187}$$

The local oscillator is at disposal of the experimenter and its characteristics are known. So, we can subtract from the output of the counter the known signal $\frac{1}{2}\ell^2|f_0(t)|^2$; this generates a phase factor $-\frac{i}{2}\ell^2\int_0^t k(s)|f_0(s)|ds$ in the characteristic operator. Moreover, we rescale the output of the counter by a factor $2/\ell$: this amounts to replace $k(s)$ by $2k(s)/\ell$ everywhere in the characteristic functional. At the end, the output current is

$$I(t) = \int_0^t F(t-s)\left[\frac{2}{\ell}\,\mathrm{d}\tilde{N}_+(s) - \ell|f_0(s)|^2\,\mathrm{d}s\right], \tag{188}$$

which corresponds to the characteristic operator

$$\widehat{\Phi}_t^\ell(k) = \exp\left\{-i\ell\int_0^t k(s)|f_0(s)|^2\mathrm{d}s + \frac{2i}{\ell}\int_0^t k(s)\mathrm{d}N(P_+; s)\right\}. \tag{189}$$

In other terms the characteristic operator has the general form discussed in Theorem 3.1 with $d = 1$, $h^\alpha = 0$, $b = 0$, $r(k; t) = 0$, $c(t) = -\ell|f_0(t)|^2$, $S(k(t)) - \mathbb{1} = (e^{2ik(t)/\ell} - 1)\,P_+$.

By applying Proposition 3.2, we obtain that the reduced characteristic operator $\mathcal{G}_t^\ell(f; k)$ satisfies eq. (135) with generator

$$\mathcal{K}_t^\ell(f; \kappa)[\varrho] = \mathcal{L}(f(t))[\varrho] - i\ell\kappa|f_0(t)|^2\,\varrho + \frac{1}{2}\left(e^{2i\kappa/\ell} - 1\right)\left\{\left(R_1 + \sum_j S_{1j}f_j(t)\right)\right.$$
$$\times \varrho\left(R_1^* + \sum_i \overline{f_i(t)}S_{1i}^*\right) + \ell\overline{f_0(t)}\left(R_1 + \sum_j S_{1j}f_j(t)\right)\varrho$$
$$\left. + \varrho\left(R_1^* + \sum_i \overline{f_i(t)}S_{1i}^*\right)\ell f_0(t) + \ell^2|f_0(t)|^2\,\varrho\right\}. \tag{190}$$

Now, we consider a very intense laser field for the local oscillator: $\ell \to \infty$. In this limit the generator of the reduced characteristic operator becomes

$$\mathcal{K}_t(f; \kappa)[\varrho] = \mathcal{L}(f(t))[\varrho] - \kappa|f_0(t)|^2\,\varrho$$
$$+ i\kappa\left\{\overline{f_0(t)}\left(R_1 + \sum_j S_{1j}f_j(t)\right)\varrho + \varrho\left(R_1^* + \sum_i \overline{f_i(t)}S_{1i}^*\right)f_0(t)\right\}. \tag{191}$$

By comparing (191) with the general expression (136), we see that the reduced characteristic operator of heterodyne detection is given by a factor corresponding to the measurement of the compatible observables

$$Q(z_1 \otimes f_0; t) = \int_0^t \left[f_0(s) \, \mathrm{d}A_1^\dagger(s) + \overline{f_0(s)} \, \mathrm{d}A_1(s) \right], \qquad t \geq 0, \qquad (192)$$

times the expression

$$\exp\left\{ -\frac{1}{2} \int_0^t \left(k(s) \right)^2 |f_0(s)|^2 \, \mathrm{d}s \right\}; \qquad (193)$$

this term represents a classical Gaussian noise to be added to the quantum noise intrinsic to the measurement of $Q(z_1 \otimes f_0; t)$. As we now see, it is possible to eliminate this extra noise.

Balanced Heterodyne Detection

The noise in the output current can be reduced by the measurement scheme called *balanced heterodyne detection* [88]. Now two identical photoelectron counters are used for detecting the photons in both the fields $B_+(t)$ and $B_-(t)$ and the difference $I_1(t) - I_2(t)$ of the two output currents is measured; the scheme of balanced heterodyne detection is given in Fig. 3.

We scale again the final output current by a factor ℓ^{-1}, so that we have

$$I(t) = \frac{1}{\ell} \int_0^t F(t-s) \left[\mathrm{d}\tilde{N}_+(s) - \mathrm{d}\tilde{N}_-(s) \right], \qquad (194)$$

which corresponds to a continual measurement of the commuting operators

$$X_\ell(t) = \frac{1}{\ell} \left[N(P_+; t) - N(P_-; t) \right]. \qquad (195)$$

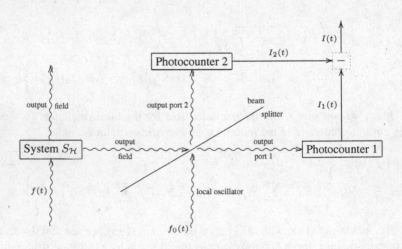

Fig. 3. Balanced heterodyne detection

Then, the characteristic operator

$$\exp\left\{i\int_0^t k(s)\,\mathrm{d}X_\ell(s)\right\}$$

has the structure given in Theorem 3.1 with $d=1$, $c=0$, $h^\alpha=0$, $r_t(k)=0$, $B^\alpha \equiv B = \frac{1}{\ell}\left(P_+ - P_-\right)$,

$$S(k(t)) - \mathbb{1} = \left(e^{ik(t)/\ell} - 1\right)P_+ + \left(e^{-ik(t)/\ell} - 1\right)P_- .$$

Now we construct the reduced characteristic operator in three steps: first we eliminate the field A_0, then we take the limit of a very intense local oscillator $\ell \to \infty$, and finally we eliminate the other fields. By recalling that the field A_0 does not interact with the system $S_\mathcal{H}$, we can do the first two steps by considering the matrix elements on the coherent vectors:

$$\langle \psi(g^1)|\widehat{\Phi}_t^\ell(k)\psi(g^2)\rangle_\Gamma :=$$

$$\left\langle \psi(\ell f_0)\otimes\psi(g^1)\Big|\exp\left\{i\int_0^t k(s)\,\mathrm{d}X_\ell(s)\right\}\psi(\ell f_0)\otimes\psi(g^2)\right\rangle_{\circ\Gamma\otimes\Gamma}. \quad (196)$$

Then, by eq. (119) we obtain

$$\langle \psi(g^1)|\widehat{\Phi}_t^\ell(k)\psi(g^2)\rangle = \langle \psi(g^1)|\psi(y^2)\rangle\exp\left\{\int_0^t \mathrm{d}s\left[\left(\ell^2\,|f_0(s)|^2 + \overline{g_1^1(s)}g_1^2(s)\right)\right.\right.$$

$$\times\left(\cos\left(k(s)/\ell\right) - 1\right) + i\ell\left(\overline{f_0(s)}g_1^2(s) + \overline{g_1^1(s)}f_0(s)\right)\sin\left(k(s)/\ell\right)\Big]\Big\}. \quad (197)$$

In the limit of a strong local oscillator we get

$$\langle \psi(g^1)|\widehat{\Phi}_t(k)\psi(g^2)\rangle = \lim_{\ell\to\infty}\langle \psi(g^1)|\widehat{\Phi}_t^\ell(k)\psi(g^2)\rangle = \langle \psi(g^1)|\psi(g^2)\rangle$$

$$\times \exp\left\{\int_0^t \mathrm{d}s\left[-\frac{1}{2}\left(k(s)\right)^2|f_0(s)|^2 + ik(s)\left(\overline{f_0(s)}g_1^2(s) + \overline{g_1^1(s)}f_0(s)\right)\right]\right\}, \quad (198)$$

which is the characteristic operator of the observables $Q(z_1 \otimes f_0; t)$ (192). With the notations of the Theorem 3.1 we have

$$d=1, \qquad B^\alpha = 0, \qquad c=0, \qquad b=0,$$
$$h^\alpha(t) \equiv h(t) = f_0(t)z_1, \qquad S_t(k) = \mathbb{1}, \qquad r(k;t) = ik(t)f_0(t)z_1. \qquad (199)$$

Then, the generator of the reduced characteristic operator becomes

$$\mathcal{K}_t(f;\kappa)[\varrho] = \mathcal{L}(f(t))[\varrho] - \frac{1}{2}\kappa^2\,|f_0(t)|^2\,\varrho$$

$$+ i\kappa\left\{\overline{f_0(t)}\left(R_1 + \sum_j S_{1j}f_j(t)\right)\varrho + \varrho\left(R_1^* + \sum_i \overline{f_i(t)}S_{1i}^*\right)f_0(t)\right\}, \quad (200)$$

which is the similar to the expression (191), a part from a smaller noise due to the factor $1/2$ in the second term.

Moments

By using $h(t) = f_0(t)z_1$, we can say that the balanced heterodyne scheme realizes a continual measurement of the compatible quantum observables

$$Q(h;t) = \sum_j \int_0^t \left[\langle z_j | h(s) \rangle \, dA_j^\dagger(s) + \langle h(s) | z_j \rangle \, dA_j(s) \right], \qquad t \geq 0; \quad (201)$$

by denoting by $\tilde{Q}(h;t)$ the associated stochastic process, the output current of the balanced heterodyne detection scheme is

$$I(t) = \int_0^t F(t-s) \, d\tilde{Q}(h;s). \quad (202)$$

The generator of the reduced characteristic operator associated to the process $\tilde{Q}(h;t)$ is

$$\mathcal{K}_t(f;\kappa)[\varrho] = \mathcal{L}(f(t))[\varrho] - \frac{\kappa^2}{2} \|h(t)\|^2 \varrho + i\kappa \left[Z(f;t)\varrho + \varrho Z(f;t)^* \right], \quad (203)$$

$$Z(f;t) = C(h(t))^* \left[R + (S - \mathbb{1})C(f(t)) \right] + \langle h(t) | f(t) \rangle. \quad (204)$$

$$(205)$$

Typical choices are

$$F(t) = k \exp\left(-\frac{\tilde{\gamma}}{2} t \right), \qquad \tilde{\gamma} > 0, \quad (206)$$

$$h(t) = e^{-i\nu t} \hat{h}, \qquad \hat{h} \in \mathcal{Z}, \qquad \nu \in \mathbb{R}, \quad (207)$$

$$\langle \hat{h} | f(t) \rangle = 0, \qquad \forall t. \quad (208)$$

These choices give

$$\mathcal{K}_t(f;\kappa)[\varrho] = \mathcal{L}(f(t))[\varrho] - \frac{\kappa^2}{2} \|\hat{h}\|^2 \varrho + i\kappa \left[Z(f;t)\varrho + \varrho Z(f;t)^* \right], \quad (209)$$

$$Z(f;t) = e^{i\nu t} C(\hat{h})^* \left[R + (S - \mathbb{1})C(f(t)) \right]. \quad (210)$$

By changing ν, which means to change the measuring apparatus, the whole spectrum of $S_\mathcal{H}$ can be explored. The condition (208) means that the light of the laser stimulating $S_\mathcal{H}$ does not hit the detector directly.

The first and second moments turn out to be given by

$$\mathbb{E}_{\rho_0}\left[\tilde{Q}(h;t) \right] = \int_0^t ds \, \mathbb{E}^c \left[\text{Tr}_\mathcal{H} \left\{ Y(f;s)\rho(f;s) \right\} \right], \quad (211)$$

$$\mathbb{E}_{\rho_0}\left[\tilde{Q}(h;t_1)\tilde{Q}(h;t_2)\right] = \int_0^{t_1 \wedge t_2} \|h(t)\|^2 \, dt$$

$$+ \int_{t_1 \wedge t_2}^{t_1 \vee t_2} dt \int_0^{t_1 \wedge t_2} ds \, \mathbb{E}^c \left[\mathrm{Tr}_{\mathcal{H}}\{Y(f;t)\Upsilon(f;t,s) \circ \mathcal{Y}(f;s)[\rho(f;s)]\}\right]$$

$$+ 2 \int_0^{t_1 \wedge t_2} dt \int_0^t ds \, \mathbb{E}^c \left[\mathrm{Tr}_{\mathcal{H}}\{Y(f;t)\Upsilon(f;t,s) \circ \mathcal{Y}(f;s)[\rho(f;s)]\}\right], \quad (212)$$

where

$$\mathcal{Y}(f;t)[\varrho] = Z(f;t)\varrho + \varrho Z(f;t)^*, \tag{213}$$

$$Y(f;t) = Z(f;t) + Z(f;t)^*. \tag{214}$$

By these equations, the measurement scheme we are discussing here can be interpreted as an imprecise, indirect, continual measurement of the system observables $Y(f;t)$.

Then, for the output current (202) we get

$$\mathbb{E}_{\rho_0}\left[I(t)\right] = \mathsf{k} \int_0^t d\tau \, e^{-\tilde{\gamma}(t-\tau)/2} \, \mathbb{E}^c \left[\mathrm{Tr}_{\mathcal{H}}\{Y(f;\tau)\rho(f;\tau)\}\right], \tag{215}$$

$$\mathbb{E}_{\rho_0}\left[I(t)I(s)\right] = \mathsf{k}^2 \int_0^{t \wedge s} e^{-\tilde{\gamma}\left(\frac{t+s}{2}-\tau\right)} \|h(\tau)\|^2 \, d\tau$$

$$+ \mathsf{k}^2 \left[\int_{t \wedge s}^{t \vee s} d\tau_1 \int_0^{t \wedge s} d\tau_2 + 2 \int_0^{t \wedge s} d\tau_1 \int_0^{\tau_1} d\tau_2\right] e^{-\tilde{\gamma}(t+s-\tau_1-\tau_2)/2}$$

$$\times \mathbb{E}^c \left[\mathrm{Tr}_{\mathcal{H}}\{Y(f;\tau_1)\Upsilon(f;\tau_1,\tau_2) \circ \mathcal{Y}(f;\tau_2)[\rho(f;\tau_2)]\}\right]. \tag{216}$$

3.6 Physical Models

In the next sections we want to present two concrete applications in quantum optics of the whole theory of quantum continual measurements.

An interesting phenomenon is the so called *shelving effect*: a three-level system with a peculiar configuration of permitted transitions and suitably stimulated by lasers exhibits bright and dark periods in its fluorescence light. This phenomenon can have a nice mathematical treatment by using QSC and the theory of continual measurements (direct detection) [16, 20, 42].

Also the simplest quantum system, a two-level atom, presents interesting features. A noteworthy characteristic of its fluorescence spectrum is a three peaked shape in the case of a very intense stimulating laser (Mollow spectrum). Again QSC and the theory of continual measurements give a way of modelling and studying this system [26–28], giving a unified treatment to known results and allowing modifications of the known results by using an Hudson-Parthasarathy equation with $S \neq 1$. It is in this problem that it is used a mixture of coherent states of type **S3** as initial state [28].

4 A Three–Level Atom and the Shelving Effect

Today experimental techniques allow to observe the fluorescent light emitted by a single atom or ion; therefore, it is possible to observe effects which are completely masked when many emitters are involved. One of these phenomena, the *electron shelving effect*, was proposed by Dehmelt as a very sensitive scheme for detecting very weak transitions in single ions [45, 46] and it was observed when single ion spectroscopy became feasible [30, 83, 85].

Consider a three–level atom with states $|g\rangle$ (ground), $|b\rangle$ (blue) and $|r\rangle$ (red); assume that the blue transition $b \leftrightarrow g$ is very strong and the red one $r \leftrightarrow g$ is very weak, while the transition $b \leftrightarrow r$ is prohibited. When both transitions are driven by two suitably tuned lasers, we expect the atom to emit blue fluorescent light. But sometimes, when the atom absorbs a red photon, the electron goes into the red state, which has a long lifetime (a fraction of a second), and the fluorescence light stops until the red state decays. Thus we expect to observe bright and dark periods, randomly distributed. In a pictorial language, we say that during a dark period the electron is 'shelved' in the red state. The energy–level scheme we have described, when indeed $|g\rangle$ is the lowest state, is called the V configuration and it is given in Fig. 4; the simple arrows represent spontaneous decay and the double arrows represent absorption/stimulated emission.

The same considerations apply to the so called Λ configuration when $|g\rangle$ is the highest state (Fig. 4). Usually the electron jumps between the $|g\rangle$ and the $|b\rangle$ states emitting blue light, but sometimes the $|g\rangle$ state decays into the $|r\rangle$ one and it stays 'shelved' there until it absorbs a red photon. A more realistic model could be obtained by adding a weak $|b\rangle \leftrightarrow |r\rangle$ transition [91].

However, this discussion is of a semi-classical character and does not takes into account that the atom-field interaction gives rise to quantum coherent superpositions of the atomic states. So, to explain the shelving effect we need a full quantum mechanical treatment. A first good quantum–mechanical explanation of this effect was given in [39].

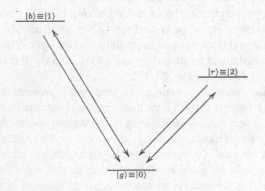

Fig. 4. Energy–level scheme for the V configuration.

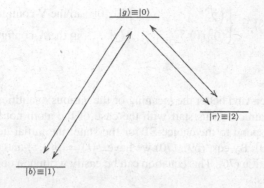

Fig. 5. Energy–level scheme for the Λ configuration.

4.1 The Atom–Field Dynamics

Let us concretize the previous discussion in the choice of the operators H, R, S appearing in the Hudson-Parthasarathy equation (39) and of the photon space $\mathcal{Z}$; this choice fixes the atom-field dynamics.

We denote by $|j\rangle$, $j = 0, 1, 2$ the three states: $|g\rangle \equiv |0\rangle$, $|b\rangle \equiv |1\rangle$, $|r\rangle \equiv |2\rangle$. The level scheme of Figs. 4 and 5 implies that the free atomic Hamiltonian is

$$H = \sum_{j=1}^{2} \omega_j |j\rangle\langle j|, \qquad \begin{cases} \omega_j > 0 & \text{in the V-configuration,} \\ \omega_j < 0 & \text{in the } \Lambda\text{-configuration.} \end{cases} \tag{217}$$

Moreover, we consider the simplest case: only the absorption/emission process is relevant and there is not direct scattering; so we take

$$S = \mathbb{1}. \tag{218}$$

According to the discussion on the quasi–monochromatic approximation of Section 2.5 we need a different field for any possible atomic transition, if well separated. We have two well separated transitions, the $g \leftrightarrow b$ one and the $g \leftrightarrow r$ one; so, we take: $\mathcal{Z} = \mathcal{Z}^1 \oplus \mathcal{Z}^2$. Then, in the rotating wave–approximation, the operator R must give the two possible decays with emission of blue photons (in $\mathcal{Z}^1$) and red photons (in $\mathcal{Z}^2$); this implies

$$R = R^1 + R^2, \qquad R^j = \begin{cases} |0 \otimes \alpha^j\rangle\langle j| & \text{in the V-configuration,} \\ |j \otimes \alpha^j\rangle\langle 0| & \text{in the } \Lambda\text{-configuration,} \end{cases} \tag{219}$$

$$\alpha^j \in \mathcal{Z}^j, \qquad \gamma_j := \left\| \alpha^j \right\|^2 > 0. \tag{220}$$

The two following quantities constructed from R appear in the master equation (74):

$$R^*R = \begin{cases} \sum_{j=1}^{2} \gamma_j |j\rangle\langle j| & \text{in the V-configuration,} \\ (\gamma_1 + \gamma_2)|0\rangle\langle 0| & \text{in the } \Lambda\text{-configuration,} \end{cases} \tag{221}$$

$$\mathrm{Tr}_{\mathcal{Z}}\{R\varrho R^*\} = \begin{cases} \left(\sum_{j=1}^2 \gamma_j \langle j|\varrho|j\rangle\right) |0\rangle\langle 0| & \text{in the V-configuration,} \\ \langle 0|\varrho|0\rangle \sum_{j=1}^2 \gamma_j |j\rangle\langle j| & \text{in the } \Lambda\text{-configuration,} \end{cases} \qquad (222)$$

Pure Decay

In order to understand better the meaning of the various quantities appearing in the atom–field dynamics, let us start with the case of the atom not stimulated in any way; this correspond to the choice **S1** for the state: the initial field is in the Fock vacuum ($f \equiv 0$). By eqs. (69), (70) we have $\rho(t) = \rho(0;t)$ and this state satisfies the master equation (76). This equation can be easily written in our case and solved; the final result is:

1. In the case of the V-configuration: $i, j = 1, 2,$

$$\langle j|\rho(0;t)|i\rangle = \exp\left\{-\tfrac{1}{2}(\gamma_i + \gamma_j)t + i(\omega_i - \omega_j)t\right\}\langle j|\rho_0|i\rangle, \qquad (223a)$$

$$\langle j|\rho(0;t)|0\rangle = \overline{\langle 0|\rho(0;t)|j\rangle} = \exp\left\{-\tfrac{1}{2}\gamma_j t - i\omega_j t\right\}\langle j|\rho_0|0\rangle, \qquad (223b)$$

$$\langle 0|\rho(0;t)|0\rangle = 1 - \sum_{j=1}^2 e^{-\gamma_j t}\langle j|\rho_0|j\rangle. \qquad (223c)$$

2. In the case of the Λ-configuration: $i, j = 1, 2,$

$$\langle 0|\rho(0;t)|0\rangle = e^{-(\gamma_1+\gamma_2)t}\langle 0|\rho_0|0\rangle, \qquad (224a)$$

$$\langle j|\rho(0;t)|0\rangle = \overline{\langle 0|\rho(0;t)|j\rangle} = \exp\left\{-\tfrac{1}{2}(\gamma_1 + \gamma_2)t - i\omega_j t\right\}\langle j|\rho_0|0\rangle, \qquad (224b)$$

$$\langle j|\rho(0;t)|i\rangle = e^{i(\omega_i-\omega_j)t}\langle j|\rho_0|i\rangle + \delta_{ij}\frac{\gamma_j}{\gamma_1 + \gamma_2}\left(1 - e^{-(\gamma_1+\gamma_2)t}\right)\langle 0|\rho_0|0\rangle. \qquad (224c)$$

These equations confirm that ω_1, ω_2 are the atomic frequencies and show that γ_1, γ_2 are the spontaneous emission rates. The assumption of a weak red transition and a strong blue one gives $\gamma_1 \gg \gamma_2 > 0$.

Stimulating Lasers

Let us consider now the case of interest, when we have two lasers stimulating the blue and the red transitions. This is the case of a state of type **S2**, with the field in a coherent state with f of the form

$$f(t) = \mathbb{1}_{[0,T]}(t)\sum_{j=1}^2 e^{-i\nu_j t}\lambda^j, \qquad \nu_j > 0, \quad \lambda^j \in \mathcal{Z}^j. \qquad (225)$$

In all the following formulae we take $T \to +\infty$.

By eqs. (69), (70), we have $\rho(t) \equiv \rho(f;t)$; this reduced state satisfies the master equation (74), with Liouvillian (75)

$$\mathcal{L}(f(t))[\varrho] = -\left[\frac{1}{2} R^* R + iH(f(t))\right] \varrho - \varrho \left[\frac{1}{2} R^* R - iH(f(t))\right]$$
$$+ \mathrm{Tr}_{\mathcal{Z}}\{R\varrho R^*\}, \quad (226)$$

where (cf eq. (73c))

$$H(f(t)) = H + iC(f(t))^* R - iR^* C(f(t))$$
$$= \sum_{j=1}^{2} \left\{\omega_j |j\rangle\langle j| + ie^{i\nu_j t}\langle\lambda^j|\alpha^j\rangle\sigma_-^j - ie^{-i\nu_j t}\langle\alpha^j|\lambda^j\rangle\sigma_+^j\right\}, \quad (227)$$

$$\sigma_-^j := |0\rangle\langle j|, \qquad \sigma_+^j := |j\rangle\langle 0|, \qquad \text{in the V-configuration,}$$
$$\sigma_-^j := |j\rangle\langle 0|, \qquad \sigma_+^j := |0\rangle\langle j|, \qquad \text{in the } \Lambda\text{-configuration.} \quad (228)$$

The explicit time dependence due to the lasers can be removed; in the physical parlance one says that the *rotating frame* is used. Let us set

$$\epsilon := \begin{cases} +1 & \text{in the V-configuration,} \\ -1 & \text{in the } \Lambda\text{-configuration,} \end{cases} \quad (229)$$

$$\Delta_j := \nu_j - |\omega_j|, \qquad \Omega_j := 2\,|\langle\alpha^j|\lambda^j\rangle|, \qquad \beta_j := \epsilon\arg\{-i\langle\alpha^j|\lambda^j\rangle\}; \quad (230)$$

note that one has

$$\langle\alpha^j|\lambda^j\rangle = \frac{i}{2}\,e^{\epsilon i\beta_j}\,\Omega_j. \quad (231)$$

The Δ_j are called the *detuning* parameters and the Ω_j are called *Rabi frequencies*. To have a strong blue transition, strongly stimulated, and a weak red one we assume:

$$\gamma_1 \gg \gamma_2 > 0, \qquad \Omega_1 \gg \Omega_2 > 0,$$
$$\gamma_1 \gg \Omega_2, \qquad \Omega_1^2 \gg \gamma_1\gamma_2. \quad (232)$$

The approximation of taking two field types, one interacting with the blue transition and one with the red one, is reasonable only if the two lasers are not too much out of resonance; moreover, we want the two transitions to be not overdamped. These two conditions give the further assumptions

$$|\Delta_j| \lesssim 2\gamma_j, \qquad 2\Omega_j > \gamma_j, \quad j = 1, 2. \quad (233)$$

Then, the new reduced state

$$\tilde{\rho}(t) := \exp\left\{\epsilon it \sum_{j=1}^{2} \nu_j |j\rangle\langle j|\right\} \rho(t) \exp\left\{-\epsilon it \sum_{j=1}^{2} \nu_j |j\rangle\langle j|\right\} \quad (234)$$

satisfies the master equation

$$\frac{d}{dt}\,\tilde{\rho}(t) = \mathcal{L}_\Delta\,[\tilde{\rho}(t)], \quad (235)$$

where

$$\mathcal{L}_\Delta[\varrho] := K\varrho + \varrho K^* + \mathrm{Tr}_{\mathcal{Z}}\left\{ R\varrho R^* \right\}, \tag{236}$$

$$K := -\frac{1}{2} R^* R - \mathrm{i} \sum_{j=1}^{2} \left[\frac{\Omega_j}{2} \left(\mathrm{e}^{\mathrm{i}\beta_j} |j\rangle\langle 0| + \mathrm{e}^{-\mathrm{i}\beta_j} |0\rangle\langle j| \right) - \epsilon \Delta_j |j\rangle\langle j| \right]. \tag{237}$$

4.2 The Detection Process

Now we assume to have a detector able to count photons flying through a solid angle S_d not containing the direction of propagation of the lasers, so that the lasers do not send light directly to the counter and only fluorescence light is detected; the efficiency of the counter can be taken into account by choosing S_d smaller than the geometrical solid angle spanned by the detector. We can formalize this setup by saying that the detector performs a continual measurement of the observable $N(P; t)$ defined by eq. (85) with

1. $P\lambda^j = 0$, $j = 1, 2$: the laser light does not impinge directly on the detector;
2. $P\mathcal{Z}^j \subset \mathcal{Z}^j$, $j = 1, 2$: the detector does not mix up blue and red photons.

Let us set

$$\eta_j := \left\| P\alpha^j \right\|^2 ; \tag{238}$$

by their definitions, we have $0 \leq \eta_j \leq \gamma_j$. We assume that "many" blue photons are detected, so we have

$$0 \leq \eta_2 \leq \gamma_2 \mathbb{L}\eta_1 \leq \gamma_1 . \tag{239}$$

The whole information on the counting probabilities is contained in the characteristic functional (129), (185) or in the probabilities of no counts (183a) and the exclusive probability densities (183b). To compute these probabilities we need to particularize to our case the quantities of eqs. (167); we get

$$\mathcal{Y}(f; t)[\varrho] = \mathrm{Tr}_{\mathcal{Z}}\left\{ (\mathbb{1} \otimes P) R\varrho R^* \right\} =: \mathcal{Y}[\varrho] , \tag{240a}$$

$$\tilde{\mathcal{L}}(f(t)) = \mathcal{L}(f(t)) - \mathcal{Y} \tag{240b}$$

From the expression of the probabilities (183a), (183b) we see that we can make everywhere the transformation (234) without changing such probabilities. This simply means that everywhere we have to make the substitutions $\mathcal{L}(f(t)) \to \mathcal{L}_\Delta$ and $\tilde{\mathcal{L}}(f(t)) \to \tilde{\mathcal{L}}_\Delta$, where

$$\tilde{\mathcal{L}}_\Delta := \mathcal{L}_\Delta - \mathcal{Y} . \tag{241}$$

In this way we get

$$P_t(0|\rho_0) = \mathrm{Tr}_{\mathcal{H}}\left\{ \mathrm{e}^{\tilde{\mathcal{L}}_\Delta t}[\rho_0] \right\} \tag{242}$$

for the probability of no counts up to time t and

$$p_t(t_m, t_{m-1}, \dots, t_1|\rho_0) = \mathrm{Tr}_{\mathcal{H}}\Big\{ \mathrm{e}^{\tilde{\mathcal{L}}_\Delta(t-t_m)} \circ \mathcal{Y} \circ \mathrm{e}^{\tilde{\mathcal{L}}_\Delta(t_m-t_{m-1})}$$
$$\circ \mathcal{Y} \circ \cdots \circ \mathrm{e}^{\tilde{\mathcal{L}}_\Delta(t_2-t_1)} \circ \mathcal{Y} \circ \mathrm{e}^{\tilde{\mathcal{L}}_\Delta t_1}[\rho_0] \Big\} \tag{243}$$

for the exclusive probability densities.

More concretely we have

$$\tilde{\mathcal{L}}_\Delta[\varrho] = K\varrho + \varrho K^*$$

$$+ \begin{cases} \left(\sum_{j=1}^2 (\gamma_j - \eta_j) \langle j|\varrho|j\rangle\right) |0\rangle\langle 0| & \text{in the V-configuration,} \\ \langle 0|\varrho|0\rangle \sum_{j=1}^2 (\gamma_j - \eta_j) |j\rangle\langle j| & \text{in the } \Lambda\text{-configuration,} \end{cases} \quad (244)$$

$$K = -\frac{\mathrm{i}}{2} \sum_{j=1}^2 \Omega_j \left(\mathrm{e}^{\mathrm{i}\beta_j} |j\rangle\langle 0| + \mathrm{e}^{-\mathrm{i}\beta_j} |0\rangle\langle j| \right)$$

$$+ \begin{cases} \sum_{j=1}^2 \left(\mathrm{i}\Delta_j - \frac{\gamma_j}{2}\right) |j\rangle\langle j| & \text{in the V-configuration,} \\ -\mathrm{i}\sum_{j=1}^2 \Delta_j |j\rangle\langle j| - \frac{\gamma_1+\gamma_2}{2} |0\rangle\langle 0| & \text{in the } \Lambda\text{-configuration,} \end{cases} \quad (245)$$

$$\mathcal{Y}[\varrho] = \varkappa(\varrho)\rho_{\mathrm{jump}}, \quad (246\text{a})$$

$$\varkappa(\varrho) := \begin{cases} \sum_{j=1}^2 \eta_j \langle j|\varrho|j\rangle & \text{in the V-configuration,} \\ (\eta_1 + \eta_2) \langle 0|\varrho|0\rangle & \text{in the } \Lambda\text{-configuration,} \end{cases} \quad (246\text{b})$$

$$\rho_{\mathrm{jump}} := \begin{cases} |0\rangle\langle 0| & \text{in the V-configuration,} \\ \sum_{j=1}^2 \frac{\eta_j}{\eta_1+\eta_2} |j\rangle\langle j| & \text{in the } \Lambda\text{-configuration.} \end{cases} \quad (246\text{c})$$

By inserting these expressions into eq. (243) and setting

$$w_0(t) := \varkappa\left(\mathrm{e}^{\tilde{\mathcal{L}}_\Delta t}[\rho_0]\right), \qquad w(t) := \varkappa\left(\mathrm{e}^{\tilde{\mathcal{L}}_\Delta t}[\rho_{\mathrm{jump}}]\right), \quad (247)$$

we obtain

$$p_t(t_m,\dots,t_1|\rho_0) = P_{t-t_m}(0|\rho_{\mathrm{jump}})\, w(t_m - t_{m-1}) \cdots w(t_2 - t_1)\, w_0(t_1). \quad (248)$$

Moreover, one can check immediately that

$$\frac{\mathrm{d}}{\mathrm{d}t} P_t(0|\varrho) = -\varkappa\left(\mathrm{e}^{\tilde{\mathcal{L}}_\Delta t}[\varrho]\right) \quad (249)$$

and, then, one has

$$P_t(0|\rho_{\mathrm{jump}}) = 1 - \int_0^t w(s)\mathrm{d}s, \qquad P_t(0|\rho_0) = 1 - \int_0^t w_0(s)\mathrm{d}s. \quad (250)$$

These equations say that our detection process is a *delayed renewal counting process* (an usual renewal counting process when $w_0(t) \equiv w(t)$); $w(t)$ is the interarrival waiting–time density. By construction $\int_0^{+\infty} w(t)\,\mathrm{d}t \le 1$, but it is possible to have $\int_0^{+\infty} w(t)\,\mathrm{d}t < 1$: this means that there is a non–zero probability that the detected fluorescence stops.

Note that the probability density $w(t)$ of the interarrival times is a continuous function and that $w(0) = 0$; this says that just after a count the atom cannot emit, but it needs some time to be excited again: this is the so called *antibunching effect*.

From now on we assume that all the fluorescence photons are detected; we can think to have an array of perfect detectors spanning the whole solid angle around the atom, with the exception of a small angle containing the direction of propagation of the lasers. In other terms we give the process which counts all the emitted photons. Mathematically this amounts in taking

$$P = \mathbb{1} \qquad \text{or} \qquad \eta_j = \gamma_j, \quad j = 1, 2. \tag{251}$$

This simplifies all the computations, because we get

$$\tilde{\mathcal{L}}_\Delta[\varrho] = K\varrho + \varrho K^*, \qquad e^{\tilde{\mathcal{L}}_\Delta t}[\varrho] = e^{Kt}\varrho e^{K^*t}, \tag{252}$$

$$P_t(0|\rho_{\text{jump}}) = 1 - \int_0^t w(s)\mathrm{d}s = \mathrm{Tr}\left\{ e^{Kt}\rho_{\text{jump}}e^{K^*t} \right\}, \tag{253a}$$

$$P_t(0|\rho_0) = 1 - \int_0^t w_0(s)\mathrm{d}s = \mathrm{Tr}\left\{ e^{Kt}\rho_0 e^{K^*t} \right\}. \tag{253b}$$

As far as many blue photons are detected and (239) holds, the assumption (251) should not alter the statistical properties of the counting process in an essential way.

4.3 Bright and Dark Periods: The V-Configuration

As we have seen, all the probabilities depend on the waiting time densities $w_0(t)$ and $w(t)$. By particularizing eqs. (245)–(247) to the V case, we get

$$K = \sum_{j=1}^2 \left[\left(\mathrm{i}\Delta_j - \frac{\gamma_j}{2} \right) |j\rangle\langle j| - \frac{\mathrm{i}}{2}\,\Omega_j \left(e^{\mathrm{i}\beta_j}|j\rangle\langle 0| + e^{-\mathrm{i}\beta_j}|0\rangle\langle j| \right) \right], \tag{254}$$

$$\varkappa(\varrho) = \sum_{j=1}^2 \gamma_j \langle j|\varrho|j\rangle, \qquad \rho_{\text{jump}} = |0\rangle\langle 0|, \tag{255}$$

$$w_0(t) = \sum_{j=1}^2 \gamma_j \left\langle j \left| e^{Kt}\rho_0 e^{K^*t} \right| j \right\rangle, \tag{256a}$$

$$w(t) = \sum_{j=1}^2 \gamma_j \left| \left\langle j \left| e^{Kt} \right| 0 \right\rangle \right|^2. \tag{256b}$$

To obtain the expression of the interarrival time density $w(t)$, we have to compute $e^{Kt}|0\rangle$. By setting

$$a_0(t) := \langle 0 | e^{Kt} | 0 \rangle, \qquad a_j(t) := e^{-i\beta_j} \langle j | e^{Kt} | 0 \rangle, \qquad j = 1, 2, \qquad (257)$$

we can write

$$w(t) = \sum_{j=1}^{2} \gamma_j \, |a_j(t)|^2, \qquad (258)$$

$$P_t(0 | \rho_{\text{jump}}) = 1 - \int_0^t w(s) \mathrm{d}s = \sum_{j=0}^{2} |a_j(t)|^2. \qquad (259)$$

By taking the time derivative, we get the system of linear differential equations

$$\frac{\mathrm{d}}{\mathrm{d}t} \, \boldsymbol{a}(t) = \boldsymbol{G} \, \boldsymbol{a}(t), \qquad \boldsymbol{a}(0) = \begin{pmatrix} 1 \\ 0 \\ 0 \end{pmatrix}, \qquad (260)$$

where

$$\boldsymbol{G} := \begin{pmatrix} 0 & -i\Omega_1/2 & -i\Omega_2/2 \\ -i\Omega_1/2 & -\xi_1/2 & 0 \\ -i\Omega_2/2 & 0 & -\xi_2/2 \end{pmatrix}, \qquad (261)$$

$$\xi_j := \gamma_j - 2i\Delta_j, \qquad j = 1, 2. \qquad (262)$$

The solution of the system (260) can be obtained by Laplace transform; we get

$$\begin{aligned}
a_0(t) = & \frac{(z_0 + \xi_1/2)(z_0 + \xi_2/2)}{(z_0 - z_1)(z_0 - z_2)} \, e^{z_0 t} \\
& + \frac{(z_1 + \xi_1/2)(z_1 + \xi_2/2)}{(z_1 - z_0)(z_1 - z_2)} \, e^{z_1 t} + \frac{(z_2 + \xi_1/2)(z_2 + \xi_2/2)}{(z_2 - z_0)(z_2 - z_1)} \, e^{z_2 t}, \quad (263a)
\end{aligned}$$

$$\begin{aligned}
a_1(t) = & -\frac{i\Omega_1}{2} \left(\frac{z_0 + \xi_2/2}{(z_0 - z_1)(z_0 - z_2)} \, e^{z_0 t} \right. \\
& \left. + \frac{z_1 + \xi_2/2}{(z_1 - z_0)(z_1 - z_2)} \, e^{z_1 t} + \frac{z_2 + \xi_2/2}{(z_2 - z_0)(z_2 - z_1)} \, e^{z_2 t} \right), \quad (263b)
\end{aligned}$$

$$\begin{aligned}
a_2(t) = & -\frac{i\Omega_2}{2} \left(\frac{z_0 + \xi_1/2}{(z_0 - z_1)(z_0 - z_2)} \, e^{z_0 t} \right. \\
& \left. + \frac{z_1 + \xi_1/2}{(z_1 - z_0)(z_1 - z_2)} \, e^{z_1 t} + \frac{z_2 + \xi_1/2}{(z_2 - z_0)(z_2 - z_1)} \, e^{z_2 t} \right), \quad (263c)
\end{aligned}$$

where z_0, z_1, z_2 are the roots (which we have assumed to be all distinct) of the characteristic polynomial of the matrix $\boldsymbol{G}$:

$$z^3 + \frac{1}{2} \left(\xi_1 + \xi_2 \right) z^2 + \frac{1}{4} \left(\Omega_1^2 + \Omega_2^2 + \xi_1 \xi_2 \right) z + \frac{1}{8} \left(\Omega_1^2 \xi_2 + \Omega_2^2 \xi_1 \right) = 0. \qquad (264)$$

By exploiting the assumptions (232) and (233), we can find an approximate expression for the three roots of this equation:

$$z_0 \simeq -\frac{\xi_2}{2} - \frac{\Omega_2^2 \xi_1}{2\Omega_1^2} \tag{265a}$$

$$z_1 \simeq -\frac{\xi_1}{4}\left(1 - \frac{\Omega_2^2}{\Omega_1^2}\right) + i\,\frac{2\Omega_1^2\left(2\Omega_1^2 + \Omega_2^2\right) - \xi_1^2\left(\Omega_1^2 + \Omega_2^2\right)}{\Omega_1^2\sqrt{4\Omega_1^2 - \xi_1^2}}, \tag{265b}$$

$$z_2 \simeq -\frac{\xi_1}{4}\left(1 - \frac{\Omega_2^2}{\Omega_1^2}\right) - i\,\frac{2\Omega_1^2\left(2\Omega_1^2 + \Omega_2^2\right) - \xi_1^2\left(\Omega_1^2 + \Omega_2^2\right)}{\Omega_1^2\sqrt{4\Omega_1^2 - \xi_1^2}}. \tag{265c}$$

Note that one has $\operatorname{Re}\left(4\Omega_1^2 - \xi_1^2\right) = 4\Omega_1^2 - \gamma_1^2 + 4\Delta_1^2 > 0$ and

$$0 > \operatorname{Re} z_0 \gg \max\left\{\operatorname{Re} z_1, \operatorname{Re} z_2\right\}. \tag{266}$$

We can see that, for the values of interest of the parameters, the three roots are indeed distinct and with strictly negative real parts. Using this in eq. (259), we get

$$\lim_{t \to +\infty} P_t(0|\rho_{\text{jump}}) = 0, \qquad \int_0^{+\infty} w(s)\mathrm{d}s = 1. \tag{267}$$

Similar considerations apply to $P_t(0|\rho_0)$ and $w_0(t)$. This implies that the fluorescence light never stops definitively.

By (258), (263), many decay times appear in the expression of $w(t)$; by (266), the longest one is $(-2\operatorname{Re} z_0)^{-1}$, while the others, $\left\{(-\operatorname{Re} z_0 - \operatorname{Re} z_j)^{-1}, (-\operatorname{Re} z_i - \operatorname{Re} z_j)^{-1}, i, j = 1, 2\right\}$ are much shorter; so, $w(t)$ has a small long living tail. Indeed, by (258) and (263), we can write

$$w(t) = w_{\text{short}}(t) + \Pi 2\left|\operatorname{Re} z_0\right| \mathrm{e}^{2\operatorname{Re} z_0 t}, \tag{268}$$

where in $w_{\text{short}}(t)$ we have grouped all the terms with a short decay time and

$$\Pi := \frac{\gamma_1 \Omega_1^2 \left|z_0 + \xi_2/2\right|^2 + \gamma_2 \Omega_2^2 \left|z_0 + \xi_1/2\right|^2}{8\left|z_0 - z_1\right|^2 \left|z_0 - z_2\right|^2 \left|\operatorname{Re} z_0\right|}; \tag{269}$$

moreover, the conditions (232) give $\Pi \ll 1$. Therefore, the interarrival times are usually very short, but sometimes (with probability Π after each emission) the interarrival time is long, with a mean of the order of $(-2\operatorname{Re} z_0)^{-1}$ and the fluorescence light stops for a detectable interval of time. In conclusion we have a sequence of bright and dark periods of random length controlled by $w(t)$; in particular, the mean length of the dark periods is approximately

$$(-2\operatorname{Re} z_0)^{-1} \simeq \frac{\Omega_1^2}{\Omega_1^2 \gamma_2 + \Omega_2^2 \gamma_1}. \tag{270}$$

4.4 Bright and Dark Periods: The Λ-Configuration

In the Λ case we can proceed in the same way. From eqs. (245)–(247) we get

$$K = -\frac{\gamma_1 + \gamma_2}{2}\,|0\rangle\langle0|$$
$$- i\sum_{j=1}^{2}\left[\Delta_j|j\rangle\langle j| + \frac{\Omega_j}{2}\left(e^{i\beta_j}|j\rangle\langle0| + e^{-i\beta_j}|0\rangle\langle j|\right)\right], \quad (271)$$

$$\varkappa(\varrho) = (\gamma_1 + \gamma_2)\,\langle0|\varrho|0\rangle, \qquad \rho_{\text{jump}} = \sum_{j=1}^{2}\frac{\gamma_j}{\gamma_1+\gamma_2}\,|j\rangle\langle j|, \quad (272)$$

$$w_0(t) = (\gamma_1 + \gamma_2)\left\langle 0\left|e^{Kt}\rho_0 e^{K^*t}\right|0\right\rangle, \quad (273a)$$

$$w(t) = \sum_{j=1}^{2}\gamma_j\left|\langle0|e^{Kt}|j\rangle\right|^2 = \sum_{j=1}^{2}\gamma_j\left|\left\langle j\left|e^{K^*t}\right|0\right\rangle\right|^2. \quad (273b)$$

The Case of Equal Detunings: The Dark State

The case $\Delta_1 = \Delta_2 =: \Delta$ is very peculiar, because in this situation K has an eigenvector with a purely imaginary eigenvalue. One can check that

$$K|\varphi_2\rangle = -i\Delta|\varphi_2\rangle, \qquad K^*|\varphi_2\rangle = i\Delta|\varphi_2\rangle, \quad (274)$$

where

$$|\varphi_2\rangle := \frac{1}{\sqrt{\Omega_1^2 + \Omega_2^2}}\left(\Omega_2 e^{i\beta_1}|1\rangle - \Omega_1 e^{i\beta_2}|2\rangle\right). \quad (275)$$

We have also

$$\mathcal{L}_\Delta[|\varphi_2\rangle\langle\varphi_2|] = 0, \qquad \tilde{\mathcal{L}}_\Delta[|\varphi_2\rangle\langle\varphi_2|] = 0; \quad (276)$$

therefore, the state φ_2 is perfectly stationary and in this state the atom cannot emit: φ_2 is a *dark state*.

It is useful to construct an orthonormal set $\{\varphi_0, \varphi_1, \varphi_2\}$ containing the stationary state φ_2; we choose

$$|\varphi_0\rangle := |0\rangle, \qquad |\varphi_1\rangle := \frac{1}{\sqrt{\Omega_1^2 + \Omega_2^2}}\left(\Omega_1 e^{i\beta_1}|1\rangle + \Omega_2 e^{i\beta_2}|2\rangle\right). \quad (277)$$

By using this c.o.n.s. and by setting

$$x(t) := \langle\varphi_0|e^{Kt}|\varphi_1\rangle, \qquad y(t) := \langle\varphi_1|e^{Kt}|\varphi_1\rangle, \quad (278)$$

we obtain from eqs. (273b), (253a)

$$w(t) = \frac{\gamma_1 \Omega_1^2 + \gamma_2 \Omega_2^2}{\Omega_1^2 + \Omega_2^2} \, |x(t)|^2 \,, \tag{279}$$

$$P_t(0|\rho_{\text{jump}}^\bullet) = \Pi + \frac{w(t)}{\gamma_1 + \gamma_2} + \frac{\gamma_1 \Omega_1^2 + \gamma_2 \Omega_2^2}{(\gamma_1 + \gamma_2)(\Omega_1^2 + \Omega_2^2)} \, |y(t)|^2 \,, \tag{280}$$

where

$$\Pi := \sum_{j=1}^{2} \frac{\gamma_j}{\gamma_1 + \gamma_2} \, |\langle j|\varphi_2\rangle|^2 = \frac{\gamma_1 \Omega_2^2 + \gamma_2 \Omega_1^2}{(\gamma_1 + \gamma_2)(\Omega_1^2 + \Omega_2^2)}. \tag{281}$$

By time differentiation we get the linear system

$$\frac{\mathrm{d}x(t)}{\mathrm{d}t} = -\frac{\gamma_1 + \gamma_2}{2} \, x(t) - \frac{\mathrm{i}}{2} \sqrt{\Omega_1^2 + \Omega_2^2} \, y(t) \,,$$
$$\frac{\mathrm{d}y(t)}{\mathrm{d}t} = -\frac{\mathrm{i}}{2} \sqrt{\Omega_1^2 + \Omega_2^2} \, x(t) - \mathrm{i}\Delta y(t) \,; \tag{282}$$

the initial conditions are $x(0) = 0$, $y(0) = 1$. The roots of its characteristic equation are

$$z_\pm = \frac{1}{2} \left[-\mathrm{i}\Delta - \frac{\gamma_1 + \gamma_2}{2} \pm \mathrm{i}\sqrt{\Omega_1^2 + \Omega_2^2 - \left(\frac{\gamma_1 + \gamma_2}{2} - \mathrm{i}\Delta \right)^2} \right] \tag{283}$$

and its solution is

$$x(t) = \frac{\mathrm{i}\sqrt{\Omega_1^2 + \Omega_2^2}}{2\,(z_+ - z_-)} \left(\mathrm{e}^{z_- t} - \mathrm{e}^{z_+ t} \right), \tag{284a}$$

$$y(t) = \frac{1}{z_+ - z_-} \left[(z_+ + \mathrm{i}\Delta)\, \mathrm{e}^{z_- t} - (z_- + \mathrm{i}\Delta)\, \mathrm{e}^{z_+ t} \right]. \tag{284b}$$

From these results we have $\lim_{t \to +\infty} P_t(0|\rho_{\text{jump}}) = \Pi$ and so, after each emission, there is a probability Π that the fluorescence stops and a probability $1 - \Pi$ of a new emission. By (253a) we have also

$$\int_0^{+\infty} w(t)\mathrm{d}t = 1 - \Pi = \frac{\gamma_1 \Omega_1^2 + \gamma_2 \Omega_2^2}{(\gamma_1 + \gamma_2)(\Omega_1^2 + \Omega_2^2)}. \tag{285}$$

Let us consider for simplicity the case $\rho_0 = \rho_{\text{jump}}$, which gives $w_0(t) = w(t)$. If N_{tot} denotes the total number of emissions before the light stops, by the renewal structure we have $P[N_{\text{tot}} = n] = (1 - \Pi)^n \Pi$, $n = 0, 1, 2, \ldots$. Moreover, the inter-emission times T_i are independent and identically distributed random variables with probability density

$$\frac{w(t)}{1 - \Pi} = \frac{(\gamma_1 + \gamma_2)(\Omega_1^2 + \Omega_2^2)}{4\,|z_+ - z_-|^2} \, \left| \mathrm{e}^{z_- t} - \mathrm{e}^{z_+ t} \right|^2 . \tag{286}$$

Then, $T := \sum_{i=1}^{N_{\text{tot}}} T_i$ is the duration of the initial bright period and its mean is

$$\mathbb{E}[T] = \mathbb{E}[N_{\text{tot}}]\,\mathbb{E}[T_1] = \frac{1 - \Pi}{\Pi} \int_0^{+\infty} t\,\frac{w(t)}{1 - \Pi}\,\mathrm{d}t = \frac{1}{\Pi} \int_0^{+\infty} t w(t)\mathrm{d}t \,. \tag{287}$$

The Case of Different Detunings

In the generic case $\Delta_1 \neq \Delta_2$ we can write the expression (258) for $w(t)$ by setting

$$a_0(t) := \langle 0|e^{K^*t}|0\rangle, \qquad a_j(t) := e^{-i\beta_j}\langle j|e^{K^*t}|0\rangle, \quad j = 1, 2. \qquad (288)$$

Then, we have to solve the system of linear differential equations (260) where now

$$G := \begin{pmatrix} -(\gamma_1 + \gamma_2)/2 & i\Omega_1/2 & i\Omega_2/2 \\ i\Omega_1/2 & i\Delta_1 & 0 \\ i\Omega_2/2 & 0 & i\Delta_2 \end{pmatrix}. \qquad (289)$$

One can show that, when $\Delta_1 \neq \Delta_2$, all the eigenvalues of G have not vanishing negative real parts and that, under conditions (232), one of the decay times in $w(t)$ is very long and the other ones very short; then, the discussion on bright and dark periods goes on in a similar way as in the V-system case [42].

5 A Two–Level Atom and the Spectrum of the Fluorescence Light

The simplest, non trivial matter-field system is a two-level atom stimulated by a laser. The fluorescence spectrum of such a system, in the case of a perfectly mono-chromatic laser, was obtained by Mollow [81]; then, the case of a laser with a Lorentzian spectrum was developed by Kimble and Mandel [68, 75]. The Hudson-Parthasarathy equation (39) gives not only the way of giving an unified treatment of such a system and of connecting it to the photon-detection theory, but it allows also to explore possible corrections to the dynamics and to obtain modifications of the Mollow spectrum. In the usual treatments, the scattering of the light by the atom is described by the absorption/emission process due to the electric dipole interaction involving only two states of the atom. But even if the absorption/emission process would be forbidden and the atom would be frozen in the up or down level, some scattering of light would remain, due to the response of the atom as a whole; we can call it "direct scattering". For instance, in a perturbative development in Feynman diagrams, scattering processes would be generated also by virtual transitions starting and ending in one of the two states left in the description, but involving as intermediate states the other ones, which have been eliminated in the final description. The use of Hudson-Parthasarathy equation allows, even when the atom is approximated by a two-level system, for the introduction of both the "absorption/emission channel", through the terms containing the R operator, and the "direct scattering channel", through the terms containing the S operator. This modified model, in the case of a perfectly monochromatic laser, has been presented in [26] and developed in [27]; the results for a non monochromatic laser have been obtained in [28].

Being a relatively simple situation, we want to take the opportunity of showing also how the polarization of light can be introduced in QSC; to be consistent, also the electromagnetic selection rules have to be taken into account and the up and down levels have to be degenerate.

5.1 The Dynamical Model

The Two-Level Atom

The experiments [41, 49, 59, 60, 86] involve or the hyperfine component of the D_2 line of sodium with up level $2P_{3/2}$, $F = 3$ and down level $2S_{1/2}$, $F = 2$, or the levels $6s6p\ ^1P_1$, $F = 1$, and $6s^2\ S_0$, $F = 0$ of ^{138}Ba; F is the total angular momentum.

In order to describe an atom with two degenerate levels as in the experimental situation, we take

$$\mathcal{H} = \mathbb{C}^{2F_- + 1} \oplus \mathbb{C}^{2F_+ + 1}, \qquad F_+ = F_- + 1, \tag{290}$$

where F_- is integer or semi-integer. We denote by

$$|F_\pm, M\rangle, \qquad M = -F_\pm, \ldots, F_\pm,$$

the angular momentum basis in $\mathcal{H}$; the parities of the states of the two levels must be opposed, let us say they are $\epsilon_\pm$, with $\epsilon_+ \epsilon_- = -1$. Let us denote by

$$P_\pm = \sum_{M=-F_\pm}^{F_\pm} |F_\pm, M\rangle\langle F_\pm, M| \tag{291}$$

the two projection on the up or down states. Then, the energy H of the free atom, contained in the quantity K (37e), must be given by

$$H = \frac{1}{2}\,\omega_0(P_+ - P_-), \qquad \omega_0 > 0, \tag{292}$$

where the atomic frequency ω_0 must already include the Lamb shift.

The Photon Space

In the approximations we are considering [18, 89], the fields behave as monodimensional waves, so that a change of position is equivalent to a change of time and viceversa. Then, the space $\mathcal{Z}$ has to contain only the degrees of freedom linked to the direction of propagation and to the polarization. To describe a spin-1 0-mass particle we use the conventions of Messiah [79] pp. 550, 1032–1037.

The space $\mathcal{Z}$ is spanned by the c.o.n.s.

$$|j, m; \varpi\rangle \equiv \vec{\Theta}_{jm}^{\varpi}, \qquad j = 1, 2, \ldots, \quad m = -j, -j+1, \ldots, j, \quad \varpi = \pm 1;$$

jm is the total angular momentum, $\varpi = +1$ denotes the electrical multipoles, $\varpi = -1$ denotes the magnetic multipoles and $(-1)^j \varpi$ is the parity. By using the spherical harmonics $Y_{lm}(\theta, \phi)$ and the orbital angular momentum operator $\vec{\ell}$, one has

$$\vec{\Theta}_{jm}^{-1} = \frac{1}{\sqrt{j(j+1)}}\, \vec{\ell}\, Y_{jm}\,, \qquad \vec{\Theta}_{jm}^{+1} = \mathrm{i} \vec{p} \times \vec{\Theta}_{jm}^{-1}\,, \tag{293}$$

where $\vec{p}$ is the direction versor given by

$$p_1 = \sin\theta\, \cos\phi\,, \qquad p_2 = \sin\theta\, \sin\phi\,, \qquad p_3 = \cos\theta\,. \tag{294}$$

The Interaction

Let us consider now the terms with the creation and annihilation operators in the dynamical equation (39); they must describe the absorption/emission process. By asking spherical symmetry, parity conservation and only electrical dipole contribution in R, we must have

$$R = \alpha \sum_{M=-F_+}^{F_+} |\epsilon_+; F_-; 1; F_+, M\rangle\langle F_+, M|\,, \qquad \alpha \in \mathbb{C}\,, \quad \alpha \neq 0\,, \tag{295}$$

$$|\epsilon_+; F_-; 1; F_+, M\rangle$$
$$= \sum_{m_1=-F_-}^{F_-} \sum_{m_2=-1}^{1} |F_-, m_1\rangle \otimes |1, m_2; +1\rangle\langle F_-, m_1; 1, m_2|F_+, M\rangle\,, \tag{296}$$

where by $\langle F_-, m_1; 1, m_2|F_+, M\rangle$ we denote the Clebsch-Gordan coefficients ([79] pp. 560–563). The interaction terms containing R are responsible for the spontaneous decay of the atom and, as we shall see, $|\alpha|^2$ turns out to be the natural line width. By eqs. (37) and (291) we have

$$R_k = \alpha \sum_{m=-1}^{1} \langle z_k|1, m; +1\rangle Q_m\,, \tag{297}$$

$$Q_m = \sum_{m_1=-F_-}^{F_-} |F_-, m_1\rangle\langle F_-, m_1; 1, m|F_+, m_1 + m\rangle\langle F_+, m_1 + m|\,, \tag{298}$$

$$R_k = P_- R_k P_+\,, \qquad R^* R = |\alpha|^2 P_+\,, \tag{299}$$

$$\sum_k R_k \rho R_k^* = |\alpha|^2 \sum_{m=-1}^{1} Q_m \rho Q_m^*\,, \tag{300}$$

$$K - \frac{\mathrm{i}}{2}\, \omega_0 \mathbb{1} = -\left(\mathrm{i}\omega_0 + \frac{|\alpha|^2}{2}\right) P_+\,. \tag{301}$$

The interaction term containing the Λ-process must give the residual scattering when the atom is frozen in the up or down level, so we take the unitary operator S of the form

$$S = P_+ \otimes S^+ + P_- \otimes S^- , \qquad S^\pm \in \mathcal{U}(\mathcal{Z}). \tag{302}$$

Then, by spherical symmetry and parity conservation we must have

$$S^\pm = \sum_{j=1}^{\infty} \sum_{m=-j}^{j} \sum_{\varpi=\pm 1} |j, m; \varpi\rangle \exp\{2i\delta_\pm(j; \varpi)\}\langle j, m; \varpi|, \tag{303}$$

$0 \le \delta_\pm(j; \varpi) < 2\pi$. The unitary operators S^+ and S^- represent the scattering operators for a photon impinging on the atom frozen in the up or in the down state; the δ's are the phase shifts for these scattering processes. Quantities like $\omega_0, \alpha, \delta_\pm(j; \varpi)$ are phenomenological parameters, or, better, they have to be computed from some more fundamental theory, such as some approximation to quantum electrodynamics.

Summarizing, the possible processes are: *absorption*, through the term containing R^*, *emission*, through R, *photon scattering with the atom in the up level*, through $P_+ \otimes S^+$, *photon scattering with the atom in the down level*, through $P_- \otimes S^-$. An illustration of these processes is given in Fig. 6.

The Balance Equation for the Number of Photons

Let us introduce the observables, which we have already considered in Section 3.1 eqs. (85) and (96), "total number of photons in the time interval $[0, t]$" before and after the interaction with the atom,

$$N_{\text{tot}}^{\text{in}}(t) := N(\mathbb{1}; t) = \sum_k \Lambda_{kk}(t) ,$$
$$N_{\text{tot}}^{\text{out}}(t) := N^{\text{out}}(\mathbb{1}; t) = U(t)^* N_{\text{tot}}^{\text{in}}(t) U(t) . \tag{304}$$

By eq. (96) we obtain in the present model the balance equation

$$N_{\text{tot}}^{\text{in}}(t) + \frac{1}{2}(P_+ - P_-) = N_{\text{tot}}^{\text{out}}(t) + \frac{1}{2}U(t)^*(P_+ - P_-)U(t). \tag{305}$$

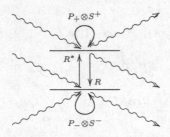

Fig. 6. The dynamical processes: absorption, emission and direct photon scattering

Such an equation has a very important meaning: it says that the number of photons entering the system up to time t plus the photons stored in the atom at time 0 is equal to the number of photons leaving the system up to time t plus the photons stored in the atom at time t.

The Phase Diffusion Model for the Laser

In [68, 75] a laser model is considered which, translated in our setup, amounts in taking as the state of the laser a mixture of coherent vectors of the type **S3**, (65). Therefore, the initial state is taken to be

$$\mathfrak{s} = \rho_0 \otimes \mathbb{E}^c \left[\eta(f) \right], \qquad \rho_0 \in \mathcal{S}(\mathcal{H}), \tag{306}$$

$$f(t) = e^{-i\left(\omega t + \sqrt{B}\, W(t)\right)} 1_{(0,T)}(t)\,\lambda, \quad \lambda \in \mathcal{Z}, \quad \omega > 0, \quad B \geq 0, \tag{307}$$

$W(t)$ is a real standard Wiener process canonically realized in the Wiener probability space $(\Omega^c, \mathcal{F}^c, P^c)$ (**S3'**); T is a large time and $T \to +\infty$ in the final results is always understood. This is the simplest model for a laser which is not perfectly monochromatic nor perfectly coherent. Only the phase fluctuates, not the intensity; moreover, the laser spectrum has a Lorentzian shape with bandwidth B:

$$\frac{\hbar\omega}{2\pi} \int_{-\infty}^{+\infty} e^{i\nu\tau} \, \mathbb{E}^c [\langle f(t)| f(t+\tau)\rangle] \, d\tau = \hbar\omega \|\lambda\|^2 \, \frac{B/(2\pi)}{(\nu - \omega)^2 + B^2/4}; \tag{308}$$

$\hbar\omega \|\lambda\|^2$ is the power of the laser. The whole model is meaningful only for ω not too "far" from ω_0.

In some experiments the laser light is taken to be circularly polarized, because in this way, in the long run, only two states are involved in the dynamics ([60] p. 206) and the usual theory has been developed for systems with only two states [68, 81]. Let the incoming light have right circular polarization and propagate along the z axis; then, the electromagnetic selection rules imply that the atomic transitions which survive are: spontaneous emission with $\Delta M = 0, \pm 1$, stimulated emission with $\Delta M = -1$, absorption with $\Delta M = 1$; the situation is summarized in the Fig. 7.

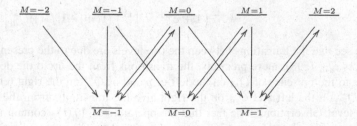

Fig. 7. Allowed atomic transitions (case $F_+ = 2$, $F_- = 1$)

In order to describe a well collimated laser beam propagating along the direction z ($\theta = 0$) and with right circular polarization, we take

$$\lambda = \alpha \Omega \, e^{i\delta} \lambda_+ \,, \qquad \Omega > 0 \,, \quad \delta \in [0, 2\pi) \tag{309}$$

$$\vec{\lambda}_+(\theta, \phi) = \frac{1_{[0, \Delta\theta]}(\theta)}{\Delta\theta\sqrt{3\pi(1 - \cos\Delta\theta)}} \left(-\frac{1}{\sqrt{2}}\right) \left(\vec{i} + i\vec{j}\right) ; \tag{310}$$

in all the physical quantities the limit $\Delta\theta \downarrow 0$ will be taken. Note that the power of the laser $\hbar\omega\|\lambda\|^2 = \frac{2}{3}\hbar\omega|\alpha|^2\Omega^2/(\Delta\theta)^2$ diverges for $\Delta\theta \downarrow 0$, because we need a not vanishing atom-field interaction in the limit.

In the following we shall need the relation

$$\langle j, m; \varpi | \lambda_+ \rangle = -\varpi \sqrt{\frac{2j + 1}{12}} \, \delta_{m,1} \,. \tag{311}$$

5.2 The Master Equation and the Equilibrium State

Let us start by considering the (not averaged) reduced statistical operator $\rho(f; t) = \mathrm{Tr}_\Gamma \{U(t)\,(\rho_0 \otimes \eta(f))\,U(t)^*\}$ (70), which satisfies the master equation (72) or (74); the time-dependent Liouvillian (73) turns out to be

$$\mathcal{L}(f(t))[\varrho] = -i\left[H(f(t)), \varrho\right]$$
$$+ \frac{1}{2}\sum_{k=1}^{\infty} \left(\left[R_k(f(t))\varrho, R_k(f(t))^*\right] + \left[R_k(f(t)), \varrho R_k(f(t))^*\right]\right), \tag{312}$$

$$R_k(f(t)) = \alpha \sum_{m=-1}^{1} \langle z_k|1, m; +1\rangle Q_m + \sum_{\epsilon=\pm} \langle z_k|(S^\epsilon - \mathbb{1})f(t)\rangle P_\epsilon \,, \tag{313}$$

$$H(f(t)) = \frac{1}{2}\left[\omega_0 - \mathrm{Im}\,\langle f(t)|\,(S^+ - S^-)\,f(t)\rangle\right](P_+ - P_-)$$
$$+ \frac{i}{2}\sum_{m=-1}^{1}\left[\alpha\left(1 + e^{-2i\delta - (1;+1)}\right)\langle f(t)|1, m; +1\rangle Q_m\right.$$
$$\left. - \overline{\alpha}\left(1 + e^{2i\delta - (1;+1)}\right)\langle 1, m; +1|f(t)\rangle Q_m^*\right]. \tag{314}$$

We see that the transitions between the two levels are due to the presence of the operators Q_m (298); more precisely, the transitions from the up to the down level are due to the presence of Q_m on the left of ϱ and of Q_m^* on the right (emission), while Q_m^* on the left and Q_m on the right give the transitions from the down to the up level (absorption). The fact that the operators $R_k(f(t))$ contain a sum of terms proportional to Q_m and of terms proportional to $P_\pm$ is an indication of an interference effect between absorption/emission and direct scattering.

Note that, when the field is in the Fock vacuum, only spontaneous emission is present; indeed, for $f(t) = 0$, the Liouvillian (312) reduces to

$$\mathcal{L}(0)[\varrho] = -i\omega_0 [P_+ , \varrho] - \frac{1}{2}|\alpha|^2 (P_+\varrho + \varrho P_+) + |\alpha|^2 \sum_{m=-1}^{1} Q_m\varrho Q_m^* , \quad (315)$$

which describes the atomic decay according to the usual selection rules for electric dipole and fixes the meaning of $|\alpha|^2$ as spontaneous decay rate or natural line width.

By inserting the expression (307), (309), (310) of $f(t)$ into eqs. (313), (314) we get

$$R_k\big(f(t)\big) = e^{-i\left(\omega t + \sqrt{B}\,W(t)\right)} U_W(t)^* D(z_k) U_W(t), \quad (316)$$

$$H\big(f(t)\big) = U_W(t)^* \tilde{H} U_W(t) + \frac{1}{2}\omega\, (P_+ - P_-), \quad (317)$$

$$U_W(t) := \exp\left\{ \frac{i}{2}\left[\omega t + \sqrt{B}\,W(t)\right](P_+ - P_-)\right\}, \quad (318)$$

$$D(h) := \alpha \sum_{m=-1}^{1} \langle h|1,m;+1\rangle Q_m + \alpha\Omega e^{i\delta} \sum_{\epsilon=\pm} \langle h|\,(S^\epsilon - \mathbb{1})\,\lambda_+\rangle P_\epsilon , \quad (319)$$

$$h \in \mathcal{Z},$$

$$\tilde{H} = \frac{1}{2}\left[\omega_0 - \omega - |\alpha|^2\,\Omega^2\,\mathrm{Im}\langle\lambda_+|\,(S^+ - S^-)\,\lambda_+\rangle\right](P_+ - P_-) \quad (320)$$
$$+ \frac{i}{4}|\alpha|^2\,\Omega\left[e^{i\delta}\left(1 + e^{2i\delta_-(1;+1)}\right) Q_1^* - e^{-i\delta}\left(1 + e^{-2i\delta_-(1;+1)}\right) Q_1\right].$$

This result suggests to consider the random atomic state in the "rotating frame"

$$\tilde{\rho}(f;t) := U_W(t)\rho(f;t)U_W(t)^* . \quad (321)$$

By classical stochastic calculus we obtain the stochastic equations of Ito type

$$dU_W(t) = \left[\frac{i}{2}\left(\omega dt + \sqrt{B}\,dW(t)\right)(P_+ - P_-) - \frac{1}{8}B dt\right] U_W(t) \quad (322)$$

and

$$d\tilde{\rho}(f;t) = \frac{i}{2}\sqrt{B}\,[P_+ - P_-,\,\tilde{\rho}(f;t)]\,dW(t) + \tilde{\mathcal{L}}_B\,[\tilde{\rho}(f;t)]\,dt, \quad (323)$$

where

$$\tilde{\mathcal{L}}_B[\varrho] := \frac{B}{8}\left([(P_+ - P_-)\varrho, P_+ - P_-] + [P_+ - P_-, \varrho(P_+ - P_-)]\right)$$
$$- i\left[\tilde{H}, \varrho\right] + \frac{1}{2}\sum_k \left([D(z_k)\varrho, D(z_k)^*] + [D(z_k), \varrho D(z_k)^*]\right). \quad (324)$$

The quantum dynamical semigroup $\exp\left\{\tilde{\mathcal{L}}_B t\right\}$ will be one of the main ingredients in the computations of the fluorescence spectrum; therefore, we need to study its properties and, in particular, its equilibrium state.

The Parameters

First of all we need to summarize all the parameters which enter the model and to introduce some shorthand notations.

We already started to give all quantities in units of the natural linewidth $|\alpha|^2$; for instance, we wrote $\lambda = \alpha\Omega\, e^{i\delta}\lambda_+$ in (309), so that Ω^2 becomes an adimensional measure of the *laser intensity*. Then, we define

$$x := \frac{\nu - \omega_0}{|\alpha|^2}\,, \qquad\qquad \text{reduced frequency,} \tag{325a}$$

$$y := B/|\alpha|^2\,, \qquad\qquad \text{reduced laser bandwidth,} \tag{325b}$$

$$z := (\omega - \omega_0)/|\alpha|^2\,, \qquad \text{reduced detuning,} \tag{325c}$$

$$\gamma := \frac{\tilde{\gamma}}{|\alpha|^2}\,, \qquad\qquad \text{reduced instrumental width,} \tag{325d}$$

It is also useful to introduce the shorthand notation

$$q := \frac{\gamma + y}{2} + i(x - z) \tag{326}$$

and the *shifted detuning parameter*

$$\tilde{z} := z - \Omega^2\varepsilon\,, \tag{327}$$

where

$$\varepsilon := \text{Im}\langle S^+ \lambda_+ | P_\perp S^- \lambda_+\rangle\,, \tag{328}$$

$$P_\perp := \mathbb{1} - |1,1;+1\rangle\langle 1,1;+1|\,; \tag{329}$$

here $\Omega^2\varepsilon$ plays the role of an *intensity dependent light shift*. It is also useful to define

$$\delta_\pm := \delta_\pm(1;+1)\,, \qquad s := \delta_+ - \delta_-\,, \tag{330}$$

$$g_\pm := P_\perp (S^\pm - \mathbb{1})\lambda_+\,, \qquad \Delta g := g_+ - g_-\,, \tag{331}$$

$$b := \frac{1}{2}\left[1 + y + \Omega^2\left(\|\Delta g\|^2 + \sin^2 s\right)\right] - i\left(\tilde{z} + \frac{1}{4}\Omega^2\sin 2s\right)\,, \tag{332}$$

$$\Gamma^2 := \left(1 + y + \Omega^2\|\Delta g\|^2\right)^2 + \Omega^2\left(2 + 2y + 2\Omega^2\|\Delta g\|^2 + \Omega^2\sin^2 s\right)\,. \tag{333}$$

We have to ask $|\varepsilon| < +\infty$, $\|g_\pm\| < +\infty$, for $\Delta\theta \downarrow 0$; roughly speaking, $S^\pm$ must introduce small corrections even when the norm of λ_+ diverges.

The Equilibrium State

By inserting the explicit expressions of all the quantities appearing in the time-independent Liouvillian (324) and by using the shorthand notations just introduced, after some computations we obtain

$$|\alpha|^{-2}\,\tilde{\mathcal{L}}_B[\varrho] = -P_+\varrho P_+ - bP_+\varrho P_- - \bar{b}P_-\varrho P_+ + \sum_{m=-1}^{1} Q_m\varrho Q_m^*$$

$$+ \frac{1}{2}\,\Omega\left[e^{i(\delta+2\delta_-)}Q_1^*\varrho + e^{-i(\delta+2\delta_-)}\varrho Q_1\right]$$

$$- \frac{1}{2}\,\Omega\sum_{\epsilon=\pm}\left[e^{i(\delta+2\delta_\epsilon)}P_\epsilon\varrho Q_1^* + e^{-i(\delta+2\delta_\epsilon)}Q_1\varrho P_\epsilon\right]. \quad (334)$$

For $|\alpha|^2 > 0$, $\Omega > 0$, the quantum dynamical semigroup $\exp\left\{\tilde{\mathcal{L}}t\right\}$ has a unique equilibrium state with support in the linear span of $|F_+, F_+\rangle$, $|F_-, F_-\rangle$: for any initial state ρ_0, $\lim_{t\to+\infty}\exp\left\{\tilde{\mathcal{L}}_B t\right\}[\rho_0] = \rho_\infty$. We skip proofs and computations and give the final expression for ρ_∞:

ρ_∞

$$\langle F_+, F_+|\rho_\infty|F_+, F_+\rangle = \Omega^2\,\mathrm{Re}\,d\,, \qquad \langle F_-, F_-|\rho_\infty|F_-, F_-\rangle = 1 - \Omega^2\,\mathrm{Re}\,d\,, \quad (335a)$$

$$\langle F_+, F_+|\rho_\infty|F_-, F_-\rangle = \overline{\langle F_-, F_-|\rho_\infty|F_+, F_+\rangle} = \Omega\exp\left[i\left(\delta+2\delta_-\right)\right]d\,, \quad (335b)$$

$$d = \frac{1 + y + \Omega^2\left(\|\Delta g\|^2 + \sin^2 s\right) + i\left(2\tilde{z} - \Omega^2\sin s\cos s\right)}{4\tilde{z}^2 + \Gamma^2}\,. \quad (335c)$$

5.3 The Detection Scheme

Heterodyne Detection

The spectrum of our two-level atom can be scanned by using the balanced heterodyne scheme (Sect. 3.5 and Fig. 3). We take a monochromatic laser of frequency ν as local oscillator and consider a measuring apparatus which spans a small solid angle with the vertex in the atom and not containing the forward direction in order that the light of the stimulating laser does not hit directly the apparatus. This means that we are in the case of eqs. (201)–(216). The measurement scheme produces an output current

$$I(\nu, \hat{h}; t) = k\int_0^t e^{-\tilde{\gamma}(t-s)/2}d\tilde{Q}(\nu, \hat{h}; s)\,, \quad (336)$$

where $\tilde{Q}(\nu, \hat{h}; s)$ is the stochastic process associated to the compatible quantum observables

$$Q(\nu, \hat{h}; t) = \sum_j\int_0^t\left[e^{-i\nu s}\langle z_j|\hat{h}\rangle\,dA_j^\dagger(s) + e^{i\nu s}\langle\hat{h}|z_j\rangle\,dA_j(s)\right], \quad (337)$$

$\nu \in \mathbb{R}$ is the frequency of the local oscillator, $\tilde{\gamma} > 0$ is an instrumental width, $k \neq 0$ is a proportionality constant and $\hat{h} \in \mathcal{Z}$, $\|\hat{h}\| = 1$. Any information on the localization and on the polarization of the detector is contained in $\hat{h}$. We assume that the detector spans a small solid angle, so that $\hat{h}$ is given by

$$\hat{h}(\theta', \phi') = \frac{1_\Xi(\theta', \phi')}{\sqrt{|\Xi|}} \, \vec{e}(\theta', \phi'), \tag{338}$$

where Ξ is a small solid angle around the direction (θ, ϕ), $\Xi \downarrow \{(\theta, \phi)\}$, $|\Xi| \simeq \sin\theta \, d\theta \, d\phi$ and $\vec{e}$ is a complex polarization vector, $|\vec{e}(\theta', \phi')| = 1$. Moreover, we assume that the transmitted wave does not reach the detector, i.e. Ξ and the laser solid angle of (310) are disjoint: $\Xi \cap \{\theta \in [0, \Delta\theta], \phi \in [0, 2\pi)\} = \emptyset$; in the limit of infinitesimal angles we have simply that Ξ is a small solid angle around the direction (θ, ϕ) with $\theta > 0$. In particular this gives

$$\langle \hat{h} | \lambda_+ \rangle = 0. \tag{339}$$

Then, the generator (209) of the characteristic operator associated to the process $\tilde{Q}(\nu, \hat{h}; s)$ becomes

$$\mathcal{K}_t(f; \kappa)[\varrho] = \mathcal{L}(f(t))[\varrho] - \frac{\kappa^2}{2} \varrho + i\kappa \left[Z(f; t)\varrho + \varrho Z(f; t)^* \right], \tag{340}$$

where, by eqs. (143), (147), $\mathcal{L}(f(t))$ is the generator of the dynamics $\Upsilon(f; t, s)$ and is given by eqs. (312)–(314), while $Z(f; t)$ is given by eq. (210), which now becomes

$$Z(f; t) = e^{i\nu t} \left\{ \alpha \sum_{m=-1}^{1} \langle \hat{h} | 1, m; +1 \rangle Q_m + \sum_{\epsilon=\pm} \langle \hat{h} | (S^\epsilon - \mathbb{1}) f(t) \rangle P_\epsilon \right\}. \tag{341}$$

By using the explicit expression (307) of $f(t)$ and the stochastic unitary operators $U_W(t)$ (318) we get

$$Z(f; t) = \exp\left\{ -i \left[(\omega - \nu) t + \sqrt{B} W(t) \right] \right\} U_W(t)^* D(\hat{h}) U_W(t), \tag{342}$$

where $D(\cdot)$ is defined in eq. (319). By defining

$$\tilde{\Upsilon}(t, s)[\varrho] := U_W(t) \Upsilon(f; t, s) \left[U_W(s)^* \varrho U_W(s) \right] U_W(t)^*, \tag{343}$$

eqs. (321), (323) give

$$d\tilde{\Upsilon}(t, s)[\varrho] = \frac{i}{2} \sqrt{B} \left[P_+ - P_-, \tilde{\Upsilon}(t, s)[\varrho] \right] dW(t) + \tilde{\mathcal{L}}_B \left[\tilde{\Upsilon}(t, s)[\varrho] \right] dt, \tag{344}$$

with $\tilde{\mathcal{L}}_B$ given by eq. (324).

The Power Spectrum

As the power of a current is proportional to the square of the current itself, the expression

$$P(\nu, \hat{h}) = \lim_{T \to +\infty} \frac{k_1}{T} \int_0^T \mathbb{E}_{\rho_0} \left[I(\nu, \hat{h}; t)^2 \right] dt \tag{345}$$

is the *mean output power* in the long run; $k_1 > 0$ is a suitable constat with the dimensions of a resistance, it is independent of ν, but it can depend on the other features of the detection apparatus. As a function of ν, $P(\nu, \hat{h})$ gives the *power spectrum* observed in the "channel $\hat{h}$"; in the case of the choice (338) it is the spectrum observed around the direction (θ, ϕ) and with polarization $\vec{e}$.

Let us pospone the computations of the mean power and let us start by giving and discussing its final expression:

$$P(\nu, \hat{h}) = \frac{k^2 k_1}{\tilde{\gamma}} + \frac{4\pi k^2 k_1}{\tilde{\gamma}} \Sigma(\nu; \hat{h}), \qquad (346)$$

$$\Sigma(\nu; \hat{h}) = \frac{1}{2\pi} \int_0^{+\infty} e^{-[(\tilde{\gamma}+B)/2+i(\nu-\omega)]t} \, \mathrm{Tr}\left\{ D(\hat{h})^* \, e^{\mathcal{K}_1 t} \left[D(\hat{h})\rho_\infty \right] \right\} \mathrm{d}t$$
$$+ \text{c.c.}, \quad (347)$$

$$\mathcal{K}_1[\varrho] = \tilde{\mathcal{L}}_B[\varrho] - BP_+\varrho P_- + BP_-\varrho P_+, \qquad (348)$$

c.c. means "complex conjugated".

- The term $k^2 k_1/\tilde{\gamma}$ is independent of ν and, for this reason, can be seen as a white noise contribution to the power spectrum. It is due to the detection scheme, but it cannot be eliminated; its origin can be traced to the canonical commutation relations of the fields and, so, it is of a quantum origin. It is known as *shot noise*.
- The mean power $P(\nu, \hat{h})$ is positive because it is the expectation of the square of a real quantity, but it can be shown that also $\Sigma(\nu; \hat{h})$ is positive. Therefore, $\Sigma(\nu; \hat{h})$ can be separated from the shot noise and can be interpreted as the *fluorescence spectrum* in the channel $\hat{h}$. To prove the positivity of $\Sigma(\nu; \hat{h})$ one needs to go back to expressions in which the quantum fields appear explicitly; see Section 3.1.3 of [28].
- The normalization of $\Sigma(\nu; h)$ in (346) has been chosen in such a way that

$$\int_{-\infty}^{+\infty} \Sigma(\nu; \hat{h}) \, \mathrm{d}\nu = \mathrm{Tr}\left\{ D(\hat{h})^* D(\hat{h})\rho_\infty \right\}, \qquad (349)$$

With this choice, the total strength of the spectrum is the asymptotic rate of emission of photons in the "channel $\hat{h}$"; indeed, we have

$$\int_{-\infty}^{+\infty} \Sigma(\nu; \hat{h}) \, \mathrm{d}\nu = \lim_{T \to +\infty} \frac{1}{T} \, \mathrm{Tr}_{\mathcal{H}\otimes\Gamma}\left\{ N^{\mathrm{out}}(P_{\hat{h}}; T)\mathfrak{s} \right\}, \qquad (350)$$

where $P_{\hat{h}}$ is the orthogonal projection on $\hat{h}$. Also for this result we refer to Section 3.1.3 of [28].
- The state ρ_∞ is the unique equilibrium state of the quantum dynamical semigroup $\exp(\tilde{\mathcal{L}}_B t)$ and it is given in eqs. (335).

– In quantum optics it is often stated that the emission spectrum is the Fourier transform of the two-times quantum correlation function of the dipole operator. This is indeed the structure appearing in eq. (347) if we interpret the operator $D(\cdot)$, defined in eq. (319), as an *effective dipole operator*. This is reasonable because a dipole operator has to take into account not only the two levels remained in the description, but also the full structure of the atom and the operator $S - \mathbb{1}$ is indeed a track of this structure. Let us note that the effective dipole operator appears also in the Liouvillian $\tilde{\mathcal{L}}_B$ and, so it contributes to the spectrum also through ρ_∞ and $\mathcal{K}_1$.

– The semigroup $\exp[\mathcal{K}_1 t]$ is trace preserving, but not positivity preserving, while $\tilde{\mathcal{L}}_B$ is a bona-fide Liouvillian, because it can be written in the Lindblad form; this peculiar structure of the quantum correlation function appearing in (347), while not explicitly formulated, was already found in [68]. By putting in evidence the terms with B we can write

$$\mathcal{K}_1[\varrho] - \frac{B}{2}\varrho = \tilde{\mathcal{L}}_0[\varrho] - \frac{B}{2}P_+\varrho P_+ - \frac{B}{2}P_-\varrho P_- + BP_+\varrho P_-. \qquad (351)$$

Let us sketch now the computations which bring to eqs. (346)–(348).

First step. By particularizing the formula for the second moments (216) to our case we get

$$\mathbb{E}_{\rho_0}\left[I(\nu,\hat{h};t)^2\right] = \frac{k^2}{\tilde{\gamma}}\left(1 - e^{-\tilde{\gamma}t}\right) + 2k^2 \int_0^t ds_1 \int_0^{s_2} ds_2\, e^{-\tilde{\gamma}(2t-s_1-s_2)/2}$$

$$\times \mathbb{E}^c\left[e^{i[(\omega-\nu)(s_1+s_2)+\sqrt{B}(W(s_1)+W(s_2))]}\operatorname{Tr}\left\{D(\hat{h})^*\tilde{\Upsilon}(s_1,s_2)\left[\tilde{\rho}(f;s_2)D(\hat{h})^*\right]\right\}\right.$$

$$\left. + e^{i[(\omega-\nu)(s_1-s_2)+\sqrt{B}(W(s_1)-W(s_2))]}\operatorname{Tr}\left\{D(\hat{h})^*\tilde{\Upsilon}(s_1,s_2)\left[D(\hat{h})\tilde{\rho}(f;s_2)\right]\right\}\right.$$

$$\left. + \text{c.c.}\right]. \qquad (352)$$

Second step. The dynamics $\tilde{\Upsilon}(t,s)$ depends on the Wiener process only through the increments $W(\tau) - W(s)$, $s \leq \tau \leq t$, and, so, the conditional expectation $\mathbb{E}^c\left[e^{i\sqrt{B}(W(t)-W(s))}\tilde{\Upsilon}(t,s)\big|\mathcal{F}_s^c\right]$ is non random and coincides with the expectation $\mathbb{E}^c\left[e^{i\sqrt{B}(W(t)-W(s))}\tilde{\Upsilon}(t,s)\right]$. Recall that $\mathcal{F}_s^c$ is the σ-algebra generated by $W(\tau)$, $\tau \in [0,s]$; see **S3'** for this notation. By eq. (344) and the classical Itô's formula we get

$$d\left(e^{i\sqrt{B}(W(t)-W(s))}\tilde{\Upsilon}(t,s)[\varrho]\right) = \left(i\sqrt{B}\left(\frac{1}{2}[P_+ - P_-,\cdot] + \mathbb{1}\right)dW(t)\right.$$

$$\left. + \left(\mathcal{K}_1 - \frac{B}{2}\mathbb{1}\right)dt\right)\left[e^{i\sqrt{B}(W(t)-W(s))}\tilde{\Upsilon}(t,s)[\varrho]\right]; \qquad (353)$$

this gives

$$\mathbb{E}^c\left[e^{i\sqrt{B}(W(t)-W(s))}\tilde{\Upsilon}(t,s)\big|\mathcal{F}_s^c\right] = e^{-B(t-s)/2}\,e^{\mathcal{K}_1(t-s)}. \qquad (354)$$

Similarly, we get

$$\mathbb{E}^c \left[e^{2i\sqrt{B}W(s)} \tilde{\rho}(f;s) \right] = e^{-2Bs} e^{\mathcal{K}_2 s}[\rho_0], \tag{355}$$

with

$$\mathcal{K}_2 = \mathcal{K}_1 - \frac{B}{2}\, [P_+ - P_-, \cdot], \tag{356}$$

or, more explicitly,

$$\mathcal{K}_2[\varrho] - 2B\varrho = \tilde{\mathcal{L}}_0[\varrho] - 2B \left(P_+ \varrho P_+ + P_- \varrho P_- + \frac{1}{4} P_- \varrho P_+ + \frac{3}{4} P_+ \varrho P_- \right). \tag{357}$$

By inserting these results into eq. (352) we get

$$\mathbb{E}_{\rho_0} \left[I(\nu, \hat{h}; t)^2 \right] = \frac{k^2}{\tilde{\gamma}} \left(1 - e^{-\tilde{\gamma}t} \right) + 2k^2 \int_0^t ds_1 \int_0^{s_2} ds_2\, e^{-\tilde{\gamma}(2t-s_1-s_2)/2}$$
$$\times \left\{ e^{i(\omega-\nu)(s_1+s_2) - \frac{B}{2}(s_1+3s_2)} \operatorname{Tr} \left\{ D(\hat{h})^* e^{\mathcal{K}_1(s_1-s_2)} \left[e^{\mathcal{K}_2 s_2}[\rho_0] D(\hat{h})^* \right] \right\} \right.$$
$$\left. + e^{i(\omega-\nu)(s_1-s_2) - \frac{B}{2}(s_1-s_2)} \operatorname{Tr} \left\{ D(\hat{h})^* e^{\mathcal{K}_1(s_1-s_2)} \left[D(\hat{h}) e^{\tilde{\mathcal{L}}_B s_2}[\rho_0] \right] \right\} \right.$$
$$\left. + \text{c.c.} \right\}. \tag{358}$$

Third step. By using the new variables of integration $\tau = s_1 - s_2$, $s = s_2$, we get

$$\mathbb{E}_{\rho_0} \left[I(\nu, \hat{h}; t)^2 \right] = \frac{k^2}{\tilde{\gamma}} \left(1 - e^{-\tilde{\gamma}t} \right) + 2k^2 \left[a_1(t) + a_2(t) + a_3(t) + \text{c.c.} \right], \tag{359}$$

$$a_1(t) := \int_0^t d\tau \int_0^{t-\tau} ds\, e^{-\tilde{\gamma}(t-s-\frac{\tau}{2}) + [i(\omega-\nu) - \frac{B}{2}]\tau} \operatorname{Tr} \left\{ D(\hat{h})^* e^{\mathcal{K}_1 \tau} \left[D(\hat{h}) \rho_\infty \right] \right\}, \tag{360a}$$

$$a_2(t) := \int_0^t d\tau \int_0^{t-\tau} ds\, e^{-\tilde{\gamma}(t-s-\frac{\tau}{2}) + [i(\omega-\nu) - \frac{B}{2}]\tau}$$
$$\times \operatorname{Tr} \left\{ D(\hat{h})^* e^{\mathcal{K}_1 \tau} \left[D(\hat{h}) e^{\tilde{\mathcal{L}}_B s}[\rho_0 - \rho_\infty] \right] \right\}, \tag{360b}$$

$$a_3(t) := \int_0^t d\tau \int_0^{t-\tau} ds\, e^{-\tilde{\gamma}(t-s-\frac{\tau}{2}) + [i(\omega-\nu) - \frac{B}{2}]\tau}$$
$$\times \operatorname{Tr} \left\{ D(\hat{h})^* e^{\mathcal{K}_1 \tau} \left[e^{\mathcal{K}_3(\nu)s}[\rho_0] D(\hat{h})^* \right] \right\}, \tag{360c}$$

$$\mathcal{K}_3(\nu) := 2[i(\omega - \nu) - B] + \mathcal{K}_2. \tag{360d}$$

282   **Alberto Barchielli**

The a_1 term becomes

$$a_1(t) = \frac{1}{\tilde{\gamma}} \int_0^t d\tau \left[e^{-\frac{\tilde{\gamma}}{2}\tau} - e^{-\frac{\tilde{\gamma}}{2}t} e^{-\frac{\tilde{\gamma}}{2}(t-\tau)} \right] e^{[i(\omega-\nu)-\frac{B}{2}]\tau}$$

$$\times \operatorname{Tr}\left\{ D(\hat{h})^* e^{\mathcal{K}_1 \tau} \left[D(\hat{h})\rho_\infty \right] \right\} \quad (361)$$

and in the limit it gives

$$\lim_{t\to+\infty} a_1(t) = \frac{1}{\tilde{\gamma}} \int_0^{+\infty} d\tau\, e^{[i(\omega-\nu)-\frac{B+\tilde{\gamma}}{2}]\tau} \operatorname{Tr}\left\{ D(\hat{h})^* e^{\mathcal{K}_1 \tau} \left[D(\hat{h})\rho_\infty \right] \right\}. \quad (362)$$

The a_2 term becomes

$$a_2(t) = \int_0^t d\tau\, e^{[i(\omega-\nu)-\frac{B}{2}]\tau} \operatorname{Tr}\left\{ D(\hat{h})^* \right.$$

$$\left. \times e^{\mathcal{K}_1 \tau} \left[D(\hat{h}) \frac{e^{\tilde{\mathcal{L}}_B(t-\tau)-\frac{\tilde{\gamma}}{2}\tau} - e^{-\frac{\tilde{\gamma}}{2}t} e^{-\frac{\tilde{\gamma}}{2}(t-\tau)}}{\tilde{\gamma}+\tilde{\mathcal{L}}_B} [\rho_0 - \rho_\infty] \right] \right\}; \quad (363)$$

because $\lim_{t\to+\infty} e^{\tilde{\mathcal{L}}_B t}[\rho_0 - \rho_\infty] = 0$, we get

$$\lim_{t\to+\infty} a_2(t) = 0. \quad (364)$$

Similarly, the a_3 term becomes

$$a_3(t) = \int_0^t d\tau\, e^{[i(\omega-\nu)-\frac{B}{2}]\tau} \operatorname{Tr}\left\{ D(\hat{h})^* \right.$$

$$\left. \times e^{\mathcal{K}_1 \tau} \left[\frac{e^{\mathcal{K}_3(\nu)(t-\tau)-\frac{\tilde{\gamma}}{2}\tau} - e^{-\frac{\tilde{\gamma}}{2}t} e^{-\frac{\tilde{\gamma}}{2}(t-\tau)}}{\tilde{\gamma}+\mathcal{K}_3(\nu)} [\rho_0] D(\hat{h})^* \right] \right\}; \quad (365)$$

then, at least almost everywhere in ν,

$$\lim_{t\to+\infty} \frac{1}{T} \int_0^T a_3(t)dt = \lim_{t\to+\infty} \frac{1}{T} \int_0^T d\tau\, e^{[i(\omega-\nu)-\frac{B+\tilde{\gamma}}{2}]\tau} \operatorname{Tr}\left\{ D(\hat{h})^* \right.$$

$$\left. \times e^{\mathcal{K}_1 \tau} \left[\frac{1}{\tilde{\gamma}+\mathcal{K}_3(\nu)} \circ \frac{e^{\mathcal{K}_3(\nu)(T-\tau)} - \mathbb{1}}{\mathcal{K}_3(\nu)} [\rho_0] D(\hat{h})^* \right] \right\} = 0. \quad (366)$$

By inserting all these results into (345) we get the expression (346)–(348) for the power spectrum.

5.4 The Fluorescence Spectrum

In order to get an explicit expression for the spectrum (347), we need to solve the pseudo–master equation $\dot{\sigma}(t) = \mathcal{K}_1[\sigma(t)]$ with the initial condition $\sigma(0) = D(\hat{h})\rho_\infty$. We already said that the equilibrium state ρ_∞ is supported by the two extreme states

$$|1\rangle = |F_+, F_+\rangle, \quad |2\rangle = |F_-, F_-\rangle \tag{367}$$

and one can check that all the operations involved in (347) leave the span of $|1\rangle$, $|2\rangle$ invariant; so, we can forget all the other states in $\mathcal{H}$ and we are left with formulae involving 2×2 matrices. Let us use the Pauli matrices

$$\sigma_+ = \begin{pmatrix} 0 & 1 \\ 0 & 0 \end{pmatrix}, \qquad \sigma_- = \begin{pmatrix} 0 & 0 \\ 1 & 0 \end{pmatrix}, \qquad \sigma_z = \begin{pmatrix} 1 & 0 \\ 0 & -1 \end{pmatrix}; \tag{368}$$

with this notation we have in particular $P_\pm = \frac{1}{2}(1 \pm \sigma_z)$ and

$$D(\hat{h}) = \alpha \Omega e^{i\delta} \left(\langle \hat{h}|1, 1; +1\rangle D_1 + \langle \hat{h}|g_+\rangle P_+ + \langle \hat{h}|g_-\rangle P_- \right), \tag{369}$$

$$D_1 := \frac{1}{\Omega} e^{-i\delta} \sigma_- - i e^{i\delta_+} \sin \delta_+ \, P_+ - i e^{i\delta_-} \sin \delta_- \, P_-, \tag{370}$$

$$\rho_\infty = \begin{pmatrix} \Omega^2 \operatorname{Re} d & \Omega \exp\left[i\left(\delta + 2\delta_- \right) \right] d \\ \Omega \exp\left[-i\left(\delta + 2\delta_- \right) \right] \overline{d} & 1 - \Omega^2 \operatorname{Re} d \end{pmatrix}. \tag{371}$$

The Angular Distribution of the Spectrum

To obtain the angular dependence of the spectrum, let us introduce for $\hat{h}$ the states $h_\pm$ concentrated around (θ, ϕ) and with right/left circular polarization, given by equation (338) with $\vec{e} = \vec{e}_\pm$, where

$$\vec{e}_\pm(\theta, \phi) = \frac{\exp(i\phi)}{\sqrt{2}} \begin{pmatrix} i \sin \phi \mp \cos \theta \cos \phi \\ -i \cos \phi \mp \cos \theta \sin \phi \\ \pm \sin \theta \end{pmatrix}. \tag{372}$$

Then, we can introduce the two angular spectra

$$\Sigma_\pm(x; \theta) := \frac{1}{|\Delta\Xi|} \Sigma(\nu; h_\pm), \tag{373}$$

which, by the cylindrical symmetry of the problem, do not depend on ϕ; recall that x is linked to ν by eq. (325a). Then, we have

$$\Sigma_\pm(x; \theta) = \frac{\Omega^2}{2\pi} \int_0^{+\infty} e^{-qt} \operatorname{Tr} \left\{ D_\pm(\theta)^* e^{\mathcal{K}_1 t} [D_\pm(\theta)\rho_\infty] \right\} dt + \text{c.c.}, \tag{374}$$

$$D_\pm(\theta) := \pm \frac{1}{4} \sqrt{\frac{3}{2\pi}} (1 \pm \cos \theta) D_1 + \sum_{\epsilon=\pm} g_\epsilon(\theta; \pm) P_\epsilon, \tag{375}$$

$$g_\epsilon(\theta; \pm) := \frac{\langle h_\pm|g_\epsilon\rangle}{\sqrt{|\Delta\Xi|}}. \tag{376}$$

The functions $g_\epsilon(\theta; \pm)$ depend on S^ϵ, but after all they are free parameters of the theory: they are square integrable θ-functions, satisfying the constraint

$$\int_0^\pi \sin\theta \left[(1 + \cos\theta)\, g_\epsilon(\theta; +) - (1 - \cos\theta)\, g_\epsilon(\theta; -) \right] \mathrm{d}\theta = 0 \,, \tag{377}$$

coming from the orthogonality of g_ϵ to $|1, 1; +1\rangle$, see (331).

Then, by integrating over the whole solid angle, one gets the total spectrum

$$\Sigma(x) = \int_0^\pi \mathrm{d}\theta \sin\theta \int_0^{2\pi} \mathrm{d}\phi \Big(\Sigma_+(x; \theta) + \Sigma_-(x; \theta) \Big) = \sum_k \Sigma(\nu; h_k) \,, \tag{378}$$

where $\{h_k\}$ is any c.o.n.s. in $\mathcal{Z}$. When one has $g_\epsilon = 0$, as in the usual case, one gets

$$\Sigma_\pm(x; \theta) = \frac{3}{8\pi} \left(\frac{1 \pm \cos\theta}{2} \right)^2 \Sigma(x) \,. \tag{379}$$

For $g_\epsilon \ne 0$ the x and θ dependencies do not factorize. In the experiments one measures something proportional to $\Sigma_+(x; \theta) + \Sigma_-(x; \theta)$ for θ around $\pi/2$; this quantity fails to be proportional to $\Sigma(x)$ only by the presence of some terms which we expect to be small and which are not qualitatively different from the other terms in $\Sigma(x)$. So, for simplicity, we shall study only the total spectrum (378).

The Total Spectrum

By choosing in (378) a basis with

$$h_1 = |1, 1; +1\rangle, \qquad h_2 = \|\Delta g\|^{-1} \Delta g \,, \tag{380a}$$

$$h_3 = \|\Delta g\|^{-1} \left[\|g_-\|^2 \|\Delta g\|^2 - |\langle \Delta g | g_- \rangle|^2 \right]^{-1/2} \left[\|\Delta g\|^2 g_- - \langle \Delta g | g_- \rangle \Delta g \right] \,, \tag{380b}$$

we get the final expression of the *total spectrum*

$$\begin{aligned}
\Sigma(x) = \frac{\Omega^2}{2\pi} \Bigg[&\frac{1}{q} \left(v_3 + \mathrm{i} \mathrm{e}^{\mathrm{i}\delta_-} \sin\delta_- + \mathrm{i}\Omega^2 v_1 \mathrm{e}^{-\mathrm{i}s} \sin s \right) \left(d - \mathrm{i} \mathrm{e}^{-\mathrm{i}\delta_-} \sin\delta_- \right. \\
&\left. - \mathrm{i}\Omega^2 \{\mathrm{Re}\, d\} \mathrm{e}^{\mathrm{i}s} \sin s \right) + \mathrm{e}^{-\mathrm{i}\delta_-} \sin\delta_+ \left(\Omega^2 c_1 \sin s - \mathrm{i} \mathrm{e}^{\mathrm{i}s} c_3 \right) \\
&+ \left(\mathrm{Re}\, d + \mathrm{i}\overline{d}\, \mathrm{e}^{\mathrm{i}s} \sin s \right) \left(u_3 + \mathrm{i}\Omega^2 u_1 \mathrm{e}^{-\mathrm{i}s} \sin s \right) + \frac{1}{q} \left\langle g_- | g_- + \Omega^2 \{\mathrm{Re}\, d\} \Delta g \right\rangle \\
&+ \Omega^2 \langle \Delta g | g_+ \rangle c_1 - \Omega^2 \overline{d} \, \|\Delta g\|^2 u_1 + \frac{\Omega^2}{q} \left\langle \Delta g | g_- + \Omega^2 d_1 \Delta g \right\rangle v_1 + \text{c.c.} \Bigg] \,,
\end{aligned} \tag{381}$$

$$\boldsymbol{v} = \frac{1}{2\,(\boldsymbol{K} + q)} \begin{pmatrix} 0 \\ 1 \\ 1 \end{pmatrix}, \qquad \boldsymbol{c} = \frac{1}{\boldsymbol{K} + q} \begin{pmatrix} \mathrm{Re}\, d \\ d \\ \overline{d} \end{pmatrix}, \qquad \boldsymbol{u} = \frac{1}{\boldsymbol{K} + q} \begin{pmatrix} 0 \\ 0 \\ 1 \end{pmatrix}, \tag{382}$$

$$\boldsymbol{K} = \begin{pmatrix} 1 & -1/2 & -1/2 \\ \Omega^2 e^{is} \cos s & b+y & 0 \\ \Omega^2 e^{-is} \cos s & 0 & \bar{b}-y \end{pmatrix}. \tag{383}$$

By integrating over the reduced frequency x, we get the strength of the total spectrum [cf. (349)]

$$|\alpha|^2 \int_{-\infty}^{+\infty} \Sigma(x)\, dx = |\alpha|^2 \Omega^2 \big\{ \{\mathrm{Re}\, d\} \left(1 + \Omega^2 \sin^2 \delta_+\right) + \left(1 - \Omega^2 \mathrm{Re}\, d\right)$$

$$\times \left(\sin^2 \delta_- + \|g_-\|^2\right) + \mathrm{Re}\left[d\left(e^{2i\delta_-} - 1\right)\right] + \Omega^2 \{\mathrm{Re}\, d\}\|g_+\|^2 \big\}. \tag{384}$$

Let us recall that $|\alpha|^2$ is the natural line width, Ω^2 is proportional to the laser intensity, $z = (\omega - \omega_0)/|\alpha|^2$ is the reduced detuning, $y = B/|\alpha|^2$ is the reduced laser bandwidth, $\Omega^2 |\alpha|^2 \varepsilon$ is an intensity dependent shift, $x = (\nu - \omega_0)/|\alpha|^2$ and $\gamma = \widetilde{\gamma}/|\alpha|^2$ are the reduced frequency and the reduced instrumental width, respectively, and $q = i(x-z) + (\gamma + y)/2$, $s = \delta_+ - \delta_-$, $\Delta g = g_+ - g_-$. Let us note that ε, $\delta_\pm$, $\|g_\pm\|^2$, $\langle g_+|g_-\rangle$ are parameters linked to the $S^\pm$ scattering matrices, satisfying the two constraints

$$|\langle g_+|g_-\rangle| \le \|g_+\|\,\|g_-\|, \tag{385a}$$

$$\|\Delta g\| = 0 \Rightarrow \varepsilon = 0; \tag{385b}$$

apart from this relation ε is an independent parameter of the model.

One can check that the spectrum $\Sigma(x)$ is invariant under the transformation:

$$x \to -x, \qquad z \to -z, \qquad \varepsilon \to -\varepsilon, \tag{386a}$$

$$\delta_\pm \to -\delta_\pm, \qquad \langle g_-|g_+\rangle \to \langle g_+|g_-\rangle. \tag{386b}$$

The case $S^\pm = \mathbb{1}$

Let us recall that the usual model, with only the absorption/emission process, corresponds to $\delta_\pm = 0$, $g_\pm = 0$, $\varepsilon = 0$, $z = \tilde{z}$, $s = 0$. In this case we obtain

$$\Sigma(x) = \frac{\Omega^2}{2\pi}\left[\frac{1}{q}\, v_3 d + \left(2v_3 + \frac{4\Omega^2}{N}\right) \mathrm{Re}\, d\right] + \text{c.c.}, \tag{387}$$

$$d = \frac{1 + y + 2iz}{4z^2 + \Gamma^2}, \qquad \Gamma^2 = (1+y)(1+y+2\Omega^2), \tag{388}$$

$$v_3 = [2 + \gamma + y + 2i(x-z)]\,[1 + \gamma + 4y + 2i(x-2z)]/N, \tag{389}$$

$$N = 4\Omega^2 [1 + \gamma + 2y + 2i(x-z)] + [2 + \gamma + y + 2i(x-z)]$$

$$\times [1 + \gamma + 4y + 2i(x-2z)]\,(1 + \gamma + 2ix). \tag{390}$$

Now, the spectrum $\Sigma(x)$ is invariant under the transformation:

$$x \to -x, \qquad z \to -z. \tag{391}$$

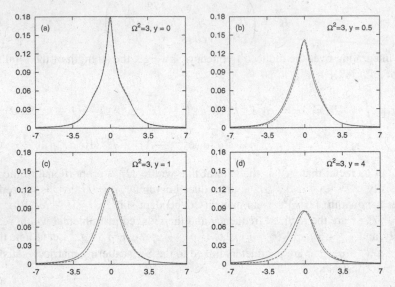

Fig. 8. The spectrum $\Sigma(x) \times 100$ for laser bandwidths $y = 0$, 0.5, 1, 4 and $\gamma = 0.6$, $\Omega^2 = 3$, $z = 0$.

Plots

Let us end by presenting some plots of the spectrum with various choices of the parameters. Let us consider only the resonant case (no detuning: $z = 0$) and let us take $\gamma = 0.6$ for the reduced instrumental width; let us recall the in our units the natural line width is 1. In all the plots we compare the spectrum predicted by the usual model (dashed lines), in which only the absorption/emission channel is present, with the spectrum predicted by the modified model (solid lines), in which both the absorption/emission channel and the direct scattering channel are present. The usual model is characterized by $\delta_{\pm} = 0$, $\|g_{\pm}\|^2 = 0$, $\varepsilon = 0$, while as an example of modified model we choose $\delta_{+} = -0.03$, $\delta_{-} = 0.13$, $\|g_{+}\|^2 = 0.0045$, $\|g_{-}\|^2 = 0.0055$, $\langle g_{+}|g_{-}\rangle = -0.004 + \mathrm{i} \times 0.002$, $\varepsilon = -0.001$.

In Figs. 8, 9, 10 we consider three laser intensities $\Omega^2 = 3$, 28, 50 and four laser bandwidths $y = 0$, 0.5, 1, 4. Our choice of the parameters for the modified model is such that for a monochromatic laser in resonance (Fig. 1(a)) the modified spectrum is not quantitatively too different from the usual one, but its asymmetry is clear. The differences between the two cases are enhanced by the presence of the bandwidth.

In Fig. 11 we have considered a stimulating laser in resonance with a very large bandwidth $y = 50$ and four levels of intensity: $\Omega^2 = 5$, 15, 30, 60. The instrumentals width is again $\gamma = 0.6$.

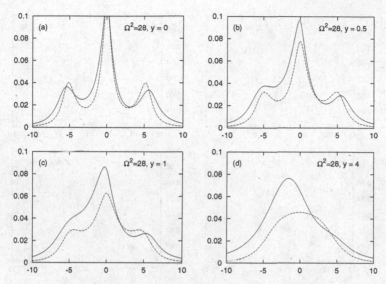

Fig. 9. The spectrum $\Sigma(x) \times 100$ for laser bandwidths $y = 0, \ 0.5, \ 1, \ 4$ and $\gamma = 0.6$, $\Omega^2 = 27$, $z = 0$.

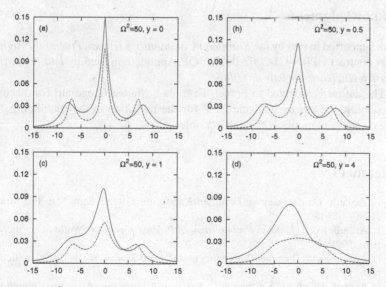

Fig. 10. The spectrum $\Sigma(x) \times 100$ for laser bandwidths $y = 0, \ 0.5, \ 1, \ 4$ and $\gamma = 0.6$, $\Omega^2 = 50$, $z = 0$.

288   Alberto Barchielli

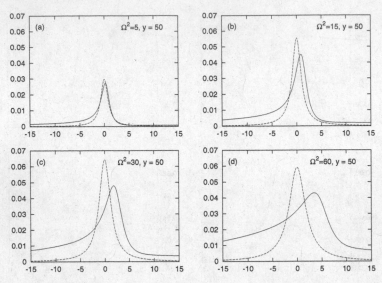

Fig. 11. The spectrum $\Sigma(x) \times 100$ for laser bandwidth $y = 50$ and $\gamma = 0.6$, $\Omega^2 = 5, 15, 30, 60, z = 0$.

Acknowledgments

Work supported in part by the *European Community's Human Potential Programme* under contract HPRN-CT-2002-00279, QP-Applications, and by *Istituto Nazionale di Fisica Nucleare, Sezione di Milano*.

The author is indebted to Franco Fagnola e Matteo Gregoratti for discussions and suggestions and to Stephane Attal for the invitation to the stimulating "École d'été de Mathématiques 2003" in Grenoble.

References

1. L. Accardi: *On the quantum Feynman-Kac formula*, Rend. Sem. Mat. Fis. Milano **48** (1980) 135-180.
2. L. Accardi (ed.): *Quantum Probability and Related Topics VI* (World Scientific, Singapore, 1991).
3. L. Accardi (ed.): *Quantum Probability and Related Topics VII* (World Scientific, Singapore, 1992).
4. L. Accardi, R. Alicki, A. Frigerio, L. Y. Gang: *An invitation to the weak coupling limit and low density limit*. In Ref. 2 pp. 3–61.
5. L. Accardi, A. Frigerio, L. Y. Gang: *On the weak coupling limit problem*. In Ref. 10 pp. 20–58.
6. L. Accardi, A. Frigerio, L. Y. Gang: *Quantum Langevin equation in the weak coupling limit*. In Ref. 11 pp. 1–16.

7. L. Accardi, A. Frigerio, J. T. Lewis: *Quantum stochastic processes*, Pub. RIMS Kyoto Univ. **18** (1982) 97-133.
8. L. Accardi, W. von Waldenfels (eds.): *Quantum Probability and Applications II*. Lecture Notes in Mathematics, Vol. 1136 (Springer, Berlin, 1985).
9. L. Accardi, W. von Waldenfels (eds.): *Quantum Probability and Applications III*. Lecture Notes in Mathematics, Vol. 1303 (Springer, Berlin, 1988).
10. L. Accardi, W. von Waldenfels (eds.): *Quantum Probability and Applications IV*. Lecture Notes in Mathematics, Vol. 1396 (Springer, Berlin, 1989).
11. L. Accardi, W. von Waldenfels (eds.): *Quantum Probability and Applications V*. Lecture Notes in Mathematics, Vol. 1442 (Springer, Berlin, 1990).
12. R. Alicki, M. Bozejko, W. A. Majewski: *Quantum Probability*, Banach Center Publications, Vol. 43 (Polish Academy of Sciences, Institute of Mathematics, Warsawa, 1998).
13. P. Alsing, G. J. Milburn, D. F. Walls: *Quantum nondemolition measurements in optical cavities*, Phys. Rev. A **37** (1988) 2970–2978.
14. A. Barchielli: *Measurement theory and stochastic differential equations in quantum mechanics*, Phys. Rev. A **34** (1986) 1642–1649.
15. A. Barchielli: *Comment on "Quantum mechanics of measurements distributed in time. A path–integral formulation"*, Phys. Rev. D **34** (1986) 2527–2530.
16. A. Barchielli: *Quantum stochastic differential equations: An application to the electron shelving effect*, J. Phys. A: Math. Gen. **20** (1987) 6341–6355.
17. A. Barchielli: *Input and output channels in quantum systems and quantum stochastic differential equations*. In Ref. 9, pp. 37–51.
18. A. Barchielli: *Direct and heterodyne detection and other applications of quantum stochastic calculus to quantum optics*, Quantum Opt. **2** (1990) 423–441.
19. A. Barchielli: *Some stochastic differential equations in quantum optics and measurement theory: the case of counting processes*. In Ref. 47 pp. 1–14.
20. A. Barchielli: *On the quantum theory of direct detection*. In O. Hirota, A. S. Holevo, C. M. Caves (eds.), *Quantum Communication, Computing, and Measurement* (Plenum Press, New York, 1997) pp. 243–252.
21. A. Barchielli, V. P. Belavkin: *Measurements continuous in time and a posteriori states in quantum mechanics*, J. Phys. A: Math. Gen. **24** (1991) 1495–1514.
22. A. Barchielli, L. Lanz, G. M. Prosperi: *A model for the macroscopic description and continual observations in quantum mechanics*, Nuovo Cimento **72B** (1982) 79–121.
23. A. Barchielli, L. Lanz, G. M. Prosperi: *Statistics of continuous trajectories in quantum mechanics: Operation valued stochastic processes*, Found. Phys. **13** (1983) 779–812.
24. A. Barchielli, G. Lupieri: *Quantum stochastic calculus, operation valued stochastic processes and continual measurements in quantum mechanics*, J. Math. Phys. **26** (1985) 2222–2230.
25. A. Barchielli, G. Lupieri: *Convolution semigroups in quantum stochastic calculus*. In Ref. 10, pp. 107–127.
26. A. Barchielli and G. Lupieri: *Photoemissive sources and quantum stochastic calculus*. In Ref. 12 pp. 53–62.
27. A. Barchielli, G. Lupieri: *Quantum stochastic models of two-level atoms and electromagnetic cross sections*, J. Math. Phys. **41** (2000) 7181–7205.
28. A. Barchielli, N. Pero: *A quantum stochastic approach to the spectrum of a two-level atom*, J. Opt. B: Quantum Semiclass. Opt. **4** (2002) 272–282.
29. V. P. Belavkin: *Nondemolition measurements, nonlinear filtering and dynamic programming of quantum stochastic processes*. In A. Blaquière (ed.), *Modelling and Control of Systems*, Lecture Notes in Control and Information Sciences, vol. 121 (Springer, Berlin, 1988) pp. 245–265.

290   **Alberto Barchielli**

30. J. C. Bergquist, R. G. Hulet, W. M. Itano, D. J. Wineland: *Observation of quantum jumps in a single atom*, Phys. Rev. Lett. **57** (1986) 1699–1702.
31. L. Bouten, H. Maassen, B. Kümmerer: *Constructing the Davies process of resonance fluorescence with quantum stochastic calculus*, Optics and Spectroscopy **94** (2003) 911–919.
32. L. Bouten, M. Guta, H. Maassen: *Stochastic Schrödinger equations*, J. Phys. A: Math. Gen. **37** (2004) 3189–3209.
33. H. J. Carmichael: *An Open System Approach to Quantum Optics*, Lect. Notes Phys., **m 18** (Springer, Berlin, 1993).
34. A. M. Chebotarev: *The quantum stochastic equation is unitarily equivalent to a symmetric boundary value problem for the Schrödinger equation*, Math. Notes **61** (1997) 510-518.
35. A. M. Chebotarev: *Quantum stochastic equation is unitarily equivalent to a symmetric boundary value problem for the Schrödinger equation*. In *Stochastic Analysis and Mathematical Physics (Viña del Mar, 1996)* (World Sci. Publishing, New York, 1998) pp. 42-54.
36. A. M. Chebotarev, D. V. Victorov: *Quantum stochastic processes arising from the strong resolvent limits of the Schrödinger evolution in Fock space*. In Ref. 12 pp. 119-133.
37. A. M. Chebotarev: *Quantum stochastic differential equation is unitary equivalent to a symmetric boundary value problem in Fock space*, Inf. Dimens. Anal. Quantum Probab. Relat. Top. **1** (1998) 175-199.
38. A. M. Chebotarev: *Lectures on Quantum Probability*. Sociedad Matemática Mexicana, Aportaciones Matemáticas, Nivel Avanzado **14**, México, 2000
39. C. Cohen–Tannoudji, J. Dalibard: *Single–atom laser spectroscopy. Looking for dark periods in fluorescence light*, Europhys. Lett. **1** (1986) 441–448.
40. M. J. Collet, D. F. Walls: *Quantum limits to light amplifiers*, Phys. Rev. Lett. **61** (1988) 2442–2444.
41. J. D. Cresser, J. Häger, G. Leuchs, M. Rateike, H. Walther: *Resonance fluorescence of atoms in strong monochromatic laser fields*. In R. Bonifacio (ed.), *Dissipative Systems in Quantum Optics*, Topics in Current Physics Vol. 27 (Springer, Berlin, 1982), pp. 21-59.
42. A. Dąbrowska : *Counting photons in the Λ-experiment*, Open Sys. & Information Dyn. **9** (2002) 381–392.
43. E. B. Davies: *Quantum stochastic processes*, Commun. Math. Phys. **15** (1969) 277–304.
44. E. B. Davies: *Quantum Theory of Open Systems*. Academic Press, London, 1976.
45. H. G. Dehmelt: Bull. Am. Phys. Soc. **20** (1975) 60.
46. H. G. Dehmelt: IEEE Trans. Instrum. Meas. **IM31** (1982) 83–87.
47. L. Diòsi, B. Lukàcs (eds.): *Stochastic Evolution of Quantum States in Open Systems and in Measurement Processes* (World Scientific, Singapore, 1994).
48. J. H. Eberly and P. Lambropoulos (eds.): *Multiphoton Processes* (Wiley, New York, 1978).
49. S. Ezekiel and F. Y. Wu: *Two-level atoms in an intense monochromatic field: A review of recent experimental investigations*. In Ref. 48 pp. 145–156.
50. A. Frigerio: *Construction of stationary quantum Markov processes through quantum stochastic calculus*. In Ref. 8 pp. 207-222.
51. A. Frigerio: *Covariant Markov dilations of quantum dynamical semigroups*, Pub. RIMS Kyoto Univ. **21** (1985) 657–675.
52. A. Frigerio, M. Ruzzier: *Relativistic transformation properties of quantum stochastic calculus*, Ann. Inst. H. Poincaré **51** (1989) 67–79.
53. C. W. Gardiner: *Inibition of atomic phase decays by squeezed light: a direct effect of squeezing*, Phys. Rev. Lett. **56** (1986) 1917–1920.

54. C. W. Gardiner, M. J. Collet: *Input and output in damped quantum systems: Quantum stochastic differential equations and the master equation*, Phys. Rev. A **31** (1985) 3761–3774.
55. C. W. Gardiner, P. Zoller: *Quantum Noise* (Springer, Berlin, 2000).
56. M. Gregoratti: *The Hamiltonian operator associated to some quantum stochastic differential equations*. PhD thesis. Università degli Studi di Milano, 2000.
57. M. Gregoratti: *On the Hamiltonian operator associated to some quantum stochastic differential equations*, Inf. Dimens. Anal. Quantum Probab. Relat. Top. **3** (2000) 483-503.
58. M. Gregoratti: *The Hamiltonian operator associated to some quantum stochastic evolutions*, Commun. Math. Phys. **222** (2001) 181–200.
59. R. E. Grove, F. Y. Wu, S. Ezekiel: *Measurement of the spectrum of resonance fluorescence from a two-level atom in a intense monochromatic field*, Phys. Rev. A **15**, (1977) 227–233.
60. W. Harting, W. Rasmussen, R. Schieder, H. Walther: *Study of the frequency distribution of the fluorescence light induced by monochromatic radiation*, Z. Physik A **278** (1976) 205–210.
61. A. S. Holevo: *A noncommutative generalization of conditionally positive definite functions*. In Ref. 9 pp. 128–148.
62. A. S. Holevo: *Limit theorems for repeated measurements and continuous measurement processes*. In Ref. 10, pp. 229–257.
63. A. S. Holevo: *Time-ordered exponentials in quantum stochastic calculus*. In Ref. 3 pp. 175–202.
64. A. S. Holevo: *Statistical Structure of Quantum Theory*, Lecture Notes in Physics m 67. Springer, Berlin, 2001.
65. R. L. Hudson and K. R. Parthasarathy: *Quantum Itô's formula and stochastic evolutions*, Commun. Math. Phys. **93** (1984) 301–323.
66. P. L. Kelley, W. H. Kleiner: *Theory of electromagnetic field measurement and photoelectron counting*, Phys. Rev. A **136** (1964) 316–334.
67. T. Kennedy, D. F. Walls: *Squeezed quantum fluctuations and macroscopic quantum coherence*, Phys. Rev. A **37** (1988) 152–157.
68. H. J. Kimble, L. Mandel: *Resonance fluorescence with excitation of finite bandwidth*, Phys. Rev. A **15** 689–699.
69. A. S. Lane, M. D. Reid, D. F. Walls: *Quantum analysis of intensity fluctuations in the nondegenerate parametric oscillator*, Phys. Rev. A **38** (1988) 788–799.
70. M. Lax: *Quantum noise. IV. Quantum theory of noise source*, Phys. Rev. **145** (1966) 110–129.
71. J. M. Lindsay, S. J. Wills: *Existence, positivity and contractivity for quantum stochastic flows with infinite dimensional noise*, Probab. Theory Relat. Fields **116** (2000) 505–543.
72. G. Lupieri: *Generalized stochastic processes and continual observations in quantum mechanics*, J. Math. Phys. **24** (1983) 2329–2339.
73. H. Maassen: *The construction of continuous dilations by solving quantum stochastic differential equations*, Semesterbericht Funktionalanalysis Tübingen Sommersemester 1984 (1984) 183-204.
74. H. Maassen: *Quantum Markov processes on Fock space described by integral kernels*. In Ref. 8 pp. 361-374.
75. L. Mandel, H. J. Kimble: *Resonance fluorescence under finite bandwidth excitation*. In Ref. 48 pp. 119–128.
76. M. A. Marte, H. Ritsch, D. F. Walls: *Squeezed-reservoir lasers*, Phys. Rev. A **38** (1988) 3577–3588.

77. M. A. Marte, D. F. Walls: *Quantum theory of a squeezed-pump laser*, Phys. Rev. A **37** (1988) 1235–1247.

78. M. B. Mensky: *Continuous quantum measurements and path integrals* (Institute of Physics, Bristol, 1993).

79. A. Messiah: *Quantum Mechanics, Vol. II* (North-Holland, Amsterdam, 1970).

80. G. J. Milburn: *Quantum measurement theory of optical heterodyne detection*, Phys. Rev. A **36** (1987) 5271–5279.

81. B. R. Mollow: *Power spectrum of light scattered by two-level systems*, Phys. Rev. **188** (1969) 1969–1975.

82. K. Mølmer, Y. Castin, J. Dalibard: *A Monte–Carlo wave function method in quantum optics*, J. Opt. Soc. Am. B **10** (1993) 524–538.

83. W. Nagourney, J. Sandberg, H. G. Dehmelt: *Shelved optical electron amplifier: observation of quantum jumps*, Phys. Rev. Lett. **56** (1986) 2797–2799.

84. K. R. Parthasarathy: *An Introduction to Quantum Stochastic Calculus* (Birkhäuser, Basel, 1992).

85. T. Sauter, W. Neuhauser, R. Blatt, P. E. Toschek: *Observation of quantum jumps*, Phys. Rev. Lett. **57** (1986) 1696–1698.

86. F. Schuda, C. R. Stroud Jr., M. Hercher: *Observation of the resonant Stark effect at optical frequencies*, J. Phys. B: Atom. Molec. Phys. **7** (1974) L198–202.

87. J. H. Shapiro, H. P. Yuen, J. A. Machado Mata: *Optical communication with two-photon coherent states — Part II: Photoemissive detection and structured receiver performance*, IEEE Trans. Inf. Theory **IT-25** (1979) 179–192.

88. H. P. Yuen, V. W. S. Chan: *Noise in homodyne and heterodyne detection*, Optics Lett. **8** (1983) 177–179.

89. H. P. Yuen, J. H. Shapiro: *Optical communication with two-photon coherent states — Part I: Quantum-state propagation and quantum-noise reduction*, IEEE Trans. Inf. Theory **IT-24** (1978) 657–668.

90. H. P. Yuen, J. H. Shapiro: *Optical communication with two-photon coherent states — Part III: Quantum measurements realizable with photoemissive detectors*, IEEE Trans. Inf. Theory **IT-26** (1980) 78–92.

91. P. Zoller, M. Marte, D. F. Walls: *Quantum jumps in atomic systems*, Phys. Rev. A **35** (1987) 198–207.

Index of Volume III

Information About the Other Two Volumes

Contents of Volume I

Index of Volume I

quantum, 142

Ensemble
  canonical, 60
  grand canonical, 63
  microcanonical, 57
Entropy
  Boltzmann, 57
Enveloping von Neumann algebra, 119
Essential support, 281
Evolution group, 29
Exponential law, 203

Factor, 118
Faithful
  representation, 80
Fermi gas, 134, 172
Fermion, 53, 186
Finite particle subspace, 192
Finite quantum system, 133
Fock space, 186
Folium, 119
Free energy, 61
Functional calculus, 16, 25, 281

G.N.S. representation, 82

Hahn decomposition theorem, 240
Hamiltonian, 290
Hamiltonian system, 43
Hardy class, 258
Harmonic oscillator, 50, 205
Heisenberg picture, 51
Heisenberg uncertainty principle, 49, 290
Helffer-Sjöstrand formula, 17
Hille-Yosida theorem, 37

Ideal
  left, 84
  right, 84
  two-sided, 84
Ideal gas, 185
Indistinguishable, 186
Individual ergodic theorem, 125
Infinitesimal generator, 35
Internal energy, 58
Invariant subspace, 22, 272
Invertible, 73
Isometric element, 75

Jensen's formula, 259

Kaplansky density theorem, 111
Kato-Rellich theorem, 285
Kato-Rosenblum theorem, 287
Koopman ergodicity criterion, 129
Koopman lemma, 128
Koopman mixing criterion, 129
Koopman operator, 128

Lebesgue-Radon-Nikodym theorem, 240
Legendre transform, 62
Liouville equation, 43
Liouville's theorem, 43
Liouvillean, 128, 143, 150, 161, 168
Lummer Phillips theorem, 38

Mean ergodic theorem, 32, 128
Measure
  absolutely continuous, 240
  complex, 239
  regular Borel, 238
  signed, 239
  space, 238
  spectral, 274, 280, 295
  support, 238
Measurement, 48
  simultaneous, 49
Measures
  equivalent, 280
  mutually singular, 240
Modular
  conjugation, 96
  operator, 96
Morphism
  *-algebra, 77
  C^*-algebra, 77

Nelson's analytic vector theorem, 32
Norm resolvent convergence, 27
Normal element, 75
Normal form, 143

Observable, 42, 46, 123, 290
Operator
  (anti-)symmetrization, 187
  closable, 5, 268
  closed, 2, 268
  core, 31, 268

Contents of Volume II

Quantum Noises

Stéphane Attal 79

**Complete Positivity and the Markov structure of Open Quantum
Systems**

Rolando Rebolledo 149

Quantum Stochastic Differential Equations and Dilation of Completely Positive Semigroups

Index of Volume II

Lecture Notes in Mathematics

For information about earlier volumes
please contact your bookseller or Springer
LNM Online archive: springerlink.com

Recent Reprints and New Editions